D0083396

Utility-Based
Learning from Data

Chapman & Hall/CRC
Machine Learning & Pattern Recognition Series

SERIES EDITORS

Ralf Herbrich and Thore Graepel
Microsoft Research Ltd.
Cambridge, UK

AIMS AND SCOPE

This series reflects the latest advances and applications in machine learning
and pattern recognition through the publication of a broad range of reference
works, textbooks, and handbooks. The inclusion of concrete examples, applications, and methods is highly encouraged. The scope of the series includes,
but is not limited to, titles in the areas of machine learning, pattern recognition, computational intelligence, robotics, computational/statistical learning
theory, natural language processing, computer vision, game AI, game theory,
neural networks, computational neuroscience, and other relevant topics, such
as machine learning applied to bioinformatics or cognitive science, which
might be proposed by potential contributors.

PUBLISHED TITLES

MACHINE LEARNING: An Algorithmic Perspective
Stephen Marsland

HANDBOOK OF NATURAL LANGUAGE PROCESSING,
Second Edition
Nitin Indurkhya and Fred J. Damerau

UTILITY-BASED LEARNING FROM DATA
Craig Friedman and Sven Sandow

Chapman & Hall/CRC
Machine Learning & Pattern Recognition Series

Utility-Based Learning from Data

Craig Friedman
Sven Sandow

CRC Press
Taylor & Francis Group
Boca Raton   London   New York

CRC Press is an imprint of the
Taylor & Francis Group, an **informa** business
A CHAPMAN & HALL BOOK

Chapman & Hall/CRC
Taylor & Francis Group
6000 Broken Sound Parkway NW, Suite 300
Boca Raton, FL 33487-2742

Printed in the United States of America on acid-free paper
10 9 8 7 6 5 4 3 2 1

International Standard Book Number: 978-1-58488-622-8 (Hardback)

Library of Congress Cataloging-in-Publication Data

Friedman, Craig.
    Utility-based learning from data / authors, Craig Friedman, Sven Sandow.
        p. cm. -- (Machine learning & pattern recognition series)
    Includes bibliographical references and index.
    ISBN 978-1-58488-622-8 (hardcover : alk. paper)
    1. Machine learning. I. Sandow, Sven. II. Title. III. Series.

Q325.5.F75 2010
006.3'1--dc22
                                                        2010023847

Visit the Taylor & Francis Web site at
http://www.taylorandfrancis.com

and the CRC Press Web site at
http://www.crcpress.com

To Donna, Michelle, and Scott – C.F.
To Emily, Jonah, Theo, and my parents – S.S.

Contents

Preface

Statistical learning — that is, learning from data — and, in particular, probabilistic model learning have become increasingly important in recent years. Advances in information technology have facilitated an explosion of available data. This explosion has been accompanied by theoretical advances, permitting new and exciting applications of statistical learning methods to bioinformatics, finance, marketing, text categorization, and other fields.

A welter of seemingly diverse techniques and methods, adopted from different fields such as statistics, information theory, and neural networks, have been proposed to handle statistical learning problems. These techniques are reviewed in a number of textbooks (see, for example, Mitchell (1997), Vapnik (1999), Witten and Frank (2005), Bishop (2007), Cherkassky and Mulier (2007), and Hastie et al. (2009)).

It is *not* our goal to provide another comprehensive discussion of all of these techniques. Rather, we hope to

(*i*) provide a pedagogical and self-contained discussion of a select set of methods for estimating probability distributions that can be approached coherently from a decision-theoretic point of view, and

(*ii*) strike a balance between rigor and intuition that allows us to convey the main ideas of this book to as wide an audience as possible.

Our point of view is motivated by the notion that probabilistic models are usually *not* learned for their own sake — rather, they are used to make decisions. We shall survey select popular approaches, and then adopt the point of view of a decision maker who

(*i*) operates in an uncertain environment where the consequences of every possible outcome are explicitly monetized,

(*ii*) bases his decisions on a probabilistic model, and

(*iii*) builds and assesses his models accordingly.

We use this point of view to shed light on certain standard statistical learning methods.

Fortunately finance and decision theory provide a language in which it is natural to express these assumptions — namely, utility theory — and formulate, from first principles, model performance measures and the notion of optimal and robust model performance. In order to present the aforementioned approach, we review utility theory — one of the pillars of modern finance and decision theory (see, for example, Berger (1985)) — and then connect various key ideas from utility theory with ideas from statistics, information theory, and statistical learning. We then discuss, using the same coherent framework, probabilistic model performance measurement and probabilistic model learning; in this framework, model performance measurement flows naturally from the economic consequences of model selection and model learning is intended to optimize such performance measures on out-of-sample data.

Bayesian decision analysis, as surveyed in Bernardo and Smith (2000), Berger (1985), and Robert (1994), is also concerned with decision making under uncertainty, and can be viewed as having a more general framework than the framework described in this book. By confining our attention to a more narrow explicit framework that characterizes real and idealized financial markets, we are able to describe results that need not hold in a more general context.

This book, which evolved from a course given by the authors for graduate students in mathematics and mathematical finance at the Courant Institute of Mathematical Sciences at New York University, is aimed at advanced undergraduates, graduate students, researchers, and practitioners from applied mathematics and machine learning as well as the broad variety of fields that make use of machine learning techniques (including, for example, bioinformatics, finance, physics, and marketing) who are interested in practical methods for estimating probability distributions as well as the theoretical underpinnings of these methods. Since the approach we take in this book is a natural extension of utility theory, some of our terminology will be familiar to those trained in finance; this book may be of particular interest to financial engineers. This book should be self-contained and accessible to readers with a working knowledge of advanced calculus, though an understanding of some notions from elementary probability is highly recommended. We make use of ideas from probability, as well as convex optimization, information theory, and utility theory, but we review these ideas in the book's second chapter.

Acknowledgments

We would like to express our gratitude to James Huang; it was both an honor and a privilege to work with him for a number of years. We would also like to express our gratitude for feedback and comments on the manuscript provided by Piotr Mirowski and our editor, Sunil Nair.

Disclaimer

This book reflects the personal opinions of the authors and does not represent those of their employers, Standard & Poors (Craig Friedman) and Morgan Stanley (Sven Sandow).

Chapter 1

Introduction

In this introduction, we informally discuss some of the basic ideas that underlie the approach we take in this book. We shall revisit these ideas, with greater precision and depth, in later chapters.

Probability models are used by human beings who make decisions. In this book we are concerned with evaluating and building models for decision makers. We do not assume that models are built for their own sake or that a single model is suitable for all potential users. Rather, we evaluate the performance of probability models and estimate such models based on the assumption that these models are to be used by a decision maker, who, informed by the models, would take actions, which have consequences.

The decision maker's perception of these consequences, and, therefore, his actions, are influenced by his risk preferences. Therefore, one would expect that these risk preferences, which vary from person to person,[1] would also affect the decision maker's evaluation of the model.

In this book, we assume that individual decision makers, with individual risk preferences, are informed by models and take actions that have associated costs, and that the consequences, which need not be deterministic, have associated payoffs. We introduce the costs and payoffs associated with the decision maker's actions in a fundamental way into our setup.

In light of this, we consider model performance and model estimation, taking into account the decision maker's own appetite for risk. To do so, we make use of one of the pillars of modern finance: utility theory, which was originally developed by von Neumann and Morgenstern (1944).[2] In fact, this book can be viewed as a natural extension of utility theory, which we discuss in Section 1.1 and Chapter 4, with the goals of

(*i*) assessing the performance of probability models, and

---

[1]Some go to great lengths to avoid risk, regardless of potential reward; others are more eager to seize opportunities, even in the presence of risk. In fact, recent studies indicate that there is a significant genetic component to an individual's appetite for risk (see Kuhnen and Chiao (2009), Zhong et al. (2009), Dreber et al. (2009), and Roe et al. (2009)).

[2]It would be possible to develop more general versions of some of the results in this book, using the more general machinery of decision theory, rather than utility theory — for such an approach, see Grünwald and Dawid (2004). By adopting the more specific utility-based approach, we are able to develop certain results that would not be available in a more general setting. Moreover, by taking this approach, we can exploit the considerable body of research on utility function estimation.

(*ii*) estimating (learning) probability models

in mind.

As we shall see, by taking this point of view, we are led naturally to

(*i*) a model performance measurement principle, discussed in Section 1.2 and Chapter 8, that we describe in the language of utility theory, and

(*ii*) model estimation principles, discussed in Section 1.3.2 and Chapter 10, under which we maximize, in a robust way, the performance of the model with respect to the aforementioned model performance principle.

Our discussion of these model estimation principles is a bit different from that of standard textbooks by virtue of

(*i*) the central role accorded to the decision maker, with general risk preferences, in a market setting, and

(*ii*) the fact that the starting point of our discussion explicitly encodes the robustness of the model to be estimated.

In more typical, related treatments, for example, treatments of the maximum entropy principle, the development of the principle is *not* cast in terms of markets or investors, and the robustness of the model is shown as a *consequence* of the principle.[3]

We shall also see, in Section 1.3.3, Chapter 7, and Chapter 10, that a number of classical information-theoretic quantities and model estimation principles are, in fact, special cases of the quantities and model estimation principles, respectively, that we discuss. We believe that by taking the aforementioned utility-based approach, we obtain access to a number of interpretations that shed additional light on various classical information-theoretic and statistical notions.

1.1   Notions from Utility Theory

Utility theory provides a way to characterize the risk preferences and the actions taken by a rational decision maker *under a known probability model*. We will review this theory more formally in Chapter 4; for now, we informally introduce a few notions. We focus on a decision maker who makes decisions in a probabilistic market setting where all decisions can be identified with

---

[3]This is consistent with the historical development of the maximum entropy principle, which was first proposed in Jaynes (1957a) and Jaynes (1957b); the robustness was only shown much later by Topsøe (1979) and generalized by Grünwald and Dawid (2004).

asset allocations. Given an allocation, a wealth level is associated with each outcome. The decision maker has a utility function that maps each potential wealth level to a utility. Each utility function must be increasing (more is preferred to less) and concave (incremental wealth results in decreasing incremental utility). We plot two utility functions in Figure 1.1. An investor (we

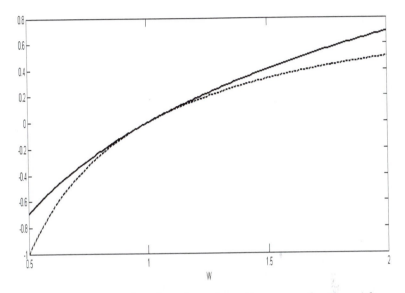

FIGURE 1.1: Two utility functions from the power family, with $\kappa = 2$ (more risk averse, depicted with a dashed curve) and $\kappa = 1$ (less risk averse, depicted with a solid curve).

use the terms decision maker and investor interchangeably) with the utility function indicated with the dashed curve is more risk averse than an investor with the utility function indicated with the solid curve, since, for the dashed curve, higher payoffs yield less utility and lower payoffs are more heavily penalized. The two utility functions that we have depicted in this figure are both members of the well-known family of power utility functions

$$U_\kappa(W) = \frac{W^{1-\kappa} - 1}{1 - \kappa} \to log(W), \text{ as } \kappa \to 1, \kappa > 0. \tag{1.1}$$

In Figure 1.1, $\kappa = 2$ (more risk averse, depicted with a dashed curve) and $\kappa = 1$ (less risk averse, depicted with a solid curve). The utility function $U_\kappa(W)$ is known to have constant relative risk aversion κ;[4] the higher the

---

[4]We shall formally define the term "relative risk aversion" later.

value of κ, the more risk averse is the investor with that utility function. Sometimes we will refer to a less risk averse investor as a more aggressive investor. For example, an investor with a logarithmic utility function is more aggressive than an investor with a power 2 utility function.

From a practical point of view, perhaps the most important conclusion of utility theory is that, *given a probability model*, a decision maker who subscribes to the axioms of utility theory acts to maximize his expected utility under that model. We illustrate these notions with Example 1.1, which we present in Section 1.6.[5]

We'd like to emphasize that, *given a probability measure*, and employing utility theory, there are no single, one-size-fits-all methods for

(*i*) allocating capital, or

(*ii*) measuring the performance of allocation strategies.

Rather, *the decision maker allocates and assesses the performance of allocation strategies based on his risk preferences*. Examples 1.1 and 1.2 in Section 1.6 illustrate these points.

1.2 Model Performance Measurement

In this book we are concerned with situations where a decision maker must select or estimate a probability model. Is there a single, one-size-fits all, best model that all individuals would prefer to use, or do risk preferences enter into the picture when assessing model performance? If risk preferences do indeed enter into model performance measurement, how can we estimate models that maximize performance, given specific risk preferences? We shall address the second question (model estimation) briefly in Section 1.3 of this introduction (and more thoroughly in Chapter 10), and the first (model performance measurement) in this section (and more thoroughly in Chapter 8).

We incorporate risk preferences into model performance measurement by means of utility theory, which, as we have seen in the previous section, allows for the quantification of these risk preferences. In order to derive explicit model performance measures, we will need two more ingredients:

(*i*) a specific setting, in which actions can be taken and a utility can be associated with the consequences, and

---

[5]Some of the examples in this introduction are a bit long and serve to carefully illustrate what we find to be very intuitive and plausible points. So, to smooth the exposition, we present our examples in the last section of this introduction. In these examples, we use notions from basic probability, which (in addition to other background material) is discussed in Chapter 2.

(*ii*) a probability measure under which we can compute the expected utility of the decision maker's actions.

Throughout most of this book, we choose as ingredient (*i*) a horse race (see Chapter 3 for a detailed discussion of this concept), in which an investor can place bets on specific outcomes that have defined payoffs. We shall also discuss a generalization of this concept to a so-called incomplete market, in which the investor can bet only on certain outcomes or combinations of outcomes. In this section we refer to both settings simply as the market setting.

As ingredient (*ii*) we choose the empirical measure (frequency distribution) associated with an out-of-sample test dataset. The term out-of-sample refers to a dataset that was not used to build the model. This aspect is important in practical situations, since it protects the model user to some extent from the perils of overfitting, i.e., from models that were built to fit a particular dataset very well, but generalize poorly. Example 1.3 in Section 1.6 illustrates how the problem of overfitting can arise.

Equipped with utility theory and the above two ingredients, we can state the following model performance measurement principle, which is depicted in Figure 1.2.

Model Performance Measurement Principle: *Given*

(*i*) *an investor with a utility function, and*

(*ii*) *a market setting in which the investor can allocate,*

the investor will allocate according to the model (so as to maximize his expected utility under the model).

We will then measure the performance of the candidate model for this investor via the average utility attained by the investor on an out-of-sample test dataset.

We note that somebody who interprets probabilities from a frequentist point of view might want to replace the test dataset with the "true" probability measure.[6] The problem with this approach is that, even if one believed in the existence of such a "true" measure, it is typically not available in practice. In this book, we do not rely on the concept of a "true" measure, although we shall use it occasionally in order to discuss certain links with the frequentist interpretation of probabilities, or to interpret certain quantities under a hypothetical "true" measure. The ideas described here are consistent with both a frequentist or a subjective interpretation of probabilities.

The examples in Section 1.6 illustrate how the above principle works in practice. It can be seen from these examples that risk preferences do indeed matter, i.e., that decision makers with different risk preferences may prefer

---

[6]One can think of the "true" measure as a theoretical construct that fits the relative frequencies of an infinitely large sample

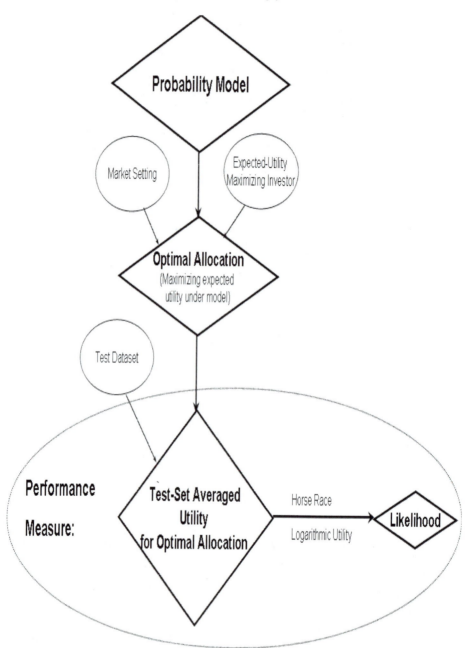

FIGURE 1.2: Model performance measurement principle (also see Section 1.2.2).

different models.[7] The intuitive reason for this is that different decision makers possess

(*i*) different levels of discomfort with unsuccessful bets, and

(*ii*) different levels of satisfaction with successful bets.

This point has important practical implications; it implies that there is no single, one-size-fits-all, best model in many practical situations.

1.2.1 Complete versus Incomplete Markets

This section is intended for readers who have a background in financial modeling, or are interested in certain connections between financial modeling and the approach that we take in this book. Financial theory makes a distinction between

(*i*) complete markets (where every conceivable payoff function can be replicated with traded instruments) — perhaps the simplest example is the horse race, where we can wager on the occurrence of each single state individually, and

(*ii*) incomplete markets.

In the real world, markets are, in general, incomplete. For example, given a particular stock, it is not, in general, possible to find a trading strategy involving one or more liquid financial instruments that pays $1 only if the stock price is exactly $100.00 in one year's time, and zero otherwise. Even though real markets are typically incomplete, much financial theory has been based on the idealized complete market case, which is typically more tractable.

As we shall see in Chapter 8, the usefulness of the distinction between the complete and incomplete market settings extends beyond financial problems — this distinction proves important with respect to measuring model performance. In horse race markets, the allocation problem can be solved via closed-form or nearly closed-form formulas, with an associated simplification of the model performance measure; in incomplete markets, it is necessary to rely to a greater extent on numerical methods to measure model performance.

1.2.2 Logarithmic Utility

We shall see in Chapter 8 that, for investors with utility functions in a logarithmic family, and only for such investors, in the horse race setting, the utility-based model performance measures are equivalent to the likelihood

---

[7]We shall show later in this book that all decision makers would agree that the "true" model is the best. However, this is of little practical relevance, since the latter model is typically not available, even to those who believe in its existence.

from classical statistics, establishing a link between our utility-based formulation and classical statistics. This link is depicted in Figure 1.2.

1.3   Model Estimation

As we have seen, different decision makers may prefer different models. This naturally leads to the notion that different decision makers may want to build different models, taking into account different performance measures. In light of this notion, we formulate the following goals:

(i) to discuss how, by starting with the model performance measurement principle of Section 1.2, we are led to robust methods for estimating models appropriate for individual decision makers, and

(ii) to establish links between some traditional information-theoretic and statistical approaches for estimating models and the approach that we take in this book, and

(iii) to briefly compare the problem settings in this book with those typically used in probability model estimation and certain types of financial modeling.

To keep things as simple as possible, we (mostly) confine the discussion in this introduction to discrete, unconditional models.[8] In the discussion that follows, before addressing the main goals of this section, we shall first review some traditional information-theoretic approaches to the probability estimation problem.

1.3.1   Review of Some Information-Theoretic Approaches

The problem of estimating a probabilistic model is often articulated in the language of information theory and solved via maximum entropy (ME), minimum relative entropy (MRE), or minimum mutual information (MMI) methods. We shall review some relevant classical information theoretic quantities, such as entropy, relative entropy, mutual information, and their properties in Chapter 2; we shall discuss modeling via the ME, MRE, and MMI principles in Chapters 9 and 10. In this introduction, we discuss a few notions informally.

Let Y be a discrete-valued random variable that can take values, y, in the finite set $\mathcal{Y}$ with probabilities p_y. The entropy of this random variable is given

---

[8]We do consider conditional models, where there are explanatory variables with known values and we seek the probability distribution of a response variable, in the chapters that follow.

by the quantity

$$H(p) \equiv -\sum_y p_y log\,(p_y)\,.$$ (1.2)

It can be shown that the entropy of a random variable can be interpreted as a measure of the uncertainty of the random variable. We note that this measure of uncertainty, unlike, for example, the variance, does not depend on the values, $y \in \mathcal{Y}$; the entropy depends only on the probabilities, p_y.

Given another probability measure on the same states, with probabilities, $\{p_1^0, \ldots, p_n^0\}$, the Kullback-Leibler relative entropy (we often refer to this quantity as, simply, relative entropy) from p to p^0 is given by

$$D(p\|p^0) \equiv \sum_y p_y log\left(\frac{p_y}{p_y^0}\right).$$ (1.3)

It can be shown that

(i) $D(p\|p^0) \geq 0$, and

(ii) $D(p\|p^0) = 0$ only if $p = p^0$.

Thus, relative entropy has some, but not all,[9] of the properties associated with a distance. We note that if the measure p^0 is uniform on the states, i.e., if there are n elements in $\mathcal{Y}$, and

$$p_y^0 = \frac{1}{n} \text{ for all } y,$$ (1.4)

then in this special case,

$$D(p\|p^0) = -H(p) - log(n),$$ (1.5)

so relative entropy can be viewed as a more general quantity than entropy. Moreover, minimizing relative entropy is equivalent, in this special case, where (1.4) holds, to maximizing entropy.

Let X be a discrete-valued random variable that can take values, x, in the finite set $\mathcal{X}$ with probabilities p_x. The mutual information between X and Y is given by

$$I(X;Y) = \sum_{x,y} p_{x,y} log \frac{p_{x,y}}{p_x p_y},$$ (1.6)

where $p_{x,y}$ denotes the joint probability that $X = x$ and $Y = y$. Thus, the mutual information is also a special case of the relative entropy for the joint random variables X and Y, where $p_{x,y}^0 = p_x p_y$. It can be shown that the mutual information can be interpreted as the reduction in the uncertainty of Y, given the knowledge of X.

---

[9]Relative entropy is not symmetric; more importantly, it does not satisfy the triangle inequality.

Armed with these information-theoretic quantities, we return to the goal of formulating methods to estimate probabilistic models from data; we discuss ME, MRE, and MMI modeling.

(i) ME modeling is governed by the maximum entropy principle, under which we would seek the probability measure that is most uncertain (has maximum entropy), given certain data-consistency constraints,

(ii) MRE modeling is governed by the minimum relative entropy principle, under which we would seek the probability measure satisfying certain data-consistency constraints that is closest (in the sense of relative entropy) to a *prior* measure, p^0; this prior measure can be thought of as a measure that one might be predisposed to use, based on prior belief, before coming into contact with data, and

(iii) MMI modeling is governed by the minimum mutual information principle, under which we would seek the probability measure satisfying certain data-consistency constraints, where X provides the least information (in the sense of mutual information) about Y. If the marginal distributions, p_x and p_y, are known, then the MMI principle becomes an instance of the MRE principle.

For ME, MRE, and MMI modeling, the idea is that the data-consistency constraints reflect the characteristics that we want to incorporate into the model, and that we want to avoid introducing additional (spurious) characteristics, with the specific means for avoiding introducing additional (spurious) characteristics described in the previous paragraph. Since entropy and mutual information are special cases of relative entropy, the principles are indeed related, though the interpretations described above might seem a bit disparate.

1.3.1.1   Features

The aforementioned data-consistency constraints are typically expressed in terms of features. Formally, a feature is a function defined on the states, for example, a polynomial feature like $f_1(y) = y^2$, or a so-called Gaussian kernel feature, with center μ and bandwidth, σ

$$f_2(y) = e^{-\frac{(y-\mu)^2}{2\sigma^2}}.$$

The model, p, can be forced to be consistent with the data, for example via a series of J constraints

$$E_p[f_j] = E_{\tilde{p}}[f_j], j = 1, \dots, J, \tag{1.7}$$

where $\tilde{p}$ denotes the empirical measure.[10] We can think of the expectation under the empirical measure on the right hand side of (1.7) as the sample average of the feature values.

Thus, by taking empirical expectations of features, we garner information about the data, and by enforcing constraints (1.7), we impose consistency of the model with the data.

1.3.1.2 The MRE Problem

The MRE problem formulation is given by

$$\text{minimize } D(p\|p^0) \text{ with respect to } p , \tag{1.8}$$

subject to data-consistency constraints, for example,

$$E_p[f_j] = E_{\tilde{p}}[f_j], j = 1, \ldots, J. \tag{1.9}$$

The solution to this problem is robust, in a sense that we make precise in Section 1.2 and Chapter 10.

1.3.1.3 The ME Problem

The ME problem formulation is given by

$$\text{maximize } H(p) \text{ with respect to } p , \tag{1.10}$$

subject to data-consistency constraints, for example,

$$E_p[f_j] = E_{\tilde{p}}[f_j], j = 1, \ldots, J. \tag{1.11}$$

As a special case of the MRE problem, the solution of the ME problem inherits the robustness of the MRE problem solution.

1.3.1.4 The MMI Problem

Under the MMI problem formulation, we seek the probability measure that minimizes the mutual information subject to certain expectation constraints.[11]

1.3.1.5 Dual Problems

Fortunately, the MRE, ME, and MMI principles all lead to convex optimization problems. We shall see that each of these problems has a corresponding *dual problem* which yields the same solution. In many cases (for example,

---

[10]Later, we shall relax the equality constraints (1.7).

[11]In this setting, the features depend on x and y; moreover, the expectation constraints can be a bit more complicated; for ease of exposition, we do not state them here. For additional details, see Globerson and Tishby (2004).

conditional probability model estimation), the dual problem is more tractable than the primal problem.

We shall see that for the MRE and ME problems,

(*i*) the solutions to the dual problem are members of a parametric exponential family, and

(*ii*) the dual problem objective function can be interpreted as the logarithm of the likelihood function.

These points sometimes, but not always (we shall elaborate in Chapter 10), apply to the MMI problem. Thus, the dual problem is typically interpreted as a search, over an exponential family, for the likelihood maximizing probability measure.[12] This establishes a connection between information theory and statistics.

1.3.2   Approach Based on the Model Performance Measurement Principle of Section 1.2

In this section, we discuss how we might develop a model estimation principle around the model performance measurement principle of Section 1.2. At first blush, it might seem natural for an investor to choose the model that maximizes the utility-based performance measures, discussed in Section 1.2, on the data available for building the model (the training data). However, it can be shown that this course of action would lead to the selection of the empirical measure (the frequency distribution of the training data) — for many interesting applications,[13] a very poor model indeed, if we want our model to generalize well on out-of-sample data; we illustrate this idea in Example 1.3 (see Section 1.6).

Though it is, generally speaking, unwise to build a model that adheres too strictly to the individual outcomes that determine the empirical measure, the observed data contain valuable statistical information that can be used for the purpose of model estimation. We incorporate statistical information from the data into a model via data-consistency constraints, expressed in terms of features, as described in Section 1.3.1.1.

---

[12] Depending on the exact choice of the data-consistency constraints, the objective function of this search may contain an additional regularization term. We shall elaborate on this in Chapters 9 and 10.

[13] For some simple applications, for example a biased coin toss with many observations, the empirical probabilities may serve well as a model. For other applications, for example, conditional probability problems where there are several real-valued explanatory variables and few observations, the empirical distribution will, generally speaking, generalize poorly out-of-sample.

1.3.2.1 Robust Outperformance Principle

Armed with the notions of features and data-consistency constraints, we return to our model estimation problem. The empirical measure typically does not generalize well because it is all too precisely attuned to the observed data. We seek a model that is consistent with the observed data, in the sense of conforming to the data-consistency constraints, yet is not too precisely attuned to the data. The question is, which data-consistent measure should we select? We want to select a model that will perform well (in the sense of the model performance measurement principle of Section 1.2), no matter which data-consistent measure might govern a potential out-of-sample test set. To address this question, we consider the following game against nature[14] (which we assume is adversarial) that occurs in a market setting.

A game against "nature" *Let Q denote the set of all probability measures, K denote the set of data-consistent probability measures, and U_q^* denote the (random) utility that is realized when allocating (so as to maximize expected utility) under the measure q in this market setting.*[15]

(i) (Our move) We choose a model, $q \in Q$; then,

(ii) (Nature's move) given our choice of a model, and, as a consequence, the allocations we would make, "nature" cruelly inflicts on us the worst (in the sense of the model performance measurement principle of Section 1.2) possible data-consistent measure; that is, "nature" chooses the measure

$$p^* = \arg\min_{p \in K} E_p[U_q^*]. \tag{1.12}$$

If we want to perform as well as possible in this game we will seek the solution of

$$q^* = \arg\max_{q \in Q} \min_{p \in K} E_p[U_q^*]. \tag{1.13}$$

By solving (1.13), we estimate a measure that (as we shall see later) conforms to the data-consistency constraints, and is *robust*, in the sense that the expected utility that we can derive from it will be attained, or surpassed, no matter which data-consistent measure "nature" chooses. The resulting estimate therefore, in particular, avoids being too precisely attuned to the individual observations in the training dataset, thereby mitigating overfitting.[16]

---

[14]This game is a special case of a game in Grünwald and Dawid (2004), which was preceded by the "log loss game" of Good (1952).

[15]We note that we are speaking informally here, since we have not specified the market setting or how to calculate U_q^*. We shall discuss these issues more precisely in the remainder of the book.

[16]This strategy does not guarantee a cure to overfitting, though! If there are too many data-consistency constraints, or the data-consistency constraints are not chosen wisely, problems

This game can be further enriched by introducing a rival, who allocates according to the measure $q^0 \in Q$.[17] In this case, we would seek the solution according to the robust outperformance principle:

Robust Outperformance Principle
We seek

$$q^* = \arg\max_{q \in Q}\min_{p \in K} E_p[U_q^* - U_{q^0}^*].\qquad(1.14)$$

Estimating q^* would allow us to to maximize the worst-case outperformance over our competitor (who allocates according to the measure $q^0 \in Q$), in the presence of a "nature" that conforms to the data-consistency constraints and tries to minimize our outperformance (in the sense of the model performance measurement principle of Section 1.2) over our rival.

Jaynes (2003), page 431, has pointed out that "this criterion concentrates attention on the worst possible case regardless of the probability of occurrence of this case, and it is thus in a sense too conservative." In our view, this may be so, given a fixed collection of features. However, by enriching the collection of features, it is always possible to go too far in the other direction, overly constraining the set of measures consistent with the data, and estimating a model that is too aggressive. We shall have more to say about ways to attempt to tune (optimally) the extent to which the data are consistent with the model in Section 1.3.5 and Chapter 10.

We note that this formulation has been cast entirely in the language of utility theory. The model that is produced is therefore specifically tailored to the risk preferences of the model user with utility function U. We also note that we have not made use of the concept of a "true" measure in this formulation.

1.3.2.2   Minimum Market Exploitability Principle

As we shall see in Chapter 10, under certain technical conditions, it is possible to reverse the order of the max and min in the robust outperformance principle. Moreover, as we shall see in Chapter 10, subject to regularity conditions, by solving the resulting minimax problem, we obtain the solution to the maxmin problem (1.14) arising from the robust outperformance principle.

By reversing the order of the max and min in (1.14), we obtain the minimum market exploitability principle:

Minimum Market Exploitability Principle

---

can arise. We shall discuss these issues, and countermeasures that can be taken to further protect against overfitting, at greater length below in this introduction, as well as in Chapters 9 and 10.

[17]Later, we shall see that this rival's allocation measure q^0 can be identified with the prior measure in an MRE problem.

We seek

$$p^* = \arg\min_{p \in K} \max_{q \in Q} E_p[U_q^* - U_{q^0}^*].$$ (1.15)

Here,

$$E_p[U_q^* - U_{q^0}^*]$$ (1.16)

can be interpreted as the gain in expected utility, for an investor who allocates according to the model q, rather than q^0, when the "true" measure is p. Under the minimum market exploitability principle, we seek the data-consistent measure, p, that minimizes the maximum gain in expected utility over an investor who uses the model q^0. After a little reflection, this principle is consistent with a desire to avoid overfitting. The intuition here is that the data-consistency constraints completely reflect the characteristics of the model that we want to incorporate, and that we want to avoid introducing additional (spurious) characteristics. Any additional characteristics (beyond the data-consistency constraints) could be exploited by an investor; so, to avoid introducing additional such characteristics, we minimize the exploitability of the market by an investor, given the data-consistency constraints.

Fortunately, as we shall see in Chapter 10, the minimum market exploitability principle leads to a convex optimization problem with an associated dual problem that can be solved robustly via efficient numerical techniques. Moreover, as we shall also see in Chapter 10, this dual problem can be interpreted as a utility maximization problem over a parametric family, and can be solved robustly via efficient numerical techniques.

By virtue of their equivalence, both the minimum market exploitability principle and the robust outperformance principle lead us down the same path; both lead to a tractable approach to estimate statistical models tailor-made to the risk preferences of the end user.

1.3.3 Information-Theoretic Approaches Revisited

As we shall see in Chapter 7, the quantity $\max_{q \in Q} E_p[U_q^* - U_{q^0}^*]$ in (1.15) is a generalization of relative entropy, with a clear economic interpretation. In particular, we shall see in Chapter 7, that the relative entropy, $D(p\|p^0)$, can be interpreted as the gain in expected utility, for a logarithmic utility investor who allocates in a horse race on the states according to the "true" measure p, rather than the measure p^0.

We shall also see in Chapter 10 that the minimum market exploitability principle, in fact, includes as special cases the maximum entropy (ME) principle, the minimum relative entropy (MRE) principle, and the minimum mutual information (MMI) principle, and that all of these principles can be expressed in economic terms.

The common intuition underlying these expressions in economic terms is that the additional characteristics (beyond the data-consistency constraints) that we want to avoid introducing could be exploited by an investor; so, to avoid introducing additional characteristics beyond the data-consistency constraints, we minimize the exploitability of the market by an investor, given the data-consistency constraints. In particular, as we shall see in Chapter 10,

(*i*) the ME principle can be viewed as the requirement that, given the data-consistency constraints, our model have as little (spurious) expected logarithmic utility as possible,

(*ii*) the MRE principle can be viewed as the requirement that, given the data-consistency constraints, our model have as little (spurious) expected logarithmic utility gain as possible over an investor who allocates to maximize his expected utility under the prior measure, and

(*iii*) the MMI principle can be viewed as the requirement that, given the data-consistency constraints, our model have as little (spurious) expected logarithmic utility gain as possible over an investor who allocates to maximize his expected utility without making use of the information given by the realizations of X.

We believe that this economic intuition provides a convincing and unifying rationale for the ME, MRE, and MMI principles.

We shall also see that

(*i*) for the ME, MRE, and certain MMI problems,[18] the objective function of the dual problem can be interpreted as the expected utility of an investor with a logarithmic utility function, so the dual problem can be formulated as the search, over an exponential family of measures, for the measure that maximizes expected (logarithmic) utility, or, equivalently, maximizes the likelihood, and that

(*ii*) for the ME, MRE, and MMI problems, by construction, the solutions possess the optimality and robustness properties discussed in Section 1.3.2.1 — they provide maximum expected utility with respect to the worst-case measures that conform to the data-consistency constraints.

For more general utility functions, we would obtain more general versions of the ME, MRE, and MMI principles; in this book, when we discuss more general utility functions, we shall concentrate on more general version of the MRE principle, rather than the ME or MMI principles.

1.3.4 Complete versus Incomplete Markets

As indicated in Section 1.2.1, there is an important distinction between the complete horse race setting and the more general incomplete market setting. In

---

[18]We shall specify these cases in Chapter 10.

the more tractable horse race setting, with data-consistency constraints under which the feature expectations under the model are related to the feature expectations under the empirical measure, the generalized relative entropy principle has an associated dual problem that can be viewed as an expected utility maximization over a parametric family. We are not aware of similar results in incomplete market settings.

1.3.5 A Data-Consistency Tuning Principle

As we have discussed, the above problem formulations bake in a robust out-performance over an investor who allocates according to the prior, or benchmark model, *given* a set of data-consistency constraints. But how, given a set of feature functions,[19] can we formulate data-consistency constraints that will prove effective?

The simplest (and most analytically tractable) way to generate data-consistency constraints from features is to require that the expectation of the features under the model be exactly the same as the expectation under the empirical measure (the frequency distribution of the training data). However, this requirement does not always lead to effective models. Two of the things that can go wrong with this approach, depending on the number and type of features and the nature of the training data, are

(i) the feature expectation constraints are not sufficiently restrictive, resulting in a model that has not "learned enough" from the data, and

(ii) the feature expectation constraints are too restrictive, resulting in a model that has learned "too much" (including noise) from the data.

In case (i), where the features are not sufficiently restrictive, we can add new features. In case (ii), where the features are too restrictive, we can relax them. By controlling the degree of relaxation in the feature expectation constraints, we can control the tradeoff between consistency with the data and the extent to which we can exploit the market, relative to the performance of our rival investor. In the end, in this case, our investor chooses the model that best balances this tradeoff, with respect to the model performance measurement principle of Section 1.2 applied to an out-of-sample dataset, as indicated in the following principle

Data-Consistency Tuning Principle
Given a family of data constraint sets indexed by the parameter α, let $q^*(\alpha)$ denote the model selected under one of the equivalent principles of Section 1.3.2 as a function of α. We tune the level of data-consistency to maximize

---

[19]In this book, we do not discuss methods to generate features — we assume that they are given. In some cases, though, we discuss ways to select a sparse set of features from some predetermined set.

(over α) the out-of-sample performance under the performance measurement principle of Section 1.2.

1.3.6 A Summary Diagram for This Model Estimation, Given a Set of Data-Consistency Constraints

We display some of the relationships discussed above in Figure 1.3, where

- (*i*) we have used a dashed arrow to signify that the MMI principle sometimes, but not always (we shall elaborate in Chapter 10), leads to a utility maximization problem over a parametric family, and

- (*ii*) we have used bi-directional arrows between the generalized MRE principle and the robust outperformance and minimum market exploitability principles, since, as we shall see in Chapter 10, all three principles are equivalent.

1.3.7 Problem Settings in Finance, Traditional Statistical Modeling, and This Book

In this section, which may be of particular interest to readers with a background in financial modeling, we compare the problem settings used in this book with problem settings used in finance and traditional statistical modeling.

Though we use methods drawn from utility theory, the problems to which we apply these methods are (statistical) probability model estimation problems, rather than more typical financial applications of utility theory. One such application — the least favorable market completion principle (discussed in Section 11.2), which is used in finance to price contingent claims[20] — is quite similar in spirit to our minimum market exploitability principle. As we shall see, (statistical) probability model estimation problems and the pricing problems from finance can be structurally similar.

In the case of contingent claim pricing problems, given the statistical measure on the system (in finance, this measure is often called the physical measure, or the real-world measure) the modeler seeks a different probability measure, a probability measure consistent with known market prices (a so-called pricing measure, or risk-neutral measure).

In the case of traditional probability model estimation problems, outside of finance, the modeler seeks a statistical (real-world) measure consistent with certain data-consistency constraints. Thus, the traditional statistical modeler

---

[20]Contingent claims are financial instruments with contractually specified payments that depend on the prices of other financial instruments. Examples include puts and calls on a stock, and interest rate futures.

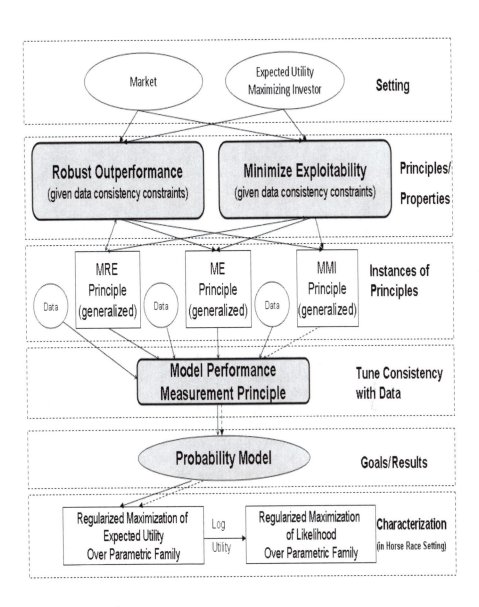

FIGURE 1.3: Model estimation approach. Note that the model performance measurement principle enters this figure twice: once as a building block (not shown) for our model estimation principles, and then later, as a means of tuning the degree of consistency with the data to maximize out-of-sample performance.

seeks a statistical (real-world) measure, and is, typically, not at all concerned with pricing measures; the contingent claim modeler assumes that the statistical (real-world) measure is known, and seeks a pricing measure.

In particular, in the horse race setting, with payoffs specified for each state,

(*i*) the contingent claim modeler essentially already has the pricing measure (which can easily be determined from the payoffs) and is able to price any contingent claim by reconstructing its payoff in terms of the horse race payoffs,

(*ii*) the traditional statistical modeler is, typically, not influenced by the payoffs, and must take whatever steps are necessary to find the statistical measure, and

(*iii*) we use the payoffs, together with utility theory, as described in the preceding sections, to evaluate model performance and estimate models (given data-consistency constraints) for expected utility maximizing investors.

1.4 The Viewpoint of This Book

As discussed in the preceding sections, we take the viewpoint of a decision maker who uses a probability model to make decisions in an uncertain environment. We believe that this viewpoint is natural, appropriate for many practical problems, and leads to intuitive, desirable, and tractable model performance measurement and model construction principles. When taking this point of view, it is relatively straightforward to construct, generalize, and shed light on some well-known principles from information theory, finance, and statistical learning. Moreover, the mindset and language adopted in this book lead to various nontraditional methods that can be brought to bear on practical problems. These nontraditional methods, some of which are discussed in this book, can be used to

(*i*) relate the performance of probability models to the risk preferences of the model user,

(*ii*) build robust probability models that are custom-tailored to the model user's risk preferences (which can, depending on the investor's risk preferences, result in relatively elegant representations of fat-tailed, yet flexible distributions),

(*iii*) measure the monetary value of a probability model,

(iv) quantify the impact of information exploitability on the performance of a probability model (measuring model performance in incomplete financial markets), and

(v) derive robust performance measures for regression models.

1.5 Organization of This Book

Chapters 8 and 10 constitute the crux of this book; each of these chapters depends on the chapters that precede it.

In Chapter 2, we review mathematical preliminaries from probability theory, convex optimization (all of the methods for building probabilistic models that we discuss in this book require solution to convex programming problems), and information theory. These are the building blocks that we use in later chapters.

In Chapter 3, we review the horse race setting, which is also known as a complete market. This is a particularly tractable and simple "market" setting — used heavily throughout this book — in which we can consider model performance and model building from a decision-theoretic point of view.

In Chapter 4, we review elements of utility theory. Utility theory provides a framework that we use to describe investor risk preferences. Expected utility maximization (von Neumann and Morgenstern (1944)) allows for plausible and practical model performance measurement and provides a goal for model construction.

In Chapter 5, we discuss an expected utility maximizing investor who bets in a horse race type market. In particular, we introduce the notion of compatibility between the utility function, the horse race market, and the probability measure. When the utility function, the market, and the probability measure are compatible, there is always an optimal allocation.

In Chapter 6, we discuss select popular methods for measuring model performance. In particular, we discuss the likelihood principle, the likelihood ratio, and the Neyman-Pearson Lemma, and draw connections between likelihood and the horse race.

In Chapter 7, we discuss information theory from a decision-theoretic point of view. This chapter starts by observing that there are decision-theoretic interpretations (in terms of an investor with a particular utility function) for fundamental information theoretic quantities. We then note that by generalizing the utility function, we obtain more general information theoretic quantities, which we explore.

In Chapter 8, we discuss model performance measurement from the point of view of an investor who would use the models to make financial decisions, and is concerned with the performance of his investments under the model.

In Chapter 9, we review select methods for learning probabilistic models from data, including maximum likelihood inference and regularized maximum likelihood inference, including the ridge and lasso models. We also discuss Bayesian inference and minimum relative entropy methods.

In Chapter 10, we develop the model learning problem in the horse race context. Based on the general principals introduced in the introduction to this book, we formulate explicit primal and dual problems.

In Chapter 11, we discuss various extensions of the material in earlier chapters. We discuss model performance measures for leveraged investors, model performance and estimation in incomplete markets, and model performance measurement for regression models.

In Chapter 12, we discuss applications to four important financial modeling problems, a breast cancer model, and a text categorization problem.

1.6   Examples

Example 1.1 *Four gamblers allocate to a coin toss*

In this example, we see that *given a probability measure*, and employing utility theory, there are no single, one-size-fits-all methods for allocating capital.

The specific setting, which is summarized in Table 1.1, is as follows.

TABLE 1.1:   Four gamblers (completely risk averse, expectation maximizing or linear utility, log utility, and power 2 utility) allocate to a coin toss. The probability of heads is .51.

	Heads Occurs	Tails Occurs
Payoffs on a $1 bet on heads:	$2	$0
Payoffs on a $1 bet on tails:	$0	$2

The payoff for a $1 bet on heads is $2 if heads occurs and zero otherwise; the payoff for a $1 bet on tails is $2 if tails occurs and zero otherwise. Each gambler must allocate his wealth to heads and/or tails. The probability of heads is known to be .51. How should a gambler allocate his capital? That depends on the gambler's risk preferences.

Suppose that our first gambler is completely averse to any risk of loss whatsoever. He can perfectly hedge, allocating $.5 to heads and $.5 to tails. In this case, his total payoff will always be exactly \$1, no matter whether the coin toss results in heads or tails.

Suppose that, at the other extreme, our second gambler will do whatever it takes to maximize the expected payoff after a single play of the game. If he allocates the fraction, b, of his wealth to heads and the fraction $1 - b$ to tails, he would allocate his wealth so as to solve the problem

$$\max_{\{b:0\leq b\leq 1\}} [0.51 * 2b + 0.49 * 2(1 - b)]. \tag{1.17}$$

This expected payoff maximizing gambler will choose $b = 1$, i.e., he will allocate his entire wealth to heads, with expected payoff $0.51 * 2 = 1.02$. We note that though this strategy maximizes the expected wealth gain on a single play, in the long run, under repeated play, the strategy of allocating all of the wealth to heads is almost surely a recipe for ruin. Eventually, almost surely, a tail will occur and the gambler will lose all of his wealth. This gambler is oblivious to that risk.

If a gambler subscribes to the axioms of utility theory, he would allocate so as to maximize his expected utility. Such a gambler, allocating fraction b of his wealth to heads and the fraction $1 - b$ to tails, would solve the problem

$$\max_{\{b:0\leq b\leq 1\}} [0.51 * U(2b) + 0.49U(2(1 - b))]. \tag{1.18}$$

We note that a gambler with the linear utility function, $U(W) = W$, would formulate precisely the optimization problem (1.17). Thus the expectation maximizing gambler can be characterized by the utility function $U(W) = W$.

Suppose that our third gambler has the utility function $U(W) = log(W)$. He would solve the problem

$$\max_{b} [0.51 \log(2b) + 0.49 \log(2(1 - b))]. \tag{1.19}$$

It is easy to verify, by calculus, that the investor with the utility $U(W) = log(W)$ will allocate the fraction $b^* = 0.51$ to heads and 0.49 to tails.

Suppose that our fourth gambler's utility function is given by the power utility with $\kappa = 2$. He would solve the problem

$$\max_{b} \left[0.51 \frac{(2b)^{1-2} - 1}{1 - 2} + 0.49 \frac{(2(1 - b))^{1-2} - 1}{1 - 2} \right]. \tag{1.20}$$

After setting the derivative to zero, solving, and checking the second derivative, we see that this investor will allocate $b^* = 0.505$ to heads and 0.495 to tails, midway between the allocations of the more aggressive logarithmic utility investor and the completely risk averse investor. These differences in allocation may seem small, but over repeated play, they can have a profound impact on the long-term experience of the gambler: the long-term wealth growth rate for the logarithmic utility investor is 0.0002, albeit with a non-negligible probability of large drawdowns (runs of "bad luck"). The long-term wealth growth rate for the completely risk averse decision maker is zero, since his wealth never changes, with no possibility of drawdowns.

Example 1.2 *Two of our gamblers rank wealth distributions*

In this example, we shall see that different investors may rank wealth distributions differently, depending on their risk preferences.

In the same setting as Example 1.1 (see Table 1.1), after repeated play, with heads occuring 51% of the time, we suppose that each of our decision makers, with log utility and power, with $\kappa = 2$, utility functions, respectively, can choose among the wealth distributions generated by the two strategies $b^* = 0.51$, and $b^* = 0.505$. We assume that these decision makers measure the success of the strategies in a manner consistent with the axioms of utility theory — by computing expected utility with respect to the probabilities actually experienced. This formulation leads to the quantities that we maximized in (1.19) and (1.20). As we have already seen from the optimization problems, the log utility investor will prefer the wealth distribution generated by the allocations $b^* = 0.51$ and the power 2 decision maker prefers the wealth distribution generated by $b^* = 0.505$.

Example 1.3 *An overfit model*

In this example, we shall see that the empirical measure can be a very poor model.

Let the random variable X denote the daily return of a stock. We observe the daily stock returns $x_1, \ldots, x_{10}$, over a two week period (10 trading days). The empirical measure is then

$$prob(X = x) = \begin{cases} \frac{1}{10}, & \text{if } x \in \{x_1, \ldots, x_{10}\}, \text{ and} \\ 0, & \text{otherwise,} \end{cases} \quad (1.21)$$

assuming that each of the returns is unique. This model reflects the training data perfectly, but will fail out-of-sample, since it only attaches nonzero probability to events that have already occurred. If this model is to be believed, then it would make sense to risk all on the bet that $x_n \in \{x_1, \ldots, x_{10}\}$, for $n > 10$, a strategy doomed to fail when a previously unobserved return (inevitably) occurs.

Example 1.4 *Two gamblers who bet on a coin toss*

In this example, we shall see that different decision makers may prefer different models and that the economic implications of the model choices can be considerable.

The specific setting, which is summarized in Table 1.2, is the following. The payoff for a \$1 bet on heads is \$100 if heads occurs and zero otherwise; the payoff for a \$1 bet on tails is \$1 if tails occurs and zero otherwise. We suppose that there are two decision makers with utility functions $U_1(W)$ and $U_2(W)$ (here we are using the notation given in (1.1)). Each decision maker can allocate to heads and/or tails. There are two probabilistic candidate models, $q^{(1)}$ and $q^{(2)}$ (these models are not associated with the investors); $q^{(1)}$ assigns the probabilities 0.004 and 0.996 to heads and tails, respectively, while $q^{(2)}$ assigns the probabilities 0.0225 and 0.9775 to heads and tails, respectively.

TABLE 1.2: Two gamblers rank model performance.

	Heads Occurs	Tails Occurs
Payoffs on a \$1 bet on heads:	\$100	\$0
Payoffs on a \$1 bet on tails:	\$0	\$1
Probabilities according to $q^{(1)}$:	0.004	0.996
Probabilities according to $q^{(2)}$:	0.0225	0.9775
Assessment dataset frequencies:	0.01	0.99

We assume that the investors would allocate under the two models by maximizing their respective utility functions under the two models. We ask the question, how would these investors rank the two models, should the frequency (empirical probability) of heads on a model assessment dataset be 0.01, which is different from either model?

(*i*) **Allocation**

Under model $q^{(1)}$, the more aggressive (log utility) investor would allocate to heads the fraction of wealth

$$0.0040 = \arg \max_{\{b:0 \leq b \leq 1\}} \left[0.004 \log(100b) + 0.996 \log(1 - b) \right]. \qquad (1.22)$$

Under model $q^{(2)}$, the more aggressive investor would allocate to heads the fraction of wealth

$$0.0225 = \arg \max_{\{b:0 \leq b \leq 1\}} \left[0.0225 \log(100b) + 0.9775 \log(1 - b) \right]. \qquad (1.23)$$

Under model $q^{(1)}$, the more risk averse ($\kappa = 2$ power utility) investor would allocate to heads the fraction of wealth

$$0.0063 = \arg \max_{\{b:0 \leq b \leq 1\}} \left[0.004 \frac{(100b)^{1-2} - 1}{1 - 2} + 0.996 \frac{(1(1 - b))^{1-2} - 1}{1 - 2} \right]. \tag{1.24}$$

Under model $q^{(2)}$, the more risk averse investor would allocate to heads the fraction of wealth

$$0.0149 = \arg \max_{\{b:0 \leq b \leq 1\}} \left[0.0225 \frac{(100b)^{1-2} - 1}{1 - 2} \right.$$
$$\left. + 0.9775 \frac{(1(1 - b))^{1-2} - 1}{1 - 2} \right]. \tag{1.25}$$

Given the models $q^{(1)}$ and $q^{(2)}$, and the associated allocation strategies for each investor, which model works best for which investor?

(*ii*) **Performance Measurement** Each decision maker, allocating according to the two models, would compute his expected utility under the empirical measure with probability of heads equal to 0.01, based on their allocations under the measures $q^{(1)}$ and $q^{(2)}$.

Under model $q^{(1)}$, the more aggressive investor's average (or expected, under the empirical measure) utility is

$$0.01 \log(100 * 0.004) + 0.99 \log(0.996) = -0.0131. \tag{1.26}$$

Under model $q^{(2)}$, the more aggressive investor's average (or expected, under the empirical measure) utility is

$$0.01 \log(100 * 0.0225) + 0.99 \log(0.9775) = -0.0144. \tag{1.27}$$

Thus, based on this performance, the more aggressive investor prefers model $q^{(1)}$.

Under model $q^{(2)}$, the more risk averse investor's average (or expected, under the empirical measure) utility is

$$0.01 \frac{(100 * 0.0063)^{1-2} - 1}{1 - 2} + 0.99 \frac{(1(1 - 0.0063))^{1-2} - 1}{1 - 2} = -0.0121. \tag{1.28}$$

Under model $q^{(1)}$, the more risk averse investor's average (or expected, under the empirical measure) utility is

$$0.01 \frac{(100 * 0.0149)^{1-2} - 1}{1 - 2} + 0.99 \frac{(1(1 - 0.0149))^{1-2} - 1}{1 - 2} = -0.0117. \tag{1.29}$$

Thus, based on this performance, the more risk averse investor prefers model $q^{(2)}$.

To give some economic perspective, we note that (as we shall see in Chapter 8) for the more aggressive investor, the difference in expected utility (0.0013) can be interpreted as a loss, for each bet, in expected wealth growth rate. The effects, after repeated play, can be quite substantial. We shall also see, in Chapter 8, that the logarithmic investor would be willing to pay the fraction $e^{0.0013} - 1 = 0.0013$ of his capital every time he makes a bet, to upgrade from model $q^{(2)}$ to model $q^{(1)}$. For models that are used frequently, the value of such a model upgrade can be considerable.

Example 1.5 *Betting on a stock*

In this example, we shall see again, this time in an incomplete market setting, that different decision makers, who allocate and assess performance in a manner consistent with utility theory, might prefer different models.

In this example, which is summarized in Table 1.3, we consider two decision makers, each of whom must select a model of the probability distribution of the logarithm of single period stock price returns. We suppose that the candidate models are t-distributed with different degrees of freedom. The univariate t-distribution with mean μ, standard deviation σ and degrees of freedom $\nu > 2$ has probability density[21]

$$f(t; \mu, \sigma, \nu) = \frac{\Gamma\left(\frac{\nu+1}{2}\right) \sqrt{\frac{\nu}{\nu-2}}}{\Gamma\left(\frac{\nu}{2}\right) \sqrt{\nu \pi} \sigma} \left(1 + \frac{1}{\nu-2}\left(\frac{t-\mu}{\sigma}\right)^2\right)^{-\frac{\nu+1}{2}}. \tag{1.30}$$

All of the stock price return distributions that we consider in this example have the same mean ($\mu = 0.1$) and standard deviation ($\sigma = 0.2$); the different degrees of freedom govern the tail fatness, with the probability distributions approaching a normal distribution as $\nu \to \infty$ and having progressively fatter tails as ν decreases.

The more aggressive decision maker has power utility with $\kappa = 2$; the more risk-averse has power utility with $\kappa = 3$. There are two stock return models that the investors weigh: one has 3 degrees of freedom, the other has 200 degrees of freedom ($\nu = 3$ and 200, respectively). We shall refer to the first distribution as the fat-tailed distribution and the second as the thin-tailed distribution. We illustrate these two models in Figure 1.4, which depicts the two distributions over much of their support, and Figure 1.5, which depicts part of the right tail of the distributions.

---

[21] This probability density is not in standard form (compare, for example, with the definition in Example 2.6 of Section 2.1.3, below), but the form described here explicitly provides the density function in terms of its mean, standard deviation, and degrees of freedom, which is convenient for this example.

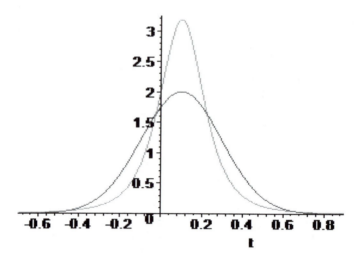

FIGURE 1.4: Two t-distributions, with degrees of freedom, $\nu = 3$ (higher center, fatter tails), and degrees of freedom, $\nu = 200$ (lower center, thinner tails).

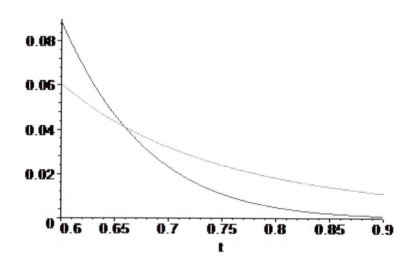

FIGURE 1.5: Two t-distribution right tails, with degrees of freedom, $\nu = 4$ (fatter tail) and degrees of freedom, $\nu = 9$ (thinner tail).

TABLE 1.3: Two power utility investors weigh two t-distributed candidate models for log stock returns. Models are assessed with respect to their expected utility under the t-distributed assessment distribution $f(t; 0.1, 0.2, 5)$.

Models:	$f(t; 0.1, 0.2, 3)$ (Fat-tailed)	$f(t; 0.1, 0.2, 200)$ (Thin-tailed)
Investors:	Aggressive utility $U_2(W)$	Risk Averse utility $U_3(W)$

The investor allocates the fraction, b, of his capital to stocks, and the fraction $1 - b$ to cash. For simplicity, we assume that here is no interest and the initial price of the stock is \$1. Let X denote the (random) logarithm of the stock price at the end of the trading period. Then, the investor's wealth at the end of the trading period is

$$W(b, X) = 1 - b + be^X. \qquad (1.31)$$

Adopting the fat-tailed model ($\nu = 3$), the more aggressive investor ($\kappa = 2$) would allocate to stocks, the fraction of wealth

$$0.9711 = \arg\max_b \int_{-\infty}^{\infty} f(t; 0.1, 0.2, 3) U_2(W(b, t)) dt. \qquad (1.32)$$

(We have solved this and the following three concave maximization problem numerically.) Adopting the thin-tailed model ($\nu = 200$), the more aggressive investor ($\kappa = 2$) would allocate to stocks, the fraction of wealth

$$0.9999 = \arg\max_b \int_{-\infty}^{\infty} f(t; 0.1, 0.2, 200) U_2(W(b, t)) dt. \qquad (1.33)$$

Adopting the fat-tailed model ($\nu = 3$), the more risk averse investor ($\kappa = 3$) would allocate to stocks, the fraction of wealth

$$0.8299 = \arg\max_b \int_{-\infty}^{\infty} f(t; 0.1, 0.2, 3) U_3(W(b, t)) dt. \qquad (1.34)$$

Adopting the thin-tailed model ($\nu = 200$), the more risk-averse investor ($\kappa = 3$) would allocate to stocks, the fraction of wealth

$$0.9989 = \arg\max_b \int_{-\infty}^{\infty} f(t; 0.1, 0.2, 200) U_3(W(b, t)) dt. \qquad (1.35)$$

We note that under the conditions of this example, the more aggressive investor always allocates more to the stock than the more risk averse investor; this is natural, since the stock affords a greater opportunity for growth, but at a greater risk. We also note that both investors allocate more to stock under the assumption of a thin-tailed return distribution than a fat-tailed distribution (which has greater probability of large losses).

We ask the question, how would these investors rank the two models, under an assessment dataset (for log returns) that is described by the t-distribution with $\nu = 5$? We see, after numerical quadrature, that

- by using and allocating according to the fat-tailed distribution model ($\nu = 3$), the more aggressive investor ($\kappa = 2$) would experience an expected utility of

$$0.0756 = \int_{-\infty}^{\infty} f(t; 0.1, 0.2, 5)U_2(W(0.9711, t))dt, \qquad (1.36)$$

- by using and allocating according to the thin-tailed distribution model ($\nu = 200$), the more aggressive investor ($\kappa = 2$) would experience an expected utility of

$$0.0763 = \int_{-\infty}^{\infty} f(t; 0.1, 0.2, 5)U_2(W(0.9999, t))dt, \qquad (1.37)$$

- by using and allocating according to the fat-tailed distribution model ($\nu = 3$), the more risk averse investor ($\kappa = 3$) would experience an expected utility of

$$0.0548 = \int_{-\infty}^{\infty} f(t; 0.1, 0.2, 5)U_3(W(0.8299, t))dt, \qquad (1.38)$$

and

- by using and allocating according to the thin-tailed distribution model ($\nu = 200$), the more risk averse investor ($\kappa = 3$) would experience an expected utility of

$$0.0324 = \int_{-\infty}^{\infty} f(t; 0.1, 0.2, 5)U_3(W(0.9989, t))dt. \qquad (1.39)$$

Thus, under a validation set consistent with the log returns being distributed as t with $\nu = 5$, the more aggressive investor would favor the performance of the thin-tailed ($\nu = 200$) model, while the more risk averse investor would prefer the performance of the fat-tailed ($\nu = 3$) model. The more aggressive investor allocates more to stock under the thin-tailed probability distribution and, under the validation measure, prefers the risk/reward profile to that of the alternative. The more risk averse investor allocates less to the risky stock under the fat-tailed probability measure, and, under the validation measure, prefers the risk/reward profile to that of the alternative.

We summarize these preferences in the Table 1.4:

TABLE 1.4: Model and allocation preferences for our investors.

Investor	Preferred Model	Allocation to Stock
More aggressive	Thin-tailed density	More
More conservative	Thick-tailed density	Less

To give some economic perspective, we note that (as we shall see in Chapter 8), under certain conditions, the more risk averse investor would be willing to pay approximately the fraction 0.0224 of his capital *every time* he makes a bet, to upgrade from the thin-tailed model ($\nu = 3$) to the fat-tailed model ($\nu = 3$). The more frequently the model is used, the more rapidly these upgrade fees would accumulate.

Chapter 2

Mathematical Preliminaries

In this chapter, we discuss some mathematical concepts that will be used in later chapters. In Section 2.1, we outline some elements of probability theory. In Section 2.2 we discuss some basics of convex optimization, and in Section 2.3 we introduce entropy and relative entropy.

2.1 Some Probabilistic Concepts

In this section, we review some probabilistic concepts, many of which are employed in this book. We refer the reader seeking a more thorough introduction to one of many textbooks on this subject, for example Ross (2005), or Bertsekas and Tsitsiklis (2002).

2.1.1 Probability Space

According to Parker (1968), "The essentials of the mathematical theory of probability were worked out in 1654 in a correspondence between the French mathematicians Pierre de Fermat (1601-65) and Blaise Pascal (1623-62)...." It was not, however, until Kolmogorov (1933) that the current formal, axiomatic, rigorous, mathematical underpinnings of probability theory were formulated. In this section we review the formal definition of a probability space as formulated by Kolmogorov.

Formally, a probability space is a triple: $(\Omega, \mathcal{F}, P)$, where

(*i*) the set $\Omega \neq \emptyset$ (the sample space) contains all possible outcomes from a random experiment,

(*ii*) $\mathcal{F}$ is a sigma algebra of subsets of Ω, i.e.,

    1. the elements of $\mathcal{F}$ (the events) are themselves *subsets* of Ω,

    2. $A \in \mathcal{F}$ implies that the complement of A in Ω, A^c, is also an element of $\mathcal{F}$, i.e.,

$$A \in \mathcal{F} \text{ implies that } A^c \in \mathcal{F}, \tag{2.1}$$

and

3. If α is an index over a countably infinite set, and A_α denotes a set indexed by alpha, then

$$A_\alpha \in \mathcal{F}, \forall \alpha \text{ implies that } \bigcup_\alpha A_\alpha \in \mathcal{F}. \qquad (2.2)$$

(*iii*) P is a probability measure, defined on $\mathcal{F}$, such that

1. if $A \in \mathcal{F}$, then $P(A) \in [0, 1]$,
2. $P(\Omega) = 1$
3. $P(\emptyset) = 0$, and
4. if $\{A_\alpha\}$ is a countably infinite set of pairwise disjoint sets indexed by α, then $P(\bigcup_\alpha A_\alpha) = \sum_\alpha P(A_\alpha)$.

To clarify this rather abstract formal definition, we introduce the following example.

Example 2.1 *Probability space generated by 3 coin tosses*

For 3 coin tosses, where each toss might result in a head (H) or tail (T),

$$\Omega = \{\omega_1, \ldots, \omega_8\}, \qquad (2.3)$$

where

$$\begin{aligned}
\omega_1 &= \{HHH\} \\
\omega_2 &= \{HHT\} \\
\omega_3 &= \{HTH\} \\
\omega_4 &= \{HTT\} \\
\omega_5 &= \{THH\} \\
\omega_6 &= \{THT\} \\
\omega_7 &= \{TTH\} \\
\omega_8 &= \{TTT\}.
\end{aligned}$$

We provide two possible sigma algebras, corresponding to the information revealed after one and three tosses, respectively:

$$\begin{aligned}
\mathcal{F}_1 &= \{\emptyset, \Omega, \{\omega_1, \ldots, \omega_4\}, \{\omega_5, \ldots, \omega_8\}\}, \text{ and} \\
\mathcal{F}_3 &= 2^\Omega \equiv \{A : A \subset \Omega\}.
\end{aligned}$$

An (natural) example of a probability measure is given by

$$P(A) = \frac{\text{the number of elements in } A}{8}. \qquad (2.4)$$

We note that even though this book takes place in a probabilistic setting, we do not make extensive, direct use of this formal definition (nor the formal definition of a random variable, described below) in the remainder of this book.

2.1.2 Random Variables

A random variable represents a particular measurement or state of the world. Formally, if Y denotes a random variable, then $Y(\omega), \omega \in \Omega$ is an $\mathcal{F}$-measurable function; i.e., the preimages, under the mapping Y, of certain "nice" sets are elements of $\mathcal{F}$. For the purposes of this book, we can assume that $Y(\omega) \in R^n$.

Example 2.2 *Random variables generated by 3 coin tosses in Example 2.1*

We consider two functions:

(i) $Y(\omega) = 1$, if the first toss is H, and 0, otherwise, is measurable (and is therefore a random variable) with respect to both $\mathcal{F}_1$ and $\mathcal{F}_3$.

(ii) $Y(\omega) = 1$, if the last toss is H, and 0, otherwise, is measurable (and is therefore a random variable) with respect to $\mathcal{F}_3$, but not $\mathcal{F}_1$.

2.1.3 Probability Distributions

We denote the probability that Y has value y (the result of a draw, a sampling, or a measurement) by $prob\{Y = y\}$. y can be discrete, continuous, or multidimensional.

If y takes discrete values, we usually denote $prob\{Y = y\}$ by p_y, with

$$0 \le p_y \le 1 \text{ and} \tag{2.5}$$

and

$$\sum_y p_y = 1. \tag{2.6}$$

At times, for clarity, it may be necessary to use alternative notation, which will be clear from the context.

Example 2.3 *Probability measure for the roll of a fair die*

Here, there are six outcomes ($y = 1, \ldots, 6$), each occurring with probability $\frac{1}{6}$. We have $p_y = \frac{1}{6}, y = 1, \ldots, 6$.

It is often natural to consider random variables that have values on a continuum in one or several dimensions. For such random variables, probabilities can be characterized by probability density functions. A function $p(y)$ with the properties

$$p(y) \ge 0, \tag{2.7}$$

$$\int p(y)dy = 1, \tag{2.8}$$

and

$$prob(Y \in A) = \int_A p(y)dy, \tag{2.9}$$

for all $A \in \mathcal{F}$ is called the probability density function (pdf) associated with the random variable Y. As we shall see below, the probabilities of discrete valued random variables can be described using pdf's, if we are willing to consider a sufficiently large class of pdf's. Sometimes, depending on the context, we may use alternative notation: $p(y), p_Y(y)$, or p_y.[1]

Another notion that is often useful is that of the cumulative distribution function. For $Y = (Y_1, \ldots, Y_n) \in R^n$, Y has cumulative distribution function

$$F(y_1, \ldots, y_n) = \int_{u_1 \leq y_1, \ldots, u_n \leq y_n} p(u)du. \tag{2.10}$$

If $Y \in R^1$ and Y has continuous distribution function $F(y)$, the qth quantile y_q is given by $y_q = F^{-1}(q)$.

We now provide examples of some important random variables:

Example 2.4 *Uniform random variable*

If Y has probability density $\frac{1}{b-a}$ on the interval (a, b), then Y is distributed uniformly on (a, b). We illustrate a uniformly distributed random variable in Figure 2.1.

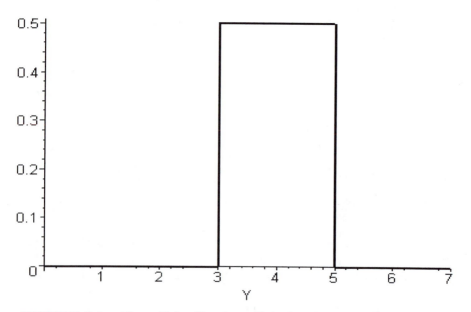

FIGURE 2.1: The pdf for Y, where Y is distributed uniformly on the interval $(3, 5)$.

---

[1]Though we sometimes abuse notation, the meaning should be clear from the context.

Example 2.5 *Standard Gaussian or standard normal random variable*

If Z has probability density

$$\frac{1}{\sqrt{2\pi}}e^{-\frac{z^2}{2}}, \text{ on } (-\infty, \infty), \tag{2.11}$$

we write $Z \sim N(0,1)$ and refer to Z as a standard Gaussian or standard normal random variable. The cumulative distribution function is given by

$$F(y) = \int_{-\infty}^{y} f(u)du = \int_{-\infty}^{y} \frac{1}{\sqrt{2\pi}}e^{-\frac{u^2}{2}}du = \frac{1}{2}erf(\sqrt{2}y) + \frac{1}{2}, \tag{2.12}$$

where the erf function is defined via

$$\mathrm{erf}(y) = \frac{2}{\sqrt{\pi}}\int_{0}^{y} e^{-t^2}dt. \tag{2.13}$$

The probability density function for a standard normal random variable is depicted in Figure 2.2. We illustrate the cumulative distribution function and the notion of a quantile in Figure 2.3.

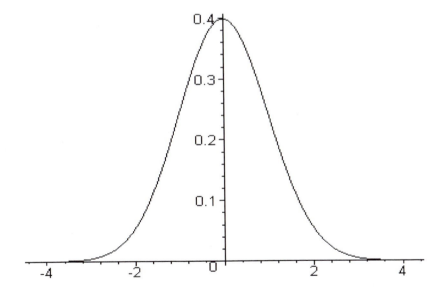

FIGURE 2.2: Probability density function for a standard normal random variable.

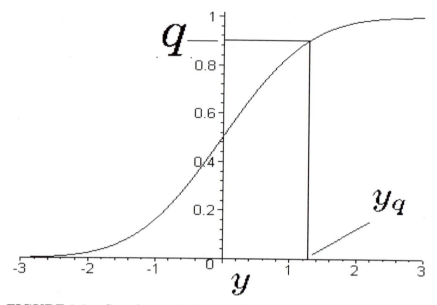

FIGURE 2.3:　Cumulative distribution function for a standard normal random variable and the quantile $y_{.9}$.

It is easy to generate a probability density function. Let A denote the set on which we want our probability density to be defined. If $\phi(y) \geq 0$ for $y \in A$ and $0 < \int_A \phi(y)dy < \infty$, then

$$f(y) = \frac{\phi(y)}{\int_A \phi(y)dy} \tag{2.14}$$

is a probability density.

A number of well known probability density functions can readily be described in terms of simple positive fucnctions, ϕ with finite integrals over a set, A:

Example 2.6 (*t- with ν degrees of freedom*) Here,

$$\phi(y) = \left(1 + \frac{x^2}{\nu}\right)^{-\left(\frac{\nu+1}{2}\right)}, \tag{2.15}$$

where $\nu \geq 1$ and $A = (-\infty, \infty)$.

Example 2.7 (*Multivariate normal distribution with mean μ and covariance matrix Σ*) Here,

$$\phi(y) = exp(-(y-\mu)^T \Sigma^{-1}(y-\mu)), \tag{2.16}$$

where Σ is positive definite, and $A = \mathbf{R}^n$.

Example 2.8 (*Exponential distribution*) Here,

$$\phi(y) = exp(-a(y-b)), \tag{2.17}$$

where $A = (b, \infty)$ and $a > 0)$.

Example 2.9 (*Uniform distribution*) Here,

$$\phi(y) = 1, \tag{2.18}$$

where $A = (a, b)$.

Example 2.10 (*Beta distribution*) Here,

$$\phi(y) = y^a (1-y)^b, \tag{2.19}$$

where $A = (0, 1)$.

Example 2.11 (*Chi-square distribution with k degrees of freedom*) Here,

$$\phi(y) = y^{\frac{k}{2}-1} e^{-\frac{y}{2}}, \tag{2.20}$$

where $A = (0, \infty)$.

Moreover, it is easy to confirm that convex combinations of probability densities are probability densities; i.e., if $p_i(y)$ is a probability density and if $\lambda_i \geq 0$ for $i \in \{1, \ldots, n\}$ and $\sum_i \lambda_i = 1$, then

$$\sum_{i=1}^{n} \lambda_i p_i(y) \tag{2.21}$$

is a probability density.

Probability density functions can accommodate real valued random variables with discrete values if the class of probability density functions is broad enough to include the Dirac delta function, $\delta(\cdot)$, which, for any function f, has the defining property

$$\int_a^b f(y)\delta(y - y_0)dy = f(y_0), \tag{2.22}$$

if $a \leq y \leq b$.

We do not provide (or require) a rigorous or thorough description of the Dirac delta function and its properties in this book; rather, we refer the interested reader to Zemanian (1987). In Exercise 3, we indicate an intuitive way of thinking about the Dirac delta function as the limit of a sequence of smooth functions.

Example 2.12 *Probability density function of a discrete valued random variable*

Suppose that the random variable Y can take the value 0, with probability .4, and the value 1, with probability .6. Then Y has the pdf

$$f(y) = .4\delta(y) + .6\delta(y-1). \tag{2.23}$$

Here, the values $y = 0$ and $y = 1$ can be interpreted as having point masses with probabilities .4 and .6, respectively, with

$$prob(Y = 0) = \lim_{\epsilon \to 0} \int_{-\epsilon}^{\epsilon} f(y)dy = .4 \tag{2.24}$$

and

$$prob(Y = 1) = \lim_{\epsilon \to 0} \int_{1-\epsilon}^{1+\epsilon} f(y)dy = .6. \tag{2.25}$$

2.1.4   Univariate Transformations of Random Variables

Suppose that we have a density function for $Y \in \mathbf{R}^1$, which is given by $f_Y(y)$ and that we seek the density function for $U = h(Y)$, where $h : \mathbf{R}^1 \to \mathbf{R}^1$ is monotone. Note that the probability that $Y \in (y, y+dy)$ is given, to leading order in dy, by $f_Y(y) \cdot dy$. By setting

$$f_U(u) \cdot |du| = f_Y(y) \cdot |dy|, \tag{2.26}$$

we obtain

Theorem 2.1 *(Density of a transformed univariate random variable) Let the random variable $Y \in \mathbf{R}^1$ have the density function $f_Y(y)$ and let $h : \mathbf{R}^1 \to \mathbf{R}^1$. If $h(y)$ is either increasing or decreasing for all y, then $U = h(Y)$ has density*

$$f_U(u) = f_Y(h^{-1}(u)) \left| \frac{d}{du} h^{-1}(u) \right|. \tag{2.27}$$

Example 2.13 *Lognormal random variable*

If $Z \sim N(0,1)$, then the density function for $Y = exp(Z)$ is given by

$$\frac{1}{y\sqrt{2\pi}} e^{-\frac{(\log(y))^2}{2}}. \tag{2.28}$$

This distribution is known as the lognormal distribution and is depicted in Figure 2.4.

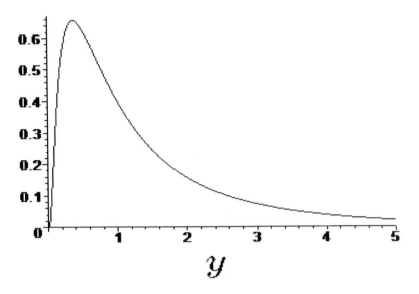

FIGURE 2.4: Probability density function for the lognormal distribution.

2.1.5 Multivariate Transformations of Random Variables

For a multivariate transformation of a vector of random variables, we make use of the following theorem from multivariate calculus:

Theorem 2.2 *(Integration by substitution) Let $f_Y : \mathbf{R}^n \to \mathbf{R}$ be integrable. Let $h : \mathbf{R}^n \to \mathbf{R}^n$ be invertible and smooth. Then*

$$\int_A f_Y(y)dy = \int_{h(A)} f_Y(h^{-1}(u))|J|du, \qquad (2.29)$$

for $A \subset \mathbf{R}^n$, where $|\cdot|$ denotes the determinant and J is the Jacobian matrix given by

$$J = \begin{pmatrix} \frac{\partial}{\partial u_1} h_1^{-1} & \cdots & \frac{\partial}{\partial u_m} h_1^{-1} \\ \vdots & \ddots & \vdots \\ \frac{\partial}{\partial u_1} h_m^{-1} & \cdots & \frac{\partial}{\partial u_m} h_m^{-1} \end{pmatrix}. \qquad (2.30)$$

We can read off the density function for the transformed random variable vector U, which leads us to the following theorem.

Theorem 2.3 *(Density of a transformed multivariate random variable) Let the random variable $Y \in \mathbf{R}^m$ have the density function $f_Y(y)$. Let $h : \mathbf{R}^m \to \mathbf{R}^m$ be invertible and smooth. Then $U = (U_1, \ldots, U_n)^T = (h_1(Y), \ldots, h_n(Y))^T$ has density*

$$f_U(u) = f_Y(h^{-1}(u))|J|. \qquad (2.31)$$

2.1.6   Expectations

In this section, we give definitions for some standard terms relating to the moments of random variables. The expectation is a measure of the central tendency of the random variable Y. For a continuous random variable with density function $p(y)$, the expected value is given by

$$E_p[Y] = \int yp(y)dy. \tag{2.32}$$

If Y is discrete-valued, with $prob\{Y = y_i\} = p_i$, then

$$E_p[Y] = \int yp(y)dy = \int y \sum_i p_i\delta(y - y_i)dy = \sum_i p_iy_i. \tag{2.33}$$

Sometimes, when the probability measure is clear, we drop the subscript p from the notation, writing

$$E[Y] = \int yp(y)dy. \tag{2.34}$$

It follows that

$$E[f(Y)] = \int f(y)p(y)dy. \tag{2.35}$$

If Y is discrete-valued, then

$$E[f(Y)] = \sum_i p_if(y_i). \tag{2.36}$$

The variance, a measure of the dispersion, of the random variable Y is given by

$$var[Y] = E[(Y - E[Y])^2]. \tag{2.37}$$

The covariance of X and Y is given by

$$cov[X, Y] = E[(X - E[X])(Y - E[Y])]. \tag{2.38}$$

The correlation of X and Y, if $var[X] > 0$ and $var[Y] > 0$, is given by

$$\rho[X, Y] = \frac{cov[X, Y]}{\sqrt{var[X]var[Y]}}. \tag{2.39}$$

Example 2.14 *Moments of linearly transformed random variables*

If the random vector $Y = (Y_1, \ldots, Y_n)^T$ has expectation vector μ (with i^{th} element μ_i), and covariance matrix $cov(Y, Y) = \Sigma$ (with ij^{th} element $cov(Y_i, Y_j)$), and if A is a matrix with n columns, then

$$E[AY] = A\mu, \tag{2.40}$$

and

$$cov(AY, AY) = A\Sigma A^T. \tag{2.41}$$

(Proof: Exercise 4.)

Not all random variables have moments, as we see from the following example.

Example 2.15 *The Cauchy distribution (Expectations need not exist)*

The Cauchy distribution

$$p(y) = \frac{1}{\pi(1+y^2)} \tag{2.42}$$

(which is a special case of the t-distribution with $\nu = 1$) has

$$E_p[Y] = \int_{-\infty}^{\infty} yp(y)dy = \int_{-\infty}^{0} yp(y)dy + \int_{0}^{\infty} yp(y)dy, \tag{2.43}$$

which is not well defined, since

$$\int_{-\infty}^{0} yp(y)dy = -\infty, \tag{2.44}$$

and

$$\int_{0}^{\infty} yp(y)dy = \infty. \tag{2.45}$$

We depict the Cauchy distribution and a standard normal distribution in Figure 2.5. Note that the Cauchy distribution has much fatter tails.

2.1.7 Some Inequalities

We list three fundamental inequalities. The first two provide bounds on the probabilities of "tail" events. The third provides a bound on the product of random variables, in terms of the second moments of each of the random variables.

- **The Markov inequality** If Y is a nonnegative random variable, then for any $c > 0$,

$$prob(Y > c) \leq \frac{E[Y]}{c}. \tag{2.46}$$

- **The Chebyshev inequality**

$$prob(|Y - \overline{Y}|^2 > c^2) \leq \frac{var(Y)}{c^2} \tag{2.47}$$

  or equivalently (when $c > 0$)

$$prob(|Y - \overline{Y}| > c) \leq \frac{var(Y)}{c^2}. \tag{2.48}$$

- **The Schwarz inequality**

$$(E[XY])^2 \leq E[X^2]E[Y^2]. \tag{2.49}$$

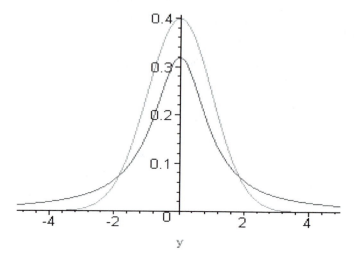

FIGURE 2.5:   Cauchy distribution (with fat tails, lower center) and Normal distribution (with thin tails, higher center).

2.1.8   Joint, Marginal, and Conditional Probabilities

Given two random variables X and Y, we use the notation $p(x, y)$ or, for discrete random variables, $p_{x,y}$, to denote the joint probability that $X = x$ and $Y = y$. We obtain marginal probabilities for continuous random variables

$$p(x) = \int p(x, y) dy, \tag{2.50}$$

or, for discrete random variables,

$$p_x = \sum_y p_{x,y}. \tag{2.51}$$

Conditional probabilities are defined via

$$p(y|x) = \frac{p(x, y)}{p(x)}. \tag{2.52}$$

For discrete random variables, we use the notation

$$p_{y|x} = \frac{p_{x,y}}{p_x}. \tag{2.53}$$

$p(y|x)$ and $p_{y|x}$ represent the probability that $Y = y$, given that $X = x$.

It is straightforward to derive Bayes' Rule. To do so, for continuous random variables, note that

$$
\begin{aligned}
p(x|y) &= \frac{p(x,y)}{p(y)} \\
&= \frac{p(y|x)p(x)}{p(y)} \\
&= \frac{p(y|x)p(x)}{\int p(x,y)dx} \\
&= \frac{p(y|x)p(x)}{\int p(y|x)p(x)dx}.
\end{aligned}
$$

Thus, Bayes' Rule reverses the conditioning. Bayes' Rule, in the discrete case, is given by

$$
p_{x|y} = \frac{p_{y|x}p_x}{\sum_x p_{y|x}p_x}.
$$

The random variables X and Y are independent if

$$
p(y|x) = p(y). \tag{2.54}
$$

That is, the probability that $Y = y$, given that $X = x$, is the same as the probability that $Y = y$ (the additional information on the value of X has no effect on the probability that $Y = y$). This leads directly to the formal definition of independence: the random variables X and Y with joint density function $p(x,y)$ and marginal density functions $p(x)$ and $p(y)$ respectively are independent if and only if

$$
p(x,y) = p(x)p(y). \tag{2.55}
$$

Example 2.16 $cov[X,Y] = 0$ *need not imply that X and Y are independent*

To see this, consider the random variables Y and Y^2 where Y is uniformly distributed on $[-1,1]$.

2.1.9 Conditional Expectations

The conditional expectation, $E[Y|x] \equiv E[Y|X = x]$, is defined as follows

$$
E[Y|x] = \int yp(y|x)dy. \tag{2.56}
$$

If Y is discrete-valued, with $prob\{Y = y_i|x\} = p_{y_i|x}$, then

$$
E[Y|x] = \int yp(y|x)dy = \int y\sum_i p_{y_i|x}\delta(y - y_i)dy = \sum_i p_{y_i|x}y_i. \tag{2.57}
$$

We shall, on occasion, encounter iterated expectations of the form $E[E[Y|X]]$. Fortunately, such iterated expectations can be simplified as follows:

$$E[E[Y|X]] = \int \left[\int yp(y|x)dy \right] p(x)dx \tag{2.58}$$

$$= \int y \left[\int p(x,y)dx \right] dy \tag{2.59}$$

$$= \int yp(y)dy \tag{2.60}$$

$$= E[Y](\text{ the tower law}). \tag{2.61}$$

We define the conditional variance, $var(Y|x)$, via

$$var(Y|x) = \int [y - E(Y|x)]^2 p(y|x)dy \tag{2.62}$$

$$= \int [y^2 - E(Y|x)^2]p(y|x)dy \tag{2.63}$$

$$= E[Y^2|x] - (E[Y|x])^2. \tag{2.64}$$

It follows that

$$var(E[Y|X]) = E\left[(E[Y|X])^2\right] - (E[E[Y|X]])^2 \tag{2.65}$$

$$= E\left[(E[Y|X])^2\right] - (E[Y])^2. \tag{2.66}$$

We now state

Theorem 2.4 *(Law of Conditional Variances)*

$$var[Y] = E[var[Y|X]] + var[E[Y|X]]. \tag{2.67}$$

Proof:

$E[var[Y|X]] + var[E[Y|X]]$
$$= E[E[Y^2|X] - (E[Y|X])^2] + E[E(Y|X)^2] - (E[Y])^2$$
$$= E[E[Y^2|X]] - E[E(Y|X)^2] + E[E(Y|X)^2] - (E[Y])^2$$
$$= E[Y^2] - (E[Y])^2$$
$$= var(Y).$$

2.1.10 Convergence

Ordinary Convergence.: A sequence of real numbers $\{y_n\}$ converges to the real number y, which we write as $\lim_{n\to\infty} y_n = y$ if

$$\forall \epsilon > 0, \exists N(\epsilon) \text{ such that } |y_n - y| < \epsilon \text{ for } n > N(\epsilon). \tag{2.68}$$

There are several generalizations for random variables. Let $Y, Y_1, Y_2, \ldots$ be jointly distributed random variables. We now list several definitions for convergence in a probabilistic context:

1. *Pointwise convergence.* $\lim_{n \to \infty} Y_n(\omega) = Y(\omega)$, $\forall \omega \in \Omega$ in the sense of ordinary convergence. This definition is the most straightforward generalization of ordinary convergence. However, for most purposes, this generalization is unnecessarily strong.

2. *Convergence with probability one (almost everywhere convergence, strong convergence).* Y_n converges to Y with probability one if

$$prob\{\omega : \lim_{n \to \infty} Y_n(\omega) = Y(\omega)\} = 1, \tag{2.69}$$

   i.e., $Y_n(\omega) \to Y(\omega)$ pointwise on a set of measure 1.

3. *Convergence in L^p, $0 < p < \infty$ (for $p = 2$, convergence in mean square).* Y_n converges to Y in L^p if and only if,

$$\lim_{n \to \infty} E[|Y_n - Y|^p] = 0. \tag{2.70}$$

4. *Convergence in probability (weak convergence).* Y_n converges to Y in probability if, for every $\epsilon > 0$,

$$\lim_{n \to \infty} prob\{|Y_n - Y| > \epsilon\} = 0. \tag{2.71}$$

   If so, we write

$$Y_n \to^p Y \tag{2.72}$$

   or

$$plim(Y_n) = Y. \tag{2.73}$$

   This definition of convergence is weaker than (is implied by) convergence with probability one or L^p convergence.

5. *Convergence in distribution.* Let $F(y) = Pr\{Y \leq y\}$ denote the *cumulative distribution function* for Y and $F_n(y) = Pr\{Y_n \leq y\}$ denote the *cumulative distribution function* for Y_n. Y_n converges to Y in distribution if

$$\lim_{n \to \infty} F_n(y) = F(y) \tag{2.74}$$

   for all y at which F is continuous. This is the weakest form of convergence discussed in this section.

2.1.11   Limit Theorems

We list below two limit theorems that govern sample averages of independent identically distributed (i.i.d.) random variables, as the number of terms becomes large.

Theorem 2.5 *(Strong Law of Large Numbers) Let Y_n be i.i.d. such that $E[Y_n] = \mu$ is finite; then*

$$\frac{1}{n} \sum_{i=1}^{n} Y_i \to \mu \text{ with probability 1.} \tag{2.75}$$

Theorem 2.6 *(Central Limit Theorem) Let Y_n be i.i.d. such that $E[Y_n] = \mu$ and $var[Y_n] = \sigma^2$ are finite. Let*

$$\overline{Y}_n = \frac{\sum_{i=1}^{n} Y_i}{n} \tag{2.76}$$

and

$$Z_n = \frac{\overline{Y} - \mu}{\frac{\sigma}{\sqrt{n}}}. \tag{2.77}$$

Let Z be a normally distributed random variable with mean 0 and variance 1. Then Z_n converges in distribution to Z, i.e., for all real z

$$\lim_{n \to \infty} prob\{Z_n \le z\} = \int_{-\infty}^{z} \frac{1}{\sqrt{2\pi}} e^{-\frac{u^2}{2}} du. \tag{2.78}$$

2.1.12   Gaussian Distributions

The Gaussian distribution, also known as the normal distribution, though not suitable for all modeling purposes, is an extremely useful distribution and is perhaps the most widely used probability distribution. Its applicability is a consequence of the Central Limit Theorem discussed in the previous section, since many outcomes spanning the natural and behavioral sciences result from the cumulative effects of many small shocks. In this section, we define general univariate Gaussian distributions with nonzero means and nonunit variances. We also define the multivariate Gaussian distribution. We then state certain facts about Gaussian distributions, sketching some proofs.

We start with the univariate standard Gaussian or standard normal distribution of Example 2.5. Suppose that Y is a *standard* normal random variable. Then $U = h(Y) = \mu + \sigma Y$ has mean μ and variance σ^2. Applying Theorem 2.3, we obtain the density for U, which is given by

$$\frac{1}{\sigma\sqrt{2\pi}} e^{-\frac{(u-\mu)^2}{2\sigma^2}}. \tag{2.79}$$

In this case we say that U has a Gaussian distribution with mean μ and variance σ^2 and write $U \sim N(\mu, \sigma^2)$.

We now consider the multivariate Gaussian distribution. If $Z_1, \ldots, Z_n$ are independent $N(0, 1)$ random variables, then their joint density function is given by

$$\prod_{i=1}^{n} \frac{1}{\sqrt{2\pi}} e^{-\frac{z_i^2}{2}} = (2\pi)^{-\frac{n}{2}} e^{-\frac{1}{2}\sum_{i=1}^{n} z_i^2} = (2\pi)^{-\frac{n}{2}} e^{-\frac{1}{2} z^T z}, \qquad (2.80)$$

where z^T denotes the transpose of z.

Let $U = CZ + \mu$, where C is an invertible $n \times n$ matrix and μ is a vector. It can be shown that U has mean μ and covariance matrix CC^T (exercise: justify this statement). According to Theorem 2.3 (exercise: fill in missing details), the density for U is given by

$$(2\pi)^{\frac{n}{2}} e^{-\frac{1}{2}(C^{-1}(u-\mu))^T (C^{-1}(u-\mu))} |C| = (2\pi)^{-\frac{n}{2}} e^{-\frac{1}{2}(u-\mu)^T (C^{-1})^T (C^{-1}(u-\mu))} |C|$$

$$= (2\pi)^{-\frac{n}{2}} e^{-\frac{1}{2}(u-\mu)^T \Sigma^{-1}(u-\mu)} |C|$$

$$= (2\pi)^{-\frac{n}{2}} |\Sigma|^{-\frac{1}{2}} e^{-\frac{1}{2}(u-\mu)^T \Sigma^{-1}(u-\mu)},$$

where $\Sigma = CC^T$. In this case we write $U \sim N(\mu, \Sigma)$.

Next, we consider affine transformations of multivariate Gaussian random variables. Suppose that $U \sim N(\mu, \Sigma)$; then it is possible to show (using Theorem 2.3) that $AU + b \sim N(\mu + b, A\Sigma A^T)$.

Finally, we consider the conditional distributions of multivariate Gaussian random variables. It is noteworthy that the conditional distributions of Gaussian random variables are themselves Gaussian random variables. Suppose that

$$Z = \begin{pmatrix} Y \\ X \end{pmatrix}, \qquad (2.81)$$

$$\mu = \begin{pmatrix} \mu_y \\ \mu_x \end{pmatrix}, \text{ and} \qquad (2.82)$$

$$\Sigma = \begin{pmatrix} \Sigma_{xx} & \Sigma_{xy} \\ \Sigma_{yx} & \Sigma_{yy} \end{pmatrix}, \qquad (2.83)$$

then, given that $X = x$, Y is distributed normally with mean

$$\mu_y + \Sigma_{yx}\Sigma_{xx}^{-1}(x - \mu_x) \qquad (2.84)$$

and covariance matrix

$$\Sigma_{yy} - \Sigma_{yx}\Sigma_{xx}^{-1}\Sigma_{xy}. \qquad (2.85)$$

2.2   Convex Optimization

In this section, we discuss some basics of convex optimization. We shall restrict ourselves to problems defined in $\mathbf{R}^n$, outlining some ideas and results and providing some but not all necessary proofs. More detailed expositions can be found in the textbooks by Rockafellar (1970), and Boyd and Vandenberghe (2004). Most of the results we discuss in this section also hold for problems defined in more general linear vector spaces. Although we shall occasionally use these generalizations later in this book, we shall not discuss them in this section, but rather refer the reader to Luenberger (1969).

The practical importance and usefulness of convexity of a optimization problem is to a large extent derived from the fact that any local minimum of a convex function on a convex set is also a global minimum. This considerably simplifies the numerical procedures that can be employed to solve these optimization problems.

A very useful concept that we shall frequently utilize in this book is the concept of convex duality. Based on this idea, a convex problem can be related to its so-called dual problem, which has, under certain conditions, the same solution as the original problem. In a later chapter, we will use this duality to relate relative entropy minimization to the maximum-likelihood method.

2.2.1   Convex Sets and Convex Functions

Definition 2.1 *A set $S \subseteq \mathbf{R}^n$ is called convex if, for any $x_1 \in S$ and $x_2 \in S$ and any $\lambda \in [0,1]$, $\lambda x_1 + (1 - \lambda)x_2 \in S$.*

Figure 2.6 illustrates this definition.

Definition 2.2 *A function $f : \mathbf{R}^n \rightarrow \mathbf{R}$ is called convex if its domain, $dom(f)$, is a convex set and if, for all $x_x, x_2 \in dom(f)$ and any $\lambda \in (0,1)$,*

$$f(\lambda x_1 + (1 - \lambda)x_2) \leq \lambda f(x_1) + (1 - \lambda)f(x_2) \,. \qquad (2.86)$$

If strict inequality holds whenever $x_1 \neq x_2$, the function is called strictly convex. A function g is said to be (strictly) concave if $-g$ is (strictly) convex.

Figure 2.7 illustrates this definition.

It is easy to see that, if a function is twice continuously differentiable on an open convex set, it is convex (strictly convex) if and only if its Hessian is positive semi-definite (positive definite). See Rockafellar (1970), Theorem 4.5, for a proof.

Convex functions have the following property, which we will find useful later.

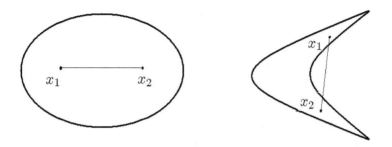

FIGURE 2.6: A convex (left) and a nonconvex (right) set.

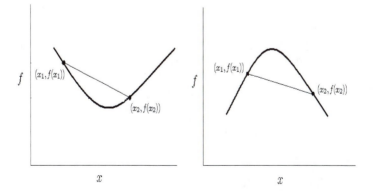

FIGURE 2.7: A convex (left) and a nonconvex (right) function.

Lemma 2.1 *(Jensen's inequality) If f is a convex function, X is a random variable with probability measure p and the expectations $E_p[X]$ and $E_p[f(X)]$ exist, then*

$$E_p[f(X)] \leq f\left(E_p[X]\right) .\tag{2.87}$$

Proof: See, for example, Feller (1966), Volume II, Chapter V.8.

Lemma 2.2 *(Supremum of a collection of convex functions) If $f_a(x)$ is convex for all $a \in A$, then*

$$g(x) = \sup_{a \in A} f_a(x)\tag{2.88}$$

is a convex function of x.

Proof: Exercise.

Lemma 2.3 *(Infimum of a collection of concave functions) If $f_a(x)$ is concave for all $a \in A$, then*

$$g(x) = \inf_{a \in A} f_a(x)\tag{2.89}$$

is a concave function of x.

Proof: Exercise.

2.2.2 Convex Conjugate Function

The following concept is very useful in the context of convex optimization problems.

Definition 2.3 *The convex conjugate function of a convex function, f, on $\mathbf{R}^n$ is defined by*

$$f^*(\lambda) = \sup_{x \in \mathbf{R}^n} \left\{\lambda^T x - f(x)\right\} ,\tag{2.90}$$

where $\lambda \in \mathbf{R}^n$.

One often calls the convex conjugate function simply the convex conjugate.
The following lemma lists some properties of the convex conjugate.

Lemma 2.4

(i) *The epigraph of the convex conjugate, f^*, is the set of all pairs (λ, μ) in $\mathbf{R}^{n+1}$ such that the affine function $h(x) = \lambda^T x - \mu$ is majorized by f.*

(ii) *$f^{**} = f$.*

(iii) *f^* is convex.*

Proof: See Rockafellar (1970), Section 12.
Later in this book, we shall make use of the following lemma.

Lemma 2.5 *Let x be an $\mathbf{R}^n$ vector, x_j denote the jth component of x, and*

$$\ell_p(x) = \left(\sum_{j=1}^{n} |x_j|^p \right)^{\frac{1}{p}} , \ 1 < p < \infty , \tag{2.91}$$

denote the ℓ_p norm of x. Then

(i) the convex conjugate of the function $\frac{1}{p}\ell_p^p$ is the function $\frac{1}{q}\ell_q^q$, where $q = \frac{p}{p-1}$, and

(ii) the convex conjugate of the function ℓ_∞, which is given by

$$\ell_\infty(x) = \max_{j=1\ldots n} |x_j| , \tag{2.92}$$

is the function ℓ_1.

Proof: The proof is straightforward; see Exercise 14 in Section 2.4.

2.2.3 Local and Global Minima

As we shall see later in this book, we often need to find a global minimum of an objective function, i.e., the minimum of the objective function over its whole domain. However, many numerical algorithms for finding a global minimum of a function focus on the search for local minima, i.e., minima of the function in small neighborhoods. This task is easier then directly finding a global minimum. For this reason, a numerical minimum search would be substantially simplified if we knew that any local minimum is also a global minimum. As the following theorem shows, this is the case for convex functions on convex sets.

Theorem 2.7 *Let f be a convex function on a convex set $S \subseteq \mathbf{R}^n$. Then, if x^* is a local minimum of f, then x^* is a global minimum of f.*

Proof: See Luenberger (1969), Proposition 1, Section 7.8.

Another issue of practical importance is the uniqueness of a minimum. The following theorem, which addresses this issue, will be useful later.

Theorem 2.8 *Let f be a strictly convex function on a convex set S. Then*

$$x^* = \arg \min_{x \in S} f(x) \tag{2.93}$$

is unique, if it exists.

In the above equations, 'arg' denotes 'argument of ...'.

Proof: We prove the theorem by contradiction. Suppose $f(x_1) = f(x_2) = \inf_{x \in S} f(x)$, where $x_1 \neq x_2 \in S$. Since S is convex, $\frac{1}{2}x_1 + \frac{1}{2}x_2 \in S$, and since f is strictly convex

$$f\left(\frac{1}{2}x_1 + \frac{1}{2}x_2\right) < \frac{1}{2}f(x_1) + \frac{1}{2}(x_2) = \inf_{x \in S} f(x), \qquad (2.94)$$

which is a contradiction. $\square$

We note that concave maximization problems can be cast as minimization problems, since

$$\max_{x \in S} f(x) = -\min_{x \in S} -f(x), \qquad (2.95)$$

with

$$\arg\max_{x \in S} f(x) = \arg-\min_{x \in S} -f(x). \qquad (2.96)$$

2.2.4 Convex Optimization Problem

A convex optimization problem is a problem of the following type.

Problem 2.1 *Find*

$$\Phi^* = \inf_{x \in \mathbf{R}^n} \Phi(x) \qquad (2.97)$$

s.t.

$$f_i(x) \leq 0, \ i = 1, ..., m \qquad (2.98)$$
$$\text{and } h_j(x) = 0, \ j = 1, ..., l, \qquad (2.99)$$

where $\Phi : \mathbf{R}^n \to \mathbf{R}$, $f_i : \mathbf{R}^n \to \mathbf{R}$ *are convex, and the* $h_j : \mathbf{R}^n \to \mathbf{R}$ *are affine.*

Below, in the context of duality, we often refer to Problem 2.1 as the *primal* problem.

We note that Problem 2.1 is not the most general class of problems that can be labeled as convex optimization problems, but is general enough for the purpose of this book.

2.2.5 Dual Problem

In this section, we will introduce the important concept of Lagrangian duality, which is also often called convex duality.

Lagrangian and Lagrange dual function

Definition 2.4 *The Lagrangian for Problem 2.1 is given by*

$$\mathcal{L}(x, \lambda, \nu) = \Phi(x) + \sum_{i=1}^{m} \lambda_i f_i(x) + \sum_{j=1}^{l} \nu_j h_j(x), \ \lambda_i \geq 0, \ \forall i = 1...m, \qquad (2.100)$$

and the λ_i, ν_i are called *Lagrange multipliers* and $\lambda = (\lambda_1, ..., \lambda_m)^T$ and $\nu = (\nu_1, ..., \nu_l)^T$. *The Lagrange dual function is given by*

$$g(\lambda, \nu) = \inf_{x \in \mathbf{R}^n} \mathcal{L}(x, \lambda, \nu). \tag{2.101}$$

In the above definition, we assume that, if the Lagrangian is unbounded below as a function of x, the Lagrange dual function takes the value $-\infty$.

The Lagrange dual function has the following properties.

Lemma 2.6 *With Φ^* defined in (2.97) and $g(\lambda, \nu)$ defined in (2.101), we must have*

(i) $g(\lambda, \nu) \leq \Phi^*$, and

(ii) g is concave.

Since the proof of this lemma is very simple and provides some insight into the working of Lagrange duality, we will provide it here.

Proof: We first prove statement (i). Let us call a point $\tilde{x}$ with $f_i(\tilde{x}) \leq 0$ and $h_i(\tilde{x}) = 0$ a feasible point. For any feasible point, we have

$$\mathcal{L}(\tilde{x}, \lambda, \nu) = \Phi(\tilde{x}) + \sum_{i=1}^{m} \lambda_i f_i(\tilde{x}) + \sum_{j=1}^{l} \nu_j h_j(\tilde{x})$$

$$\leq \Phi(\tilde{x}) \text{ (since } \lambda_i \geq 0 \text{ , } f_i(\tilde{x}) \leq 0 \text{ and } h_j(\tilde{x}) = 0 \text{ , } \forall i, j). \tag{2.102}$$

So we have

$$g(\lambda, \nu) = \inf_{x \in \mathbf{R}^n} \mathcal{L}(x, \lambda, \nu)$$

$$\leq \inf_{\tilde{x}: f_i(\tilde{x}) \leq 0 \text{ , } h_j(\tilde{x}) = 0 \text{ , } \forall i, j} \mathcal{L}(\tilde{x}, \lambda, \nu)$$

$$\leq \inf_{\tilde{x}: f_i(\tilde{x}) \leq 0 \text{ , } h_j(\tilde{x}) = 0 \text{ , } \forall i, j} \Phi(\tilde{x}) \text{ (by (2.102))}$$

$$\leq \Phi^* \text{ (by (2.97)) ,} \tag{2.103}$$

which proves statement (i).

Statement (ii) of the lemma follows directly from the fact that the dual function, g, is the pointwise infimum of a family of affine functions of (λ, ν) (see Rockafellar (1970), Theorem 5.5). $\square$

We note that Lemma 2.6 holds even for nonconvex problems.

Dual problem

We are now ready to define the so-called dual problem.

Problem 2.2 *(Dual Problem) Find*

$$g^* = \sup_{\lambda \geq 0, \nu} g(\lambda, \nu). \tag{2.104}$$

We note that, because of Lemma 2.6 (ii), Problem 2.2 amounts to the maximimization of a concave function on a convex set, which is equivalent to the minimization of a convex function on a convex set; therefore, Problem 2.2 is a convex problem.

The following theorem connects Problem 2.2 with Problem 2.1.

Theorem 2.9 *(Strong duality) If there exists some x with $h_i(x) = 0$ and $f_i(x) < 0$ (i.e., if the so-called Slater condition holds), then strong duality holds, i.e.,*

$$\Phi^* = g^* . \tag{2.105}$$

Moreover, if the minimum and the maximum in (2.106) and (2.109) exist, then the set of optimal points

$$x^* = \arg \min_{x:f_i(x)\leq 0 \,,\, h_i(x)=0 \,,\, \forall i,j} \Phi(x) \tag{2.106}$$

can be computed as

$$x^* = \hat{x}(\lambda^*, \nu^*) , \tag{2.107}$$

$$\text{where } \hat{x}(\lambda, \nu) = \arg \min_{x\in\mathbf{R}^n} \mathcal{L}(x, \lambda, \nu) , \tag{2.108}$$

$$\text{and } (\lambda^*, \nu^*) = \arg \max_{\lambda(\in\mathbf{R}^+)^m, \nu\in\mathbf{R}^l} \mathcal{L}(\hat{x}(\lambda, \nu), \lambda, \nu) . \tag{2.109}$$

In the above equations, the expressions λ^*, ν^*, x^* and $\hat{x}(\lambda, \nu)$ denote either a single point or a set, depending on whether the extrema they correspond to are uniquely attained or not.

Proof: See, for example, Boyd and Vandenberghe (2004), Chapter 5.

Theorem 2.9 has important practical implications. It states that, under the conditions of the theorem, one can solve Problem 2.2 instead of solving Problem 2.1, if one is interested in the solution of Problem 2.1. If the dimension of the dual problem, which is equal to the number of constraints in the primal problem, is smaller than the dimension, n, of the primal problem, solving the dual problem is often easier than solving the primal problem.

Later in this book, we will encounter situations where we can explicitly derive the dual problem, i.e., find an explicit expression for the Lagrange dual function, g, and where the dual problem has a much lower dimension than the primal problem. Often, an efficient strategy for solving the primal problem is to solve the dual problem instead.

Often we are interested not only in the infimum of the function Φ (under the constraints of Problem 2.1), but also in the minimizing x if such an x exists. Based on Theorem 2.9, we can also compute the latter by using (2.106) and (2.109).

2.2.6 Complementary Slackness and Karush-Kuhn-Tucker (KKT) Conditions

The Lagrange multipliers that correspond to the inequality constraint have the following useful property.

Lemma 2.7 *(Complementary slackness) For the optimal Lagrange multipliers, λ_i^*, we have either*

$$\lambda_i^* > 0 \text{ and } f_i(x^*) = 0 \tag{2.110}$$

or

$$\lambda_i^* = 0 \text{ and } f_i(x^*) < 0 . \tag{2.111}$$

Proof: See Boyd and Vandenberghe (2004), Section 5.5.2.

For a convex problem where the objective and the constraints are given in terms of differentiable functions, we can formulate the following conditions for the problem's solution.

Theorem 2.10 *(Karush-Kuhn-Tucker) Suppose the functions Φ, f_i, $i = 1, ..., m$ and g_j, $j = 1...l$ from Problem 2.1 are differentiable. Then the point x^* is primal optimal, i.e., $\Phi(x^*)$ solves Problem 2.1, and the point (λ^*, ν^*) is dual optimal, i.e., $g(\lambda^*, \nu^*)$ solves Problem 2.2, if and only if the following conditions hold.*

$$f_i(x^*) \leq 0 , \ i = 1, ..., m, \tag{2.112}$$
$$h_j(x^*) = 0 , \ j = 1, ..., l, \tag{2.113}$$
$$\lambda_i^* \geq 0, \ , \ i = 1, ..., m, \tag{2.114}$$
$$\lambda_i^* f_i(x^*) = 0, \ , \ i = 1, ..., m, \ and \tag{2.115}$$

$$0 = \nabla\Phi(x^*) + \sum_{i=1}^{m} \lambda_i^* \nabla f_i(x^*) + \sum_{j=1}^{l} \nu_j^* \nabla h_j(x^*) . \tag{2.116}$$

Proof: See Boyd and Vandenberghe (2004), Section 5.5.3.

2.2.7 Lagrange Parameters and Sensitivities

Let us consider the following perturbed version of Problem 2.1.

Problem 2.3 *Find*
$$\Phi^*(u, v) = \inf_{x \in \mathbf{R}^n} \Phi(x) \tag{2.117}$$

s.t.

$$f_i(x) \leq u_i , \ i = 1, ..., m \tag{2.118}$$
$$and \ h_j(x) = v_j , \ j = 1, ..., l , \tag{2.119}$$

where $u = (u_1, ..., u_m)^T$ and $v = (v_1, ..., v_l)^T$.

We have obtained this problem from Problem 2.1 by adding the perturbations u to the inequality constraints and the perturbations v to the equality constraints.

The following theorem relates the optimal Lagrange multipliers to the constraint perturbations.

Theorem 2.11 *Suppose that the assumptions of Theorem 2.9 hold and that Φ^* is differentiable at $u = (0, ..., 0)^T$, $v = (0, ..., 0)^T$. Then*

$$\lambda_i^* = -\left.\frac{\partial \Phi^*(u, v)}{\partial u_i}\right|_{u=(0,...,0)^T, \, v=(0,...,0)^T}, \qquad (2.120)$$

$$\text{and } \nu_j^* = -\left.\frac{\partial \Phi^*(u, v)}{\partial v_j}\right|_{u=(0,...,0)^T, \, v=(0,...,0)^T}, \qquad (2.121)$$

where the λ_i^* and the ν_j^* are the dual optimal Lagrange multipliers from Theorem 2.9.

Proof: See, for example, Boyd and Vandenberghe (2004), Section 5.6.

According to Theorem 2.11, the dual optimal Lagrange multipliers are a measure of how active a constraint is. If the absolute of a certain multiplier is small, we can modify the constraint without much effect on the solution of the optimization problem; if, on the other hand, the absolute of the multiplier is large, a small modification of the constraint leads to a big change in the solution.

2.2.8   Minimax Theorems

Suppose we have a function of two variables, which we want to minimize with respect to one of the variables and maximize with respect to the other. Can we exchange the order of the maximization and minimization? This question, which plays an important role in the theory of zero-sum games, is, as we shall see later in this book, related to the question of robustness of the minimum relative entropy method. The answer to this question is that, generally, we cannot exchange the order of the maximization and minimization. However, under certain conditions we can do so. Many such conditions are known (see, for example, von Neumann and Morgenstern (1944), Rockafellar (1970), or Frenk et al. (2002)), each of which lead to a so-called *minimax* theorem. Below, we state one of these theorems.

Before stating the theorem, we define a concave-convex function.

Definition 2.5 *Let C and D be nonempty closed convex sets in $\mathbf{R}^m$ and $\mathbf{R}^n$, respectively, and let Φ be a continuous finite function on $C \times D$. $\Phi(x, y)$ is concave-convex if for every $y \in D$, $\Phi(x, y)$ is concave in x and for every $x \in C$, $\Phi(x, y)$ is convex in y.*

Theorem 2.12 *Let C and D be nonempty closed convex sets in $\mathbf{R}^m$ and $\mathbf{R}^n$, respectively, and let Φ be a continuous finite concave-convex function on $C \times D$. If either C or D is bounded, then*

$$\inf_{y \in D} \sup_{x \in C} \Phi(x, y) = \sup_{x \in C} \inf_{y \in D} \Phi(x, y) . \tag{2.122}$$

Moreover, if both C and D are bounded, then Φ has a saddle point with respect to $C \times D$, i.e., there exists some $x^* \in C$ and $y^* \in D$ such that

$$\Phi(x, y^*) \le \Phi(x^*, y^*) \le \Phi(x^*, y) , \ \forall x \in C , \ \forall y \in D , \tag{2.123}$$

i.e.,

$$\min_{y \in D} \max_{x \in C} \Phi(x, y) = \max_{x \in C} \min_{y \in D} \Phi(x, y) = \Phi(x^*, y^*) . \tag{2.124}$$

Proof: See Rockafellar (1970), Corollaries 37.3.2 and 37.6.2.

We note that the saddle-point property from Theorem 2.12 is closely related to the duality from Theorem 2.9 (see Boyd and Vandenberghe (2004), Section 5.4).

2.2.9 Relaxation of Equality Constraints

Later in this book, when discussing the minimum relative entropy method, we shall analyze particular optimization problems under equality constraints and under relaxed equality constraints. Specifically, our starting point will be a problem of the following type.

Problem 2.4 *Find*

$$x^{(eq)} = \arg \min_{x \in \mathbf{R}^n} \Phi(x) \tag{2.125}$$

s.t.

$$f_i(x) \le 0 , \ i = 1, ..., m , \tag{2.126}$$
$$\text{and } h_j(x) = 0 , \ j = 1, ..., l , \tag{2.127}$$

where $\Phi : \mathbf{R}^n \to \mathbf{R}$ is a convex function, the functions $f_i : \mathbf{R}^n \to \mathbf{R}$ are convex, and the functions $h_j : \mathbf{R}^n \to \mathbf{R}$ are affine.

Often we are interested in the problem that is obtained from Problem 2.4 by relaxing some of the equality constraints. The following problem is a useful example for such a relaxed problem.

Problem 2.5 *Find*

$$x^{(r)} = \arg \min_{x \in \mathbf{R}^n} \Phi(x) \tag{2.128}$$

s.t.

$$f_i(x) \le 0 \; , \; i = 1, ..., m \; , \tag{2.129}$$
$$h_j(x) = c_j \; , \; j = 1, ..., \hat{l} \; , \tag{2.130}$$
$$h_j(x) = 0 \; , \; j = \hat{l} + 1, ..., l \; , \tag{2.131}$$
$$\text{and } \Psi(c) \le \alpha \; , \tag{2.132}$$

where Φ and the f_i and h_j are the same as in Problem 2.4, Ψ is a convex function of the vector $c = (c_1, ..., c_{\hat{l}})^T$ that attains its minimum, 0, at $c = (0, ..., 0)^T$, and α is a positive number.

The following theorem relates the solutions of Problems 2.4 and 2.5 to each other.

Theorem 2.13 *If the minimum and the maximum in (2.134) and (2.135) exist, then the set of optimal points of Problem 2.5 is given by*

$$x^{(r)} = \hat{x}^{(eq)}\left(\lambda^{(r)}, \nu^{(r)}\right) \; , \tag{2.133}$$

where $\hat{x}^{(eq)}(\lambda, \nu) = \arg \min_{x \in \mathbf{R}^n} \mathcal{L}^{(eq)}(x, \lambda, \nu) \; , \tag{2.134}$

$\mathcal{L}^{(eq)}(x, \lambda, \nu) = \Phi(x) + \sum_{i=1}^{m} \lambda_i f_i(x) + \sum_{j=1}^{l} \nu_j h_j(x)$ *is the Lagrangian of Problem 2.4, and*

$$\left(\lambda^{(r)}, \nu^{(r)}\right) = \arg \max_{\lambda \in (\mathbf{R}^+)^m, \; \nu \in \mathbf{R}^l} g^{(r)}(\lambda, \nu, \xi^*) \tag{2.135}$$

where $g^{(r)}(\lambda, \nu, \xi^*) = g^{(eq)}(\lambda, \nu) - \inf_{\xi \ge 0} \left\{ \xi \Psi^* \left(\frac{\nu}{\xi}\right) + \alpha\xi \right\} \; , \tag{2.136}$

$$g^{(eq)}(\lambda, \nu) = \inf_{x \in \mathbf{R}^n} \mathcal{L}^{(eq)}(x, \nu) \; , \tag{2.137}$$

$\hat{\nu}$ is the vector of the first $\hat{l}$ components of the vector ν, and Ψ^* is the convex conjugate of Ψ.

Proof: See Section 2.2.10.

This theorem states that the solution to Problem 2.5 is given by the same function of the dual optimal Lagrange parameters as the solution to Problem 2.4 is, and that the dual optimal Lagrange parameters maximize the sum of the objective function of the dual of Problem 2.4 and an additional term.

The following corollary states some explicit results for two special cases involving the ℓ_p norm given by (2.91), and the ℓ_∞-norm given by (2.92). We make use of these special cases in some important practical applications, which we shall consider later in this book.

Corollary 2.1

(i) *If the function Ψ in the relaxed constraint (2.132) in Problem 2.5 is given by*

$$\Psi(c) = \frac{1}{p}\ell_p^p(c) \,, \; 1 < p < \infty \,, \tag{2.138}$$

then the objective function of the dual problem is

$$g^{(r)}(\lambda, \nu, \xi^*) = g^{(eq)}(\lambda, \nu) - \alpha^{\frac{1}{p}} q^{\frac{1}{p}} (q-1)^{-\frac{1}{p}} \ell_q(\nu) \,, \tag{2.139}$$

where $q = \frac{p}{p-1}$.

(ii) *If the function Ψ in the relaxed constraint (2.132) in Problem 2.5 is given by*

$$\Psi(c) = \ell_\infty(c), \tag{2.140}$$

then the objective function of the dual problem is

$$g^{(r)}(\lambda, \nu, \xi^*) = g^{(eq)}(\lambda, \nu) - \alpha \ell_1(\nu) \,. \tag{2.141}$$

Proof: See Section 2.2.10.

A relaxation function that is often used is $\Psi(c) = \frac{1}{p}\ell_p^p(c)$ with $p = 2$. The objective function of the dual of Problem 2.5 is then

$$g^{(r)}(\lambda, \nu, \xi^*) = g^{(eq)}(\lambda, \nu) - \sqrt{2\alpha}\ell_2(\nu) \tag{2.142}$$

(see Corollary 2.13). This objective function is suitable for practical applications. From a theoretical perspective, however, the following corollary is useful.

Corollary 2.2 *The α-parameterized (with $\alpha \geq 0$) family of maxima of*

$$g^{(r)}(\lambda, \nu, \xi^*) = g^{(eq)}(\lambda, \nu) - \alpha^{\frac{1}{p}} q^{\frac{1}{p}} (q-1)^{-\frac{1}{p}} \ell_q(\nu) \tag{2.143}$$

is the same as the α-parameterized (with $\alpha \geq 0$) family of maxima of

$$g^{(\ell)}(\lambda, \nu) = g^{(eq)}(\lambda, \nu) - \alpha \ell_q^q(\nu) \,. \tag{2.144}$$

Proof: The corollary is a direct consequence of Theorem 2.13 and Lemma 7 in Friedman and Sandow (2003a). The proof is straightforward; it is based on the fact that a monotone transformation doesn't modify the set of Pareto-optimal values of the vector $-(g^{(eq)}(\lambda, \nu), \ell_q^q(\nu))$.

We shall apply this corollary below to demonstrate the equivalence of the ℓ_2-relaxed minimum relative entropy problem to a ℓ_2-regularized maximum-likelihood method.

We note that an alternative approach to relaxing equality constraint is discussed in Exercise 17 in Section 2.4.

2.2.10   Proofs for Section 2.2.9

Proof of Theorem 2.13: The Lagrangian of Problem 2.5 is

$$\mathcal{L}^{(r)}(x, c, \lambda, \nu, \xi) = \Phi(x) + \sum_{i=1}^{m} \lambda_i f_i(x) + \sum_{j=1}^{\hat{\imath}} \nu_j \left[h_j(x) - c_j \right] + \sum_{j=\hat{\imath}+1}^{l} \nu_j h_j(x)$$

$$+ \xi \left[\Psi(c) - \alpha \right]$$

$$= \mathcal{L}^{(eq)}(x, \lambda, \nu) - \hat{\nu}^T c + \xi \left[\Psi(c) - \alpha \right] , \qquad (2.145)$$

where $\mathcal{L}^{(eq)}$ is the Lagrangian of Problem 2.4, $\hat{\nu}$ is the vector of the first $\hat{\imath}$ components of the vector ν, and $\xi \geq 0$. This equation implies (2.133), by virtue of the fact that $\mathcal{L}^{(r)}$ depends on x only through the additive term $\mathcal{L}^{(eq)}$.

It follows further from (2.145) that the Lagrange dual function is

$$g^{(r)}(\lambda, \nu, \xi) = \inf_{x \in \mathbf{R}^n, c \in \mathbf{R}^l} \mathcal{L}^{(r)}(x, c, \lambda, \nu, \xi)$$

$$= g^{(eq)}(\lambda, \nu) + \inf_{c \in \mathbf{R}^l} \left[-\hat{\nu}^T c + \xi \Psi(c) \right] - \alpha \xi , \qquad (2.146)$$

where $g^{(eq)}(\lambda, \nu) = \inf_{x \in \mathbf{R}^n} \mathcal{L}^{(eq)}(x, \lambda, \nu)$ is the Lagrange dual function of Problem 2.4.

Next, we separately consider three cases.

(i) $\xi = 0$ and $\hat{\nu}_j \neq 0$ for some $j \in (1, ..., \hat{\imath})$. In this case,

$$\inf_{c \in \mathbf{R}^l} \left[-\hat{\nu}^T c + \xi \Psi(c) \right] = -\infty ,$$

and it follows from (2.146) that $g^{(r)}(\lambda, \nu, 0) = -\infty$.

(ii) $\xi = 0$ and $\hat{\nu}_j = 0$, $\forall j \leq \hat{\imath}$. In this case,

$$\inf_{c \in \mathbf{R}^l} \left[-\hat{\nu}^T c + \xi \Psi(c) \right] = 0 ,$$

and it follows from (2.146) that

$$g^{(r)}(\lambda, (0, ..., 0, \nu_{\hat{\imath}+1}, ..., \nu_l)^T, 0) = g^{(eq)}(\lambda, (0, ..., 0, \nu_{\hat{\imath}+1}, ..., \nu_l)^T) .$$

(iii) $\xi > 0$. Using Definition 2.3 of the convex conjugate, we obtain

$$g^{(r)}(\lambda, \nu, \xi) = g^{(eq)}(\lambda, \nu) + \xi \inf_{c \in \mathbf{R}^l} \left[-\frac{\hat{\nu}^T c}{\xi} + \Psi(c) \right] - \alpha \xi$$

$$= g^{(eq)}(\lambda, \nu) - \xi \Psi^* \left(\frac{\hat{\nu}}{\xi} \right) - \alpha \xi . \qquad (2.147)$$

Making the notational assumption that $\xi \Psi^* \left(\frac{\hat{\nu}}{\xi} \right) = 0$ if $\xi = 0$ and $\hat{\nu} = (0, ..., 0)^T$, we can collect the above three cases as follows.

$$g^{(r)}(\lambda, \nu, \xi) = \begin{cases} -\infty \text{ if } \xi = 0 \text{ and } \nu_j \neq 0 \text{ for some } j \leq \hat{l} \\ g^{(eq)}(\lambda, \nu) - \xi \Psi^* \left(\frac{\hat{\nu}}{\xi} \right) - \alpha \xi \text{ otherwise.} \end{cases} \quad (2.148)$$

(2.135)-(2.137) in Theorem 2.13 follow then from Lagrangian duality, i.e., from Theorem 2.9. This completes the proof of the theorem. □

Proof of Corollary 2.1:

(i) It follows from $\Psi = \frac{1}{p} \ell_p^p$ and Lemma 2.5 that $\Psi^* = \frac{1}{q} \ell_q^q$ where $q = \frac{p}{p-1}$. From there, it follows that

$$\inf_{\xi \geq 0} \left\{ \xi \Psi^* \left(\frac{\nu}{\xi} \right) + \alpha \xi \right\} = \alpha^{\frac{1}{p}} q^{\frac{1}{p}} (q-1)^{-\frac{1}{p}} \ell_q(\nu) . \quad (2.149)$$

This equation, in conjunction with (2.136), implies (2.139).

(ii) It follows from $\Psi = \ell_\infty$ and Lemma 2.5 that $\Psi^* = \ell_1$. Consequently,

$$\inf_{\xi \geq 0} \left\{ \xi \Psi^* \left(\frac{\nu}{\xi} \right) + \alpha \xi \right\} = \alpha \ell_1(\nu) . \quad (2.150)$$

The above equation, in conjunction with (2.136), implies (2.141). □

2.3 Entropy and Relative Entropy

In this section, we introduce two basic concepts from information theory: entropy and relative entropy. We will discuss these quantities for unconditional and conditional probabilities, on discrete and continuous state spaces. For a more detailed review we refer the reader to the textbooks by Cover and Thomas (1991), or MacKay (2003).

Entropy is a measure for the information content in a probability distribution or, alternatively, of the average length of the shortest description of a random variable. It can also be interpreted as the difference in the expected wealth growth rates between an optimal and a clairvoyant gambler in a horse race.

Relative entropy can be viewed as a measure for the discrepancy between two probability distributions. In the context of a horse race, the relative entropy between the probability measures p and q is the gain in the expected wealth growth rate experienced by an investor who bets optimally according to p as opposed to the misspecified measure q, when the horses win with the probabilities given by the measure p.

2.3.1 Entropy for Unconditional Probabilities on Discrete State Spaces

Let us consider probability measures of the random variable Y with the discrete state space $\mathcal{Y}$. We make the following definition.

Definition 2.6 *The entropy of the (unconditional) probability measure p is*

$$H(p) = -E_p[\log p] = - \sum_{y \in \mathcal{Y}} p_y \log p_y \ . \tag{2.151}$$

Here, we assign the value 0 to the expression $0 \log 0$, which is consistent with $\lim_{p \to 0} p \log p = 0$.

Some mathematical properties of the entropy

The following lemma lists some properties of the entropy.

Lemma 2.8

(i) H is a strictly concave function of p.

(ii) $H(p) \leq \log |\mathcal{Y}|$, with equality if and only if p is uniform.

(iii) $H(p) \geq 0$, with equality if and only if $p = \delta_{y,y'}$ for some y'.

Proof: Statement (i) follows from the fact that the Hessian of H is diagonal with positive entries. (ii) Exercise. Statement (iii) follows trivially from the facts that $x \log x < 0$, $\forall x \in (0,1)$ and that $H(p) = 0$ if $p = \delta_{y,y'}$ for some y'. $\square$

The concavity of the entropy turns out to be very useful; as a consequence of this property, the maximum entropy principle (see Section 9.4) leads to a convex problem.

Entropy and information

Entropy can be viewed as a measure of the information content of a probability distribution. The intuition behind this interpretation is illustrated in Fig. 2.8; probability measures that are associated with high uncertainty have a large entropy, while probability measures that are associated with little uncertainty have a low entropy. The interpretation of entropy as information content in a probability measure has been formalized by Shannon (1948), who showed that entropy is the only quantity (up to a constant) that has the following properties that one would expect an information measure to have.

(i) H is a continuous function of the probability measure, p.

(ii) If the probability measure p is uniform, $H(p)$ should be a monotone increasing function of the number of states.

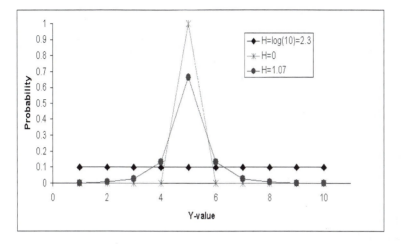

FIGURE 2.8: Three probability distributions and their entropies. The higher the uncertainty, i.e., the lower the information content, associated with a distribution, the higher is its entropy.

(*iii*) If a choice is broken down into two successive choices, the original entropy is the weighted sum of the two entropies corresponding to the two successive choice. Setting, without restriction of generality, $\mathcal{Y} = \{1, 2, ..., m\}$ (for an infinite state space, we set $m = \infty$), this means that

$$H(p_1, p_2, ..., p_m) = H(p_1 + p_2, p_3, \ldots, p_m)$$
$$+(p_1 + p_2)H\left(\frac{p_1}{p_1 + p_2}, \frac{p_2}{p_1 + p_2}\right).$$

The interpretation of entropy as the information content in a probability measure provides a motivation for the maximum entropy principle (see Section 9.4).

Entropy and minimum description length

Let us assume that our random variable Y has the probability distribution p and ask the following question: what is the expected length of the shortest description of Y? It turns out that this expected length is greater or equal to $H(p)\log_d e$, where d is the number of letters in the description alphabet (see Cover and Thomas (1991), Theorem 5.3.1).

Entropy and physics

Entropy was first introduced as a thermodynamic quantity and later related to probability distributions by Boltzmann (1877). Boltzmann identified this thermodynamic entropy with a statistical entropy, which he defined as the logarithm of the number of microstates (states of individual atoms) of a physical system that are consistent with a given macrostate (as defined by the thermodynamic properties, such as total energy) of the same physical system. Boltzmann's statistical entropy is a specialization of Definition 2.6 to a uniform probability distribution. The link between statistical and thermodynamic entropy has been generalized to nonuniform distributions and interpreted in terms of information theory by Jaynes (1957a).

One of the most fundamental laws of physics is the second law of thermodynamics, which states that, for an isolated physical system, the thermodynamic entropy always increases. This law is reflected in the following mathematical property of the entropy, H.

Lemma 2.9 *(Second law of thermodynamics) If $p^{(n)}$ denotes the probability measure for the time-n states of a Markov chain with constant transition probabilities, then*

$$H\left(p^{(n+1)}\right) \geq H\left(p^{(n)}\right) . \tag{2.152}$$

Proof: See Cover and Thomas (1991), Chapter 2.9.

Entropy and horse race

Let us identify the states of the random variable Y with horses in a horse race, and assume that horse y wins the race with probability p_y. We consider a gambler who bets on this horse race, and allocates his money so as to maximize his expected wealth growth rate (such an investor is often called a Kelly investor, after Kelly (1956)). Entropy then has the following interpretation,

$$H(p) = W_p^*(p) - W_p^{**} , \tag{2.153}$$

where $W_p^*(p)$ is the optimal wealth growth rate for an investor who knows the probability measure p, and W_p^{**} is the wealth growth rate of a clairvoyant investor, i.e., of an investor who wins every bet, if horse y wins with probability p_y. We will derive this statement below in Section 3.4, Theorem 3.2 (see also Cover and Thomas (1991), Theorem 6.1.2)).

The above interpretation of the entropy provides a financial (or a simple decision-theoretic) motivation for the maximum entropy principle (see Section 9.4).

The gambler who wants to maximize his wealth growth rate in a horse race is a simple example for a decision maker in an uncertain environment. We shall revisit this gambler later in this book, and we will provide a formal definition of the horse race in Chapter 3.

2.3.2 Relative Entropy for Unconditional Probabilities on Discrete State Spaces

We make the following definition.

Definition 2.7 *The relative entropy between the (unconditional) probability measures p and q is*

$$D(p\|q) = E_p\left[\log\left(\frac{p}{q}\right)\right] = \sum_{y\in\mathcal{Y}} p_y \log\left(\frac{p_y}{q_y}\right) . \qquad (2.154)$$

As before, we assign the value 0 to the expression $0\log 0$.

This relative entropy is also called Kullback-Leibler relative entropy or Kullback-Leibler discrepancy. It can be viewed as a measure of the discrepancy between two probability distributions. The intuitive meaning of the relative entropy as a discrepancy between two probability measures is illustrated in Fig. 2.9. We note, however, that the relative entropy doesn't exhibit all the properties of a mathematical distance; in particular, it is not symmetric in its arguments, and it does not obey the triangular inequality.

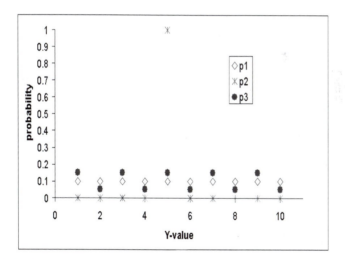

FIGURE 2.9: Three probability distributions and their relative entropies: $D(p1\|p3) = 0.14$, $D(p2\|p3) = 1.90$, $D(p2\|p1) = 2.3$.

Some mathematical properties of the relative entropy

The following lemma lists some useful properties of the relative entropy.

Lemma 2.10

   (i) $D(p\|q) = -H(p) + \log|\mathcal{Y}|$ *if q is the uniform distribution,*

   (ii) $D(p\|q)$ *is convex in* (p, q),

   (iii) $D(p\|q)$ *is strictly convex in p, and*

   (iv) $D(p\|q) \geq 0$, *with equality if and only if* $p = q$.

Proof: Statement *(i)* follows directly from Definitions 2.6 and 2.7. For statement *(ii)*, see Cover and Thomas (1991), Theorem 2.7.2. Statement *(iii)* can be easily proved by computing the Hessian, which is diagonal with strictly positive entries. Statement *(iv)* follows from Jensen's inequality and $D(p\|q) = 0$ if $p = q$ (see, Cover and Thomas (1991), Theorem 2.6.3). $\square$

    The convexity of the entropy in its first argument turns out to be very useful; as a consequence of this property, the minimum relative entropy principle (see Chapter 9.4) leads to a convex problem.

Second law of thermodynamics

Above, in Lemma 2.9, we have formulated the second law of thermodynamics in terms of the entropy, H. We can generalize this mathematical statement to relative entropies as follows.

Lemma 2.11 *If* $p^{(n)}$ *and* $q^{(n)}$ *denote the probability measures for the time-n states of two Markov-chain with constant transition probabilities, then*

$$D\left(p^{(n+1)}\|q^{(n+1)}\right) \geq D\left(p^{(n)}\|q^{(n)}\right). \tag{2.155}$$

Proof: See Cover and Thomas (1991), Chapter 2.9.

Relative entropy and horse race

Let us again identify the states of the random variable Y with horses in a horse race, and assume that horse y wins the race with probability p_y, and consider a gambler who bets on the horses. The relative entropy then has the following interpretation,

$$D(p\|q) = W_p^*(p) - W_p^*(q), \tag{2.156}$$

where $W_p^*(p)$ is the optimal wealth growth rate for a gambler who knows the probability measure p, and $W_p^*(q)$ is the optimal wealth growth rate for a gambler who believes the (misspecified) probability measure q. We have set the lower indices, p, in the above notation to indicate that horse y wins the race with probability p_y. We will prove (2.156) below in Section 3.4 (see Theorem 3.3).

The above interpretation of the relative entropy provides a financial (or a simple decision-theoretic) motivation for the minimum relative entropy principle (see Section 9.4). We will revisit this point later in this book.

2.3.3 Conditional Entropy and Relative Entropy

In the previous sections, we have introduced entropy and relative entropy for probability measures for a single random variable Y. In this section, we generalize these concepts to conditional probability measures of the random variables Y (with state space $\mathcal{Y}$) and X (with state space $\mathcal{X}$). We denote joint probability measures by $p^{X,Y}$ or $q^{X,Y}$, conditional probability measures by $p^{Y|X}$ or $q^{Y|X}$, and the marginal X-probability measures by p^X or q^X. Later in this book, when the meaning is clear from the context, we may drop the upper indices. As before, we assign the value 0 to the expression $0 \log 0$.

Definition 2.8 *The conditional entropy of the probability measure $p^{X,Y}$ is*

$$H^{Y|X}\left(p^{X,Y}\right) = -E_{p^{X,Y}}\left[\log p^{Y|X}\right] = -\sum_{x \in \mathcal{X}} p_x^X \sum_{y \in \mathcal{Y}} p_{y|x}^{Y|X} \log p_{y|x}^{Y|X} . \quad (2.157)$$

Below, we shall often call the above conditional entropy simply entropy and drop the superscript.

Definition 2.9 *The conditional relative entropy between the probability measures $p^{X,Y}$ and $q^{Y|X}$ is*

$$D^{Y|X}\left(p^{X,Y} \| q^{Y|X}\right) = E_{p^{X,Y}}\left[\log\left(\frac{p^{Y|X}}{q^{Y|X}}\right)\right]$$

$$= \sum_{x \in \mathcal{X}} p_x^X \sum_{y \in \mathcal{Y}} p_{y|x}^{Y|X} \log\left(\frac{p_{y|x}^{Y|X}}{q_{y|x}^{Y|X}}\right) .$$

Below, when the meaning is clear from the context, we shall call the above conditional relative entropy simply relative entropy and drop the superscript.

One can also define a joint entropy and a joint relative entropy; we will briefly discuss this in Exercise 20 in Section 2.4.

Some mathematical properties of the conditional entropy and the conditional relative entropy

The following lemma lists some properties.

Lemma 2.12

(i) $H^{Y|X}\left(p^{X,Y}\right) = E_{p^X}\left[H\left(p^{Y|X}|X\right)\right]$, where $H\left(p^{Y|X}|X\right)$ denotes the entropy of the probability measure $p^{Y|X}$ for a given value of X.

(ii) $H^{Y|X}$ is a strictly concave function of $p^{Y|X}$.

(iii) $H^{Y|X}\left(p^{X,Y}\right) \leq \log|\mathcal{Y}|$, with equality if and only if $p^{Y|X}$ is uniform for all $x \in \mathcal{X}$ with $p_x^X > 0$.

(iv) $H^{Y|X}\left(p^{X,Y}\right) \geq 0$, with equality if and only if, for all $x \in \mathcal{X}$ with $p_x^X > 0$, $p^{Y|X} = \delta_{y,y_x'}$ for some y'.

(v) $D^{Y|X}\left(p^{X,Y}\|q^{Y|X}\right) = E_{p^X}\left[D\left(p^{Y|X}\|q^{Y|X}|X\right)\right]$, where $D\left(p^{Y|X}\|q^{Y|X}|X\right)$ denotes the relative entropy between the probability measures $p^{Y|X}$ and $q^{Y|X}$ for a given value of X.

(vi) $D^{Y|X}\left(p^{X,Y}\|q^{Y|X}\right) = -H^{Y|X}(p^{X,Y}) + \log|\mathcal{Y}|$ if $p^{Y|X}$ is uniform for all $x \in \mathcal{X}$ with $p_x^X > 0$.

(vii) $D^{Y|X}(p^{Y|X}\|q^{Y|X})$ is convex in $(p^{Y|X}, q^{Y|X})$.

(viii) $D^{Y|X}(p^{Y|X}\|q^{Y|X})$ is strictly convex in its first argument, $p^{Y|X}$.

(ix) $D^{Y|X}\left(p^{X,Y}\|q^{Y|X}\right) \geq 0$, with equality if and only if $p^{Y|X} = q^{Y|X}$ for all $x \in \mathcal{X}$ with $p_x^X > 0$.

Proof: Statement *(i)* follows directly from Definitions 2.8 and 2.6, and statement *(v)* follows directly from Definitions 2.9 and 2.7. Statements *(ii)* and *(vii)* follow from statements *(i)* and *(v)*, the (strict) convexity of $-H$ and D (see Lemmas 2.8 and 2.10) and the fact that a nonnegatively weighted sum of (strictly) convex functions is convex itself. Statement *(viii)* follows from statement *(v)*, the strict convexity of D in its first argument (see Lemma 2.10), and the fact that a nonnegatively weighted sum of strictly convex functions is strictly convex itself. Statements *(iii)* and *(iv)* follow from statement *(i)* and Lemma 2.8. Statements *(vi)* and *(ix)* follow from statement *(v)* and Lemma 2.10. □

2.3.4 Mutual Information and Channel Capacity Theorem

We introduce the useful concept of mutual information.

Definition 2.10 *The mutual information is given by*

$$I\left(p^{X,Y}\right) = H^Y\left(p^Y\right) - H^{Y|X}\left(p^{X,Y}\right) . \tag{2.158}$$

Here we have denoted the (unconditional) entropy by H^Y as opposed to just H, in order to make the distinction between this entropy and the conditional entropy clearer. It follows from Definitions 2.10, 2.6, and 2.8 that

$$I\left(p^{X,Y}\right) = \sum_{x \in \mathcal{X}} \sum_{y \in \mathcal{Y}} p_{x,y}^{X,Y} \log \left(\frac{p_{x,y}^{X,Y}}{p_x^X \, p_y^Y} \right) . \qquad (2.159)$$

The concept of mutual information, and therefore the concept of entropy, plays an important role in the context of transmission of data through a noisy channel; this role derives from Shannon's channel capacity theorem. In order to discuss this theorem, let us consider a data transmission channel that takes the random variable X with probability distribution p^X as input and provides the random variable Y with conditional probability distribution $p^{Y|X}$ as output. We define the channel capacity, C, as

$$C = \log 2 \, \max_{p^X} I\left(p^{X,Y}\right) , \qquad (2.160)$$

where $I\left(p^{X,Y}\right)$ is the mutual information from Definition 2.10. The channel capacity theorem, which is a central result of information theory, states that the above channel capacity, C, is the highest rate in bits per channel use at which information can be sent with arbitrarily low error probability (see Shannon (1948), or Cover and Thomas (1991), Chapter 8).

2.3.5 Entropy and Relative Entropy for Probability Densities

So far, we have discussed entropy and relative entropy for random variables with a discrete state space. However, one can generalize these concepts to continuous (one-dimensional) random variables. In this section, we will explicitly discuss these generalizations for the case of unconditional probability measures; the extension to conditional measures and to higher dimensional random variables is straightforward.

Definition 2.11 *The entropy of the probability density p for the continuous random variable Y with support $\mathcal{Y}$ is*

$$H\left(p\right) = -E_p[\log p] = - \int_{\mathcal{Y}} p(y) \log p(y) dy . \qquad (2.161)$$

This entropy is sometimes called the differential entropy.

Definition 2.12 *The relative entropy between the probability densities p and q for the continuous random variable Y with support $\mathcal{Y}$ is*

$$D\left(p\|q\right) = E_p\left[\log\left(\frac{p}{q}\right)\right] = \int_{\mathcal{Y}} p(y) \log\left(\frac{p(y)}{q(y)}\right) dy . \qquad (2.162)$$

For probability densities that are Riemann integrable, we can identify the above entropy and relative entropy with the continuum limits of the corresponding quantities for a discrete random variable.

Lemma 2.13 *Let us assume that the probability densities p and q of the random variable X are Riemann integrable, and let*

$$p_k^\Delta = p(y_k)\Delta \ , \quad and \qquad (2.163)$$
$$q_k^\Delta = q(y_k)\Delta \ , \qquad (2.164)$$

where the y_k are the mid-points of size-Δ bins that partition $\mathcal{Y}$. Then

$$H(p) = \lim_{\Delta \to 0} \left[H\left(p^\Delta\right) - \log \Delta \right] \ , \quad and \qquad (2.165)$$
$$D(p\|q) = \lim_{\Delta \to 0} D\left(p^\Delta \| q^\Delta\right) \ . \qquad (2.166)$$

Proof: In order to prove (2.165), we use Definition 2.6 to write

$$H\left(p^\Delta\right) = -\sum_k p_k^\Delta \log p_k^\Delta$$
$$= -\Delta \sum_k p(y_k) \log p(y_k) - \log \Delta \ \text{(from (2.163))}.$$

(2.165) follows then directly from Definition 2.11 and the definition of the Riemann integral. Equation (2.166) follows in the same manner from Definitions 2.7 and 2.12 and (2.164); the term $\log \Delta$ drops out, since we have the logarithm of the ratio of two probabilities here. □

Some mathematical properties of the entropy and relative entropy for continuous random variables

The following lemma lists some properties.

Lemma 2.14 *The entropy, $H(p)$, of the probability distribution p of a continuous random variable*

(i) is a concave function of p, and

(ii) is nonnegative.

The relative entropy, $D(p\|q)$, of the probability distributions p and q of a continuous random variable has the following properties.

(iii) $D(p,q)$ is convex in (p,q).

(iv) If the probability densities p and q are Riemann integrable, then $D(p,q)$ is strictly convex in p.

(v) $D(p\|q) \geq 0$, with equality if and only if $p = q$ almost everywhere.

Proof: For statements *(i)* and *(iii)*, the proof of Theorem 2.7.2 in Cover and Thomas (1991) applies. Statement *(iv)* follows from the strict convexity of the discrete relative entropy in its first argument (see Lemma 2.10) and Lemma 2.13. For statement *(v)*, see Cover and Thomas (1991), Theorem 9.6.1. For statement *(ii)* the proof of Theorem 9.6.1 in Cover and Thomas (1991) can be easily modified.

2.4 Exercises

1. Suppose that $Z \sim N(0,1)$. Show that the density function for $X = Z^2$ is given by

$$\frac{1}{\sqrt{2\pi x}} e^{-\frac{x}{2}} \qquad (2.167)$$

(Chi-squared with 1 degree of freedom). Verify your result with a numerical simulation.

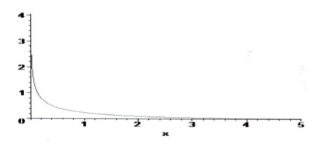

FIGURE 2.10: χ^2 distribution, plotted on the interval $[.01, 5]$.

2. (a) If X is distributed uniformly on (a, b), show that $E[X] = \frac{b+a}{2}$ and $var(X) = \frac{(b-a)^2}{12}$.

   (b) Show directly that $Z \sim N(0,1)$ indeed has mean 0 and variance 1.

   (c) Show that if $E[X] = \mu$, and $X_i, i = 1, \ldots, N$ denote repeated realizations of X, then $E[\overline{X}] = \mu$, where the sample average

$$\overline{X} = \frac{\sum_{i=1}^{N} X_i}{N}. \qquad (2.168)$$

   (d) Compute $var(\overline{X})$.

3. The Dirac delta function can be understood as the limit of a sequence of smooth functions

$$\delta(x) = \lim_{\epsilon \to 0} \delta_\epsilon(x), \tag{2.169}$$

where $\delta_\epsilon(x)$ has the property that

$$\lim_{\epsilon \to 0} \int_{-\infty}^{\infty} \delta_\epsilon(x) f(x) dx = f(0) \tag{2.170}$$

for all continuous functions, f. Functions $\delta_\epsilon(x)$ with this property are referred to as nascent delta functions. Show that the pdf of a random variable that is $N(0, \epsilon)$ is a nascent delta function.

4. Prove (2.40) and (2.41), i.e., prove that if the random vector $X = (X_1, \ldots, X_n)^T$ has expectation vector μ (with i^{th} element μ_i), and covariance matrix $cov(X, X) = \Sigma$ (with ij^{th} element $cov(X_i, X_j)$), and if A is a matrix with n columns, then

$$E[AX] = A\mu, \tag{2.171}$$

and

$$cov(AX, AX) = A\Sigma A^T. \tag{2.172}$$

5. Show that for the nonnegative discrete-valued random variable X, that takes values $0, 1, 2, \ldots$,

$$E[X] = \sum_{n \geq 0} prob\{X > n\} = \sum_{n \geq 1} prob\{X \geq n\}. \tag{2.173}$$

Hint: apply the definition of expectation, using the identity

$$n \, prob\{X = n\} = \sum_{k=1}^{n} prob\{X = k\}. \tag{2.174}$$

Verify your result for the distribution

$$prob\{X = j\} = (e - 1)e^{-j-1}, \text{ for } j \geq 0 \tag{2.175}$$

with a numerical simulation.

6. (a) Use the Markov inequality to prove the Chebyshev inequality
   (b) Using the Schwarz inequality, show that $-1 \leq \rho[X, Y] \leq 1$

7. Suppose that K is running against B in an election and p is the percentage of eligible voters who will vote for B. Using the Chebyshev inequality, estimate the number of people who should be polled to insure that the probability is .95 that the sample average differs from p by no more than .01. Support this estimate with numerical simulations.

8. If X and Y are independent random variables, then $cov[X, Y] = \rho[X, Y] = 0$.

9. The coefficients, a, b, c, of $ax^2 + bx + c$ are independent random variables and each is distributed uniformly on the interval $(0, 1)$. Give a closed-form formula for the probability that the solutions of the equation $ax^2 + bx + c = 0$ are real. Verify your result with a numerical simulation.

10. X and Y have a constant joint density, $p(x, y)$, on the region $x \geq 0, y \geq 0, x + y \leq 1$. Find $p(x, y), p(y), p(x|y), E[X|y]$, and $E[X]$. Verify your results with numerical simulations.

11. A boss leaves work at time X, which is distributed uniformly on $(0, T)$. Someone who works for the boss leaves at time Y, which is distributed uniformly on (X, T).

    Calculate $E[Y|X], E[Y], var(E[Y|X]), E[var(Y|X)]$, and $var(Y)$. Verify your results with numerical simulations.

12. Suppose that X is a standard normal random variable. Then $U = h(X) = \mu + \sigma X$ has mean μ and variance σ^2. Use Theorem 2.3 to show that (2.79) holds, i.e., that the density for U is given by

$$\frac{1}{\sigma\sqrt{2\pi}} e^{-\frac{(u-\mu)^2}{2\sigma^2}}. \tag{2.176}$$

13. We visit a random number of stores, N, and spend X_i in store $i \in \{1, \ldots, N\}$, where X_i are i.i.d. (independent and identically distributed) and independent of N, with $E[X_i] = \mu$ and $var(X_i) = \sigma^2$. Let $Y = X_1 + \cdots + X_N$. Show that

    $E[Y] = E[N]\mu,$
    $var[Y|N = n] = n\sigma^2$, and
    $var[Y] = E[N]\sigma^2 + \mu^2 var[N].$

14. Derive the convex conjugate of the function $\frac{1}{p}\ell_p^p(x)$, where

$$\ell_p(x) = \left(\sum_{j=1}^{n} |x_j|^p\right)^{\frac{1}{p}}$$

    is the ℓ_p-norm, and of the function ℓ_∞, where

$$\ell_\infty(x) = \max_{j=1\ldots n} |x_j|. \tag{2.177}$$

15. Prove Lemmas 2.2 and 2.3.

16. Let

$$x_1(\alpha) = \arg\min_{\Psi(x)\leq\alpha} \Phi(x),$$

$$\text{and } x_2(\gamma) = \arg\min_{x\in\mathbf{R}^n} \{\Phi(x)+\gamma\Psi(x)\}$$

where $\Phi : \mathbf{R}^n \to \mathbf{R}$, $\Psi : \mathbf{R}^n \to \mathbf{R}$ are strictly convex, and $\alpha, \gamma \in \mathbf{R}$. Show that, for a given α, if the Slater condition holds for the first of these optimization problems, there exists a $\gamma^*(\alpha)$ such that $x_2(\gamma^*(\alpha)) = x_1(\alpha)$.

17. Consider the following problem.

Problem 2.6 *Find*

$$F = \inf_{x\in\mathbf{R}^n, c\in\mathbf{R}^l} \{\Phi(x)+\gamma\Psi(c)\} \tag{2.178}$$

s.t.

$$f_i(x) \leq 0, \ i = 1, ..., m \tag{2.179}$$
$$\text{and } h_j(x) = c_j, \ j = 1, ..., l, \tag{2.180}$$

where $c = (c_1, ..., c_l)^T$, $\Phi : \mathbf{R}^n \to \mathbf{R}$, $f_i : \mathbf{R}^n \to \mathbf{R}$ are convex and differentiable, $\Psi : \mathbf{R}^l \to \mathbf{R}$ is a convex function of the vector $c = (c_1, ..., c_l)^T$ that attains its minimum, 0, at $c = (0, ..., 0)^T$, the $h_j : \mathbf{R}^n \to \mathbf{R}$ are affine, and $\gamma > 0$.

Let g_γ

$$g_\gamma(\lambda, \nu) = \min_{x\in\mathbf{R}^n, c\in\mathbf{R}^l} \left\{\Phi(x)+\gamma\Psi(c) + \sum_{i=1}^m \lambda_i f_i(x) + \sum_{j=1}^l \nu_j[h_j(x)-c_j]\right\} \tag{2.181}$$

be the Lagrange dual function corresponding to this problem. Show that

$$g_\gamma(\lambda, \nu) = \hat{g}(\lambda, \nu) - \gamma\Psi^*(\gamma^{-1}\nu), \tag{2.182}$$

where Ψ^* is the convex conjugate of Ψ and $\hat{g}$ is the Lagrange dual function of Problem 2.1.

18. Prove Lemma 2.8, statement (ii).

19. Provide an example which shows that the relative entropy is generally not symmetric in its arguments.

20. Prove the chain rule for the entropy:

$$H^{X,Y}\left(p^{X,Y}\right) = H^{Y|X}\left(p^{X,Y}\right) + H^X\left(p^X\right), \tag{2.183}$$

where $H^{Y|X}$ is the conditional entropy from Definition 2.8, H^X is the entropy from Definition 2.6, and

$$H^{X,Y} = -E_{p^{X,Y}}\left[\log p^{X,Y}\right] = -\sum_{x\in\mathcal{X},\, y\in\mathcal{Y}} p_{x,y}^{X,Y} \log p_{x,y}^{X,Y} \qquad (2.184)$$

is the joint entropy.

Chapter 3

The Horse Race

Probabilistic models are often used by decision makers in uncertain environments. An idealization of such a decision maker, on which we heavily rely in this book, is a gambler, or investor (we use the terms interchangeably), in a horse race. In this chapter, we introduce the notions of the horse race and the conditional horse race and discuss some simple relationships between probability measures and betting strategies, while leaving a more thorough decision-theoretic treatment for later chapters. Most of the concepts and results in this chapter can be found in the textbook by Cover and Thomas (1991) or in the original papers by Kelly (1956) and Breiman (1961).

We shall first discuss the (unconditional) horse race as a setting in which we explore unconditional probabilities, and then generalize it to the conditional horse race, which is a useful picture when we are interested in conditional probabilities. The basic ideas that we shall apply later in this book can be most easily understood in the unconditional probability context and don't have to be substantially modified in the unconditional probability context.

A horse race investor who invests so as to maximize his expected wealth growth rate — a so-called Kelly-investor — allocates money to each horse in proportion to the horse's winning probability. The expected wealth growth rate for such an investor is the difference between the expected wealth growth rate for a clairvoyant investor and the entropy of the winning-probabilities. Expected wealth growth rates are also related to the relative entropy: the latter is a difference between two expected wealth growth rates. These two relationships, which hold for the conditional and the unconditional horse race, are very important, as they provide a simple decision-theoretic interpretation for information-theoretic quantities. In Chapter 8, when we discuss decision makers with arbitrary risk preferences, we shall use these relationships as a starting point for a generalization of entropy and relative entropy.

3.1   The Basic Idea of an Investor in a Horse Race

Horse race

Definition 3.1 *(Horse race)   A horse race is characterized by the discrete random variable Y with possible states in the finite set $\mathcal{Y}$; we identify each element of $\mathcal{Y}$ with a horse. An investor can place a bet that $Y = y \in \mathcal{Y}$, which pays the odds ratio (payoff) $\mathcal{O}_y > 0$ for each dollar wagered if $Y = y$, and 0, otherwise.*

Apart from an actual horse race, the following settings are examples that meet either exactly or approximately the above definition:

- betting on a coin toss,

- investing in defaultable bonds,

- playing roulette or blackjack, and

- bringing a new product to the market.

We note that an investor who allocates \$1 of capital, investing $\frac{B}{\mathcal{O}_y}$ to state y, where

$$B = \frac{1}{\sum_{y \in \mathcal{Y}} \frac{1}{\mathcal{O}_y}},$$

receives the payoff B with certainty. This motivates the following definition:

Definition 3.2 *(Bank account)   The riskless bank account payoff, B, is given by*

$$B = \frac{1}{\sum_{y \in \mathcal{Y}} \frac{1}{\mathcal{O}_y}}, \quad . \tag{3.1}$$

We also note that $\frac{B}{\mathcal{O}_y} > 0$ and

$$\sum_{y \in \mathcal{Y}} \frac{B}{\mathcal{O}_y} = 1 ,$$

so $\frac{B}{\mathcal{O}} = \left\{ \frac{B}{\mathcal{O}_y}, y \in \mathcal{Y} \right\}$ is a probability measure on $\mathcal{Y}$. Under this measure, the expected payoff for a bet placed on a single horse, y, is always B, independent of y. So we make the following definition.

Definition 3.3 *The homogeneous expected return measure is given by*

$$p^{(h)} = \left\{ p_y^{(h)} = \frac{B}{\mathcal{O}_y}, y \in \mathcal{Y} \right\} . \tag{3.2}$$

Let us suppose the bookie was risk-neutral, i.e., demanded the same return on each horse, no matter what the associated risk is, and that there was no track take. Then, if the bookie believed in the homogeneous expected return measure, $p^{(h)}$, he would set the odds ratios $\mathcal{O}$. This provides an — albeit somewhat unrealistic — interpretation of $p^{(h)}$ as the measure that an idealized bookie believes.

Investor

The following definition makes precise what the term 'investor' shall mean throughout this book, unless indicated otherwise.

Definition 3.4 *(Investor) An investor is a gambler who invests $1 in a horse race, i.e., who allocates b_y to the event $Y = y$, where*

$$\sum_{y \in \mathcal{Y}} b_y = 1 . \tag{3.3}$$

We denote the investor's allocation by

$$b = \{b_y, \ y \in \mathcal{Y}\} . \tag{3.4}$$

We have made the assumption of $1 total investment for convenience, but without loss of generality; we may view this $1 as the investor's total wealth in some appropriate currency. In particular, we can choose the investor's initial wealth as currency.

3.2   The Expected Wealth Growth Rate

We make the following definition.

Definition 3.5 *The expected wealth growth rate corresponding to a probability measure p and a betting strategy b is given by*

$$W(b,p) = E_p[\log(b, \mathcal{O})] = \sum_{y \in \mathcal{Y}} p_y \log(b_y \mathcal{O}_y) . \tag{3.5}$$

We note that we have introduced the notation

$$E_p[f(b, \mathcal{O})] = \sum_{y \in \mathcal{Y}} p_y \log(b_y \mathcal{O}_y)$$

for the function $f = \log$ here, which we will use for more general functions below.

This definition is motivated by the following lemma, which states that, asymptotically, the gambler's wealth grows exponentially with $W(b,p)$ as growth rate.

Lemma 3.1 *The wealth, W_n, of an investor after n independent successive bets in a horse race, where the horses win with probabilities given by the measure p, is related to the expected wealth growth rate as*

$$W(b,p) = \lim_{n \to \infty} \frac{\log \left(\frac{W_n}{W_0} \right)}{n} . \tag{3.6}$$

Proof: The investor's wealth after n independent, successive bets is

$$W_n = W_0 \prod_{i=1}^{n} b_{y_i} \mathcal{O}_{y_i} , \tag{3.7}$$

where y_i is the realizations of Y in the ith bet. So we have

$$\lim_{n \to \infty} \frac{\log \left(\frac{W_n}{W_0} \right)}{n} = \lim_{n \to \infty} \frac{\sum_{i=1}^{n} \log(b_{y_i} \mathcal{O}_{y_i})}{n}$$
$$= E_p \left[\log (b, \mathcal{O}) \right] \text{ (by the law of large numbers) }. \tag{3.8}$$

The lemma follows then from Definition 3.5. $\square$

3.3   The Kelly Investor

We make the following definition, which is motivated by the work of Kelly (1956).

Definition 3.6 *(Kelly investor) A Kelly investor is an investor (in the sense of Definition 3.4) who allocates his wealth so as to maximize his expected wealth growth rate according to the model he believes.*

The Kelly investor is an investor with a particular type of risk preferences. Later in this book, we shall consider investors with different types of risk preferences.

The following theorem explicitly states the investment strategy chosen by a Kelly investor.

Theorem 3.1 *A Kelly investor who believes the probability measure p allocates his assets to the horse race according to*

$$b_y^*(p) = p_y . \tag{3.9}$$

This betting strategy is often called proportional betting, since the investor allocates to each horse proportionally to the horse's winning probability.

Proof: It follows from Definitions 3.6 and 3.5 that the optimal allocation for a Kelly investor who believes p is

$$b^*(p) = \arg \max_{\{b: \sum_{y \in \mathcal{Y}} b_y = 1\}} \sum_y p_y \log(b_y \mathcal{O}_y) . \tag{3.10}$$

In order to solve this optimization problem, which is convex, we write down its Lagrangian:

$$\mathcal{L}(b, \lambda) = \sum_y p_y \log(b_y \mathcal{O}_y) - \lambda \left(\sum_{y \in \mathcal{Y}} b_y - 1 \right) . \tag{3.11}$$

The optimal allocation, b^*, is the solution of

$$0 = \left. \frac{\partial \mathcal{L}(b, \lambda)}{\partial b_y} \right|_{b=b^*}$$

$$= \frac{p_y}{b_y^*} - \lambda ,$$

which is

$$b_y^* = \frac{p_y}{\lambda} . \tag{3.12}$$

In order to find the value, λ^*, that corresponds to the solution of our convex problem, we have to solve

$$1 = \sum_{y \in \mathcal{Y}} b_y^* \tag{3.13}$$

for λ, which, of course, is the same as solving the dual problem. We find $\lambda^* = 1$. So we have $b_y^* = p_y$. $\square$

The Kelly investor has remarkable properties. In particular, the probability that the ratio of the wealth of a Kelly investor to the wealth of a non-Kelly investor after n trials will exceed any constant can be made as close to 1 as we please, for n sufficiently large. Moreover, the expected time to double the wealth is smaller for the Kelly investor than for any other investor. For proofs of these statements, see, for example, Cover and Thomas (1991), Chapter 6, or Breiman (1961). Some of these properties also hold for Kelly investors who invest in continuous-time markets (see Merton (1971), Karatzas et al. (1991), Jamishidian (1992), and Browne and Whitt (1996)).

3.4   Entropy and Wealth Growth Rate

The following theorem relates the entropy to the expected wealth growth rate of a Kelly investor.

Theorem 3.2 *A Kelly investor who knows that the horses in a horse race win with the probabilities given by the measure p has the expected wealth growth rate*

$$W_p^*(p) = W_p^{**} - H(p) , \tag{3.14}$$

where

$$W_p^{**} = E_p[\log \mathcal{O}] \tag{3.15}$$

is the wealth growth rate of a clairvoyant investor, i.e., of an investor who wins every bet.

We have set the lower indices, p, in the above notation to indicate that horse y wins the race with probability p_y.

Proof: Theorem 3.1 states that a Kelly investor who believes the probability measure p allocates according to

$$b_y^*(p) = p_y . \tag{3.16}$$

Inserting this expression into Definition 3.5, we obtain the p-expected wealth growth rate for the Kelly investor

$$\begin{aligned} W_p^*(p) &= E_p[\log{(p, \mathcal{O})}] \\ &= E_p[\log p] + E_p[\log \mathcal{O}] \\ &= -H(p) + W_p^{**} \text{ (from (3.15) and Definition 2.6) .}\square \end{aligned}$$

As we have discussed in Section 2.3.1, Theorem 3.2 provides a financial interpretation to the entropy. In Section 7.2, where we consider investors with more general risk-preferences, we shall use Theorem 3.2 as the starting point for a generalization of entropy.

The following theorem relates the relative entropy to the expected wealth growth rate of a Kelly investor.

Theorem 3.3 *In a horse race where horses win with the probabilities given by the measure p, the difference in expected wealth growth rates between a Kelly investor who knows the probability measure p and a Kelly investor who believes the (misspecified) probability measure q is given by*

$$W_p^*(p) - W_p^*(q) = D(p\|q) . \tag{3.17}$$

Proof: Theorem 3.1 states that a Kelly investor who believes the probability measure p allocates according to

$$b_y^*(p) = p_y , \tag{3.18}$$

and a Kelly investor who believes the probability measure q allocates according to

$$b_y^*(q) = q_y . \tag{3.19}$$

Inserting the above two equations into Definition 3.5, we obtain the p-expected wealth growth rate difference

$$
\begin{aligned}
W_p^*(p) - W_p^*(q) &= E_p\left[\log\left(p, \mathcal{O}\right)\right] - E_p\left[\log\left(q, \mathcal{O}\right)\right] \\
&= E_p\left[\log p\right] - E_p\left[\log q\right] \\
&= D(p\|q) \text{ (by Definition 2.7) .}\square
\end{aligned}
$$

As we have seen in Section 2.3.2, Theorem 3.3 provides a financial interpretation to the relative entropy. In Section 7.2 we shall use this theorem as the starting point for a generalization of the relative entropy.

The following theorem states another relationship between expected wealth growth rates and relative entropy.

Theorem 3.4 *In a horse race where horses win with the probabilities given by the measure p, the expected wealth growth rate of a Kelly investor who believes the measure q is given by*

$$
W_p^*(q) = \log B + D\left(p\|p^{(h)}\right) - D\left(p\|q\right) . \tag{3.20}
$$

Proof: The expected wealth growth rate of a Kelly investor who believes the measure q is

$$
\begin{aligned}
W_p^*(q) &= W(b^*(q), p) \\
&= E_p[\log(b^*(q), \mathcal{O})] \text{ (by Definition 3.5)} \\
&= E_p[\log(q, \mathcal{O})] \text{ (by Theorem 3.1)} \\
&= E_p\left[\log\left(\frac{p}{p^{(h)}}\frac{Bq}{p}\right)\right] \text{ (by Definition 3.3)} \\
&= \log B + D\left(p\|p^{(h)}\right) - D\left(p\|q\right) \text{ (by Definition 2.7) .}\square
\end{aligned}
$$

It follows from Theorem 3.4 that the Kelly investor has an expected wealth growth rate in excess of the bank account growth rate only if the measure he believes is closer (in the relative entropy sense) to the measure p than the homogeneous measure is.

3.5 The Conditional Horse Race

So far we have discussed investors in a (unconditional) horse race as an idealized setting for a decision maker who uses a probabilistic model for a single variable. In many practical situations, however, we are interested in conditional probabilities of a variable Y given a variable X. In order to evaluate and build such probabilistic models, we introduce the notion of the conditional horse race. We shall see that most of the horse-race results from the previous sections can be generalized to the conditional horse race.

Throughout this section, we denote joint probability measures by $p^{X,Y}$ or $q^{X,Y}$, conditional probability measures by $p^{Y|X}$ or $q^{Y|X}$, and the marginal X-probability measures by p^X or q^X. Later in this book, when the meaning is clear from the context, we shall drop the upper indices.

Conditional horse race

We generalize Definition 3.1 as follows.

Definition 3.7 *(Conditional horse race)* *A conditional horse race is characterized by the discrete random variable Y with possible states in the finite set $\mathcal{Y}$ and the discrete random variable X with possible states in the finite set $\mathcal{X}$; we identify each element of $\mathcal{Y}$ with a horse and each element of $\mathcal{X}$ with a particular piece of side information. An investor can place a bet that $Y = y \in \mathcal{Y}$ after learning that $X = x \in \mathcal{X}$, which pays the odds ratio (payoff) $\mathcal{O}_{y|x} > 0$ for each dollar wagered if $Y = y$, and 0, otherwise.*

This conditional horse race (see, Friedman and Sandow (2003b)) is slightly more general than the horse race with side information from Cover and Thomas (1991), Chapter 6; the latter is restricted to payoffs that are independent of X, but is otherwise the same as the conditional horse race.

We also generalize the notions of a bank account, introduce a worst (over $x \in \mathcal{X}$) bank account, and introduce the homogeneous expected return measure.

Definition 3.8 *(Conditional bank account)* *Given that $X = x$, the riskless conditional bank account payoff, B_x, is*

$$B_x = \frac{1}{\sum_{y \in \mathcal{Y}} \frac{1}{\mathcal{O}_{y|x}}} \, . \tag{3.21}$$

Definition 3.9 *(Worst conditional bank account)* *The worst conditional bank account has payoff*

$$B = \inf_{x \in \mathcal{X}} B_x \, . \tag{3.22}$$

Definition 3.10 *The conditional homogeneous expected return measure is given by*

$$p^{Y|X(h)} = \left\{ p_{y|x}^{Y|X(h)} = \frac{B_x}{\mathcal{O}_{y|x}} \, , \, y \in \mathcal{Y}, x \in \mathcal{X} \right\} \, . \tag{3.23}$$

The above $p^{Y|X(h)}$ has the required properties of a conditional probability measure, i.e., $p_{y|x}^{Y|X(h)} > 0$, $\forall x \in \mathcal{X}, y \in \mathcal{Y}$, and $\sum_{y \in \mathcal{Y}} p_{y|x}^{Y|X(h)} = 1$, $\forall x \in \mathcal{X}$. The conditional expected return, given $X = x$, under $p^{Y|X(h)}$ is B_x; this return depends on the value of X, which is known before bets are placed, but is independent of the value of Y.

Conditional investor

We generalize Definition 3.4 as follows.

Definition 3.11 *(Conditional investor) A conditional investor is a gambler who invests \$1 in a horse race, i.e., a gambler who, after having learned that $X = x$, allocates $b_{y|x}$ to the event $Y = y$, where*

$$\sum_{y \in \mathcal{Y}} b_{y|x} = 1 , \; \forall x \in \mathcal{X} . \tag{3.24}$$

We denote the conditional investor's allocation by

$$b = \left\{ b_{y|x}, \; x \in \mathcal{X}, \; y \in \mathcal{Y} \right\} . \tag{3.25}$$

Below we shall often refer to the conditional investor simply as investor, unless we need to make a distinction between a conditional and an unconditional investor.

Expected conditional wealth growth rate

We generalize Definition 3.5 as follows.

Definition 3.12 *The expected conditional wealth growth rate corresponding to a probability measure $p^{X,Y}$ and a betting strategy b is given by*

$$W^{Y|X}\left(b, p^{X,Y}\right) = E_{p^{X,Y}}\left[\log\left(b, \mathcal{O}\right)\right] = \sum_{y \in \mathcal{Y}, x \in \mathcal{X}} p_{x,y}^{X,Y} \log(b_{y|x} \mathcal{O}_{y|x}) . \tag{3.26}$$

Below we shall often refer to the expected conditional wealth growth rate simply as expected wealth growth rate.

We generalize Lemma 3.1, which states that, asymptotically, the gambler's wealth grows exponentially with $W^{Y|X}(b, p)$ as growth rate.

Lemma 3.2 *The wealth, W_n, of an investor after n independent successive bets in a conditional horse race, where the horses win with probabilities given by $p^{X,Y}$, is related to the expected wealth growth rate as*

$$W^{Y|X}\left(b, p^{X,Y}\right) = \lim_{n \to \infty} \frac{\log\left(\frac{W_n}{W_0}\right)}{n} . \tag{3.27}$$

Proof: The investor's wealth after n independent, successive bets is

$$W_n = W_0 \prod_{i=1}^{n} b_{y_i|x_i} \mathcal{O}_{y_i|x_i} , \tag{3.28}$$

where (x_i, y_i) is the realizations of (X, Y) in the ith bet. So we have

$$\lim_{n \to \infty} \frac{\log\left(\frac{W_n}{W_0}\right)}{n} = \lim_{n \to \infty} \frac{\sum_{i=1}^{n} \log(b_{y_i|x_i} \mathcal{O}_{y_i|x_i})}{n}$$

$$= E_{p^{X,Y}}\left[\log\left(b, \mathcal{O}\right)\right] \text{ (by the law of large numbers).} \tag{3.29}$$

The lemma follows then from Definition 3.12. $\square$

Conditional Kelly investor

We generalize Definition 3.6 as follows.

Definition 3.13 *(Conditional Kelly investor) A conditional Kelly investor is a conditional investor (in the sense of Definition 3.11) who allocates his wealth so as to maximize his expected wealth growth rate according to the model he believes.*

Below we shall often, for the sake of brevity, refer to the conditional Kelly investor simply as Kelly investor.

Generalizing Theorem 3.1, the following theorem explicitly states the investment strategy chosen by a Kelly investor.

Theorem 3.5 *A conditional Kelly investor who believes the conditional probability measure $p^{Y|X}$ allocates his assets to the horse race according to*

$$b^*_{y|x}\left(p^{Y|X}\right) = p^{Y|X}_{y|x} . \tag{3.30}$$

Proof: It follows from Definition 3.13 and Definition 3.12 that

$$b^*\left(p^{Y|X}\right) = \arg \max_{\{b:\sum_{y\in\mathcal{Y}} b_{y|x}=1\}} \sum_y p^{Y|X}_{y|x} \log(b_{y|x}\mathcal{O}_{y|x}) . \tag{3.31}$$

In order to solve this optimization problem, which is convex, we write down its Lagrangian:

$$\mathcal{L}(b,\lambda) = \sum_y p^{Y|X}_{y|x} \log(b_{y|x}\mathcal{O}_{y|x}) - \lambda_x \left(\sum_y b_{y|x} - 1\right) . \tag{3.32}$$

The optimal allocation, b^*, is the solution of

$$0 = \left.\frac{\partial\mathcal{L}(b,\lambda)}{\partial b_{y|x}}\right|_{b=b^*}$$

$$= \left\{\frac{p^{Y|X}_{y|x}}{b^*_{y|x}} - \lambda_x\right\} p^X_x ,$$

which is

$$b^*_{y|x} = \frac{p_{y|x}}{\lambda_x} . \tag{3.33}$$

In order to find the value, λ^*_x, that corresponds to the solution of our convex problem, we have to solve

$$1 = \sum_y b^*_{y|x} \tag{3.34}$$

for λ_x. We find $\lambda_x^* = 1$. So we have $b_{y|x}^* = p_{y|x}^{Y|X}$. $\square$

Entropy and wealth growth rate

As we have done for the case of the (unconditional) horse race, we relate the entropy to the expected conditional wealth growth rate of a conditional Kelly investor.

Theorem 3.6 *A conditional Kelly investor who knows that the horses in a conditional horse race win with the probabilities given by the measure $p^{X,Y}$ has the expected wealth growth rate*

$$W_{p^{X,Y}}^* \left(p^{Y|X} \right) = W_{p^{X,Y}}^{**} - H^{Y|X} \left(p^{X,Y} \right) , \tag{3.35}$$

where

$$W_{p^{X,Y}}^{**} = E_{p^{X,Y}} \left[\log \mathcal{O} \right] = \sum_{y \in \mathcal{Y}, x \in \mathcal{X}} p_{x,y}^{X,Y} \log \mathcal{O}_{y|x} \tag{3.36}$$

is the wealth growth rate of a clairvoyant investor, i.e., of an investor who wins every bet.

Proof: It follows from Theorem 3.5 that a Kelly investor who believes the probability measure $p^{Y|X}$ allocates according to

$$b_{y|x}^* \left(p^{Y|X} \right) = p_{y|x}^{Y|X} . \tag{3.37}$$

Inserting this equation into Definition 3.12, we obtain the $p^{X,Y}$-expected wealth growth rate of the Kelly investor as

$$
\begin{aligned}
W_{p^{X,Y}}^* \left(p^{Y|X} \right) &= \sum_{y \in \mathcal{Y}, x \in \mathcal{X}} p_{x,y}^{X,Y} \log \left(p_{y|x}^{Y|X} \mathcal{O}_{y|x} \right) \\
&= \sum_{y \in \mathcal{Y}, x \in \mathcal{X}} p_{x,y}^{X,Y} \log p_{y|x}^{Y|X} + \sum_{y \in \mathcal{Y}, x \in \mathcal{X}} p_{x,y}^{X,Y} \log \mathcal{O}_{y|x} \\
&= -H^{Y|X} \left(p^{X,Y} \right) + W_{p^{X,Y}}^{**} \text{ (from (3.36) and Definition 2.8) .} \square
\end{aligned}
$$

In Section 7.2 we shall use Theorem 3.6 as the starting point for a generalization of the conditional entropy.

The following theorem, which is a generalization of Theorem 3.3, relates the relative entropy to the expected conditional wealth growth rate of a Kelly investor.

Theorem 3.7 *In a conditional horse race where horses win with the probabilities given by the measure $p^{X,Y}$, the difference in expected conditional wealth growth rates between a conditional Kelly investor who knows the probability measure $p^{Y|X}$ and a conditional Kelly investor who believes the (misspecified) probability measure $q^{Y|X}$ is given by*

$$W_{p^{X,Y}}^* \left(p^{Y|X} \right) - W_{p^{X,Y}}^* \left(q^{Y|X} \right) = D^{Y|X} \left(p^{X,Y} \| q^{Y|X} \right) . \tag{3.38}$$

Proof: It follows from Theorem 3.5 that a conditional Kelly investor who believes the probability measure p allocates according to

$$b^*_{y|x}\left(p^{Y|X}\right) = p^{Y|X}_{y|x} , \qquad (3.39)$$

and a conditional Kelly investor who believes the probability measure $q^{X,Y}$ allocates according to

$$b^*_{y|x}\left(q^{Y|X}\right) = q^{Y|X}_{y|x} . \qquad (3.40)$$

Inserting the above two equations into Definition 3.12, we obtain the $p^{X,Y}$- expected wealth growth rate difference

$$W^*_{p^{X,Y}}\left(p^{Y|X}\right) - W^*_{p^{X,Y}}\left(q^{Y|X}\right) = \sum_{y \in \mathcal{Y}, x \in \mathcal{X}} p^{X,Y}_{x,y} \log(p^{Y|X}_{y|x} \mathcal{O}_{y|x})$$

$$- \sum_{y \in \mathcal{Y}, x \in \mathcal{X}} p^{X,Y}_{x,y} \log(q^{Y|X}_{y|x} \mathcal{O}_{y|x})$$

$$= \sum_{y \in \mathcal{Y}, x \in \mathcal{X}} p^{X,Y}_{x,y} \log \left(\frac{p^{Y|X}_{y|x}}{q^{Y|X}_{y|x}} \right)$$

$$= D^{Y|X}\left(p^{X,Y} \| q^{Y|X}\right) \qquad (3.41)$$

(from Definition 2.9) . $\square$

In Section 7.3, where we consider investors with more general risk-preferences, we will use Theorem 3.7 as the starting point for a generalization of the conditional relative entropy.

Next, we generalize Theorem 3.4.

Theorem 3.8 *In a conditional horse race where horses win with the probabilities given by the measure $p^{X,Y}$, the expected conditional wealth growth rate of a conditional Kelly investor who believes the measure $q^{Y|X}$ can be computed as*

$$W^*_{p^{X,Y}}\left(q^{X,Y}\right) = E_{p^X}[\log B] + D^{Y|X}\left(p^{X,Y} \| p^{Y|X(h)}\right) - D^{Y|X}\left(p^{X,Y} \| q^{Y|X}\right) . \qquad (3.42)$$

Proof: The expected conditional wealth growth rate of a conditional Kelly investor who believes the measure $q^{Y|X}$ is

$$
\begin{aligned}
W_{p^{X,Y}}^* \left(q^{Y|X} \right) &= W \left(b^* \left(q^{Y|X} \right), p^{X,Y} \right) \\
&= \sum_{y \in \mathcal{Y}, x \in \mathcal{X}} p_{x,y}^{X,Y} \log \left(b_{y|x}^* \left(q^{Y|X} \right) \mathcal{O}_{y|x} \right) \\
&= \sum_{y \in \mathcal{Y}, x \in \mathcal{X}} p_{x,y}^{X,Y} \log \left(q_{y|x}^{Y|X} \mathcal{O}_{y|x} \right) \quad \text{(by Theorem 3.5)} \\
&= \sum_{y \in \mathcal{Y}, x \in \mathcal{X}} p_{x,y}^{X,Y} \log \left(\frac{p_{y|x}^{Y|X}}{p_{y|x}^{Y|X(h)}} \frac{B_x q_{y|x}^{Y|X}}{p_{y|x}^{Y|X}} \right) \quad \text{(by Definition 3.10)} \\
&= E_{p^X} \left[\log B \right] + D^{Y|X} \left(p^{X,Y} \| p^{Y|X(h)} \right) - D^{Y|X} \left(p^{X,Y} \| q^{Y|X} \right) \\
&\quad \text{(by Definition 2.9). } \square
\end{aligned}
$$

It follows from Theorem 3.8 that the conditional Kelly investor has an expected conditional wealth growth rate in excess of the bank account growth rate only if the measure he believes is closer (in the relative entropy sense) to the measure $p^{Y|X}$ than the homogeneous measure is.

Increase of expected wealth growth rate due to side information

Theorem 3.9 *The increase in the expected wealth growth rate for a conditional Kelly investor in a conditional horse race due to the information provided by the variable X is*

$$
\Delta W = I \left(p^{X,Y} \right) + W_{p^{X,Y}}^{**} - W_{p^X \times p^Y}^{**} , \tag{3.43}
$$

where $I \left(p^{X,Y} \right)$ is the mutual information, $W_{p^{X,Y}}^{**}$ is the expected wealth growth rate of a clairvoyant investor, and $W_{p^X \times p^Y}^{**}$ is the expected wealth growth rate of a clairvoyant investor in a horse race where X and Y are independent.

Proof: We compare the Kelly investor who knows the information provided by X, and therefore knows the measure $p^{Y|X}$, to an investor who doesn't know the information provided by X, and therefore believes the measure $\hat{p}^{Y|X}$ with

$$
\hat{p}_{y|x}^{Y|X} = p_y^Y , \quad \forall x \in \mathcal{X} , \tag{3.44}
$$

i.e.,

$$
\hat{p}^{X,Y} = p^X \times p^Y . \tag{3.45}
$$

The difference in expected wealth growth rate between these two investors is

$$\Delta W = W^*_{p^{X,Y}} \left(p^{Y|X} \right) - W^*_{p^{X,Y}} \left(\hat{p}^{Y|X} \right)$$

$$= -H^{Y|X} \left(p^{X,Y} \right) + W^{**}_{p^{X},Y}$$

$$+ H^{Y|X} \left(\hat{p}^{X,Y} \right) - W^{**}_{\hat{p}^{X},Y}$$

(by Theorem 3.6)

$$= I \left(p^{X,Y} \right) + W^{**}_{p^{X},Y} - W^{**}_{p^{X} \times p^{Y}}$$

(by Definitions 2.8 and 2.10, (3.44) and (3.45)) . $\Box$

As a straightforward consequence of Theorem 3.9 and (3.36), we have the following corollary.

Corollary 3.1 *In a conditional horse race where the odds ratios are independent of X, i.e., where $\mathcal{O}_{y|x} = \mathcal{O}_y$, $\forall x \in \mathcal{X}$, the increase in the expected wealth growth rate for a conditional Kelly investor in a conditional horse race due to the information provided by the variable X is*

$$\Delta W = I \left(p^{X,Y} \right) . \tag{3.46}$$

3.6   Exercises

1. A 3-horse race has win probabilities (p_1, p_2, p_3) and odds ratios $(1, 1, 1)$. A gambler places bets $(b_1, b_2, b_3), \sum_i b_i = 1$.

   (a) Calculate the gambler's expected wealth growth rate.

   (b) Find the expected wealth growth-rate-optimal gambling scheme.

   (c) Find the probabilities that cause the expected wealth to go to zero the fastest.

   (d) Calculate the expected wealth growth rate for a clairvoyant (someone who wins every bet and allocates accordingly).

2. Suppose an investor invests \$1 over one year in a financial market that offers two instruments:

   (i) a bond, which after one year pays \$1.20 (in case of no default) with a probability of 0.9 and \$0.50 (in case of default) with a probability of 0.1, and

   (ii) a bank account, which after one year pays \$1.05 with certainty.

   The investor can go long or short either position. If we were to define a horse race that corresponds to this financial market, what combinations of the bond and the bank account do the horses correspond to and what are their odds ratios?

3. A horse race has odds ratios $\mathcal{O}_y$, $y \in \mathcal{Y}$. An investor wants to bet \$1 on the horse race, and believes that horse y wins with probability p_y. How does the investor allocate his \$1 if he aims at

   (a) maximizing his expected payoff,

   (b) maximizing his best-case payoff,

   (c) maximizing his worst-case payoff, or

   (d) minimizing the uncertainty of the payoff.

   Compute, in each case, the expectation and the variance of the payoff under the assumption that the empirical probabilities are the same as the ones the investor believes.

4. A horse race has odds ratios $\mathcal{O}_y$, $y \in \mathcal{Y}$. A Kelly investor bets \$1 on the horse race, and believes that horse y wins with probability p_y. Assume that the empirical probabilities are the same as the ones the investor believes.

   (a) Compute the expectation and the variance of the investor's payoff.

   (b) Which probability measure leads to the smallest variance of the payoff?

5. Suppose an investor invests \$1 over one year in a financial market that offers two instruments:

   (i) a stock, which has an initial price of \$10, and has one of the following prices after one year: \$0.5, \$0.6 , \$0.7 , \$0.8 , \$0.9 , \$1, \$1.1, \$1.2 , \$1.3 , \$1.4 , \$1.5.

   (ii) a bank account, which after one year pays \$1.05 with certainty.

   If we were to define a horse race that corresponds to this financial market, what combinations of the bond and the bank account do the horses correspond to and what are their odds ratios? Can the investor actually place bets on any of the 'horses?' If not, what instruments are needed in order to complete the market, i.e., in order to allow the investor to place bets on any of the 'horses?'

Chapter 4

Elements of Utility Theory

Utility theory is one of the pillars of modern finance and is invoked extensively in this book. In this chapter, we briefly review elements of this theory. We refer readers interested in a more thorough introduction to J. Ingersoll, Jr. (1987) and the references therein. Utility theory can be of use when there are a number of investment alternatives and an investor is to allocate his wealth among the alternatives. Given an allocation, his wealth at the end of some time horizon is determined by the (random) outcomes for the various investment alternatives. For deterministic outcomes, it is easy to rank the choices. However, prior to the exploration of utility theory, it was not obvious how to rank such choices in a probabilistic setting. Utility theory provides a way to rank random wealth levels.

Utility theory has been criticized based on the observation that investors are not always "rational"; that is, there are investors who might not subscribe to the axioms of utility theory (stated below), or might not act in a manner consistent with those axioms. Utility theory has also been criticized because an investor's utility function cannot be observed directly. We do not address these issues; rather, we confine our attention to investors who subscribe to the axioms of utility theory and act accordingly.

In this chapter, we review the genesis of utility theory — the St. Petersburg paradox, elements of an axiomatic approach to utility theory, define risk aversion, introduce several popular utility functions, discuss field studies, and provide certain technical conditions on utility functions that we shall adopt in the remainder of the book.

4.1 Beginnings: The St. Petersburg Paradox

The beginnings of utility theory go back to the early 18th century. As described in Szekely and Richards (2004), in 1713, Nikolaus Bernoulli, in a letter to Pierre Raymond de Montmort, considered the following game:

(i) a fair coin is to be tossed until a head appears, and,

(ii) if the first head appears on the nth toss, then the game ends and the payoff from playing the game is 2^{n-1} ducats.

Thus, the payoff is always positive, and for most plays of the game, the payoff is modest, since the game typically ends when a head appears after several tosses; for example, if the first head occurs on the fourth toss, the player receives only 8 ducats. However, there is the possibility of a long string of tails, and an enormous payoff.

Montmort was intrigued by the following question: how much should one pay to play this game? The expected payoff of this game is given by

$$E[\text{payoff}] = \sum_n \frac{1}{2^n} 2^{n-1} = \frac{1}{2} + \frac{1}{2} + \frac{1}{2} + \cdots = \infty. \tag{4.1}$$

Does (4.1) suggest that this game is a bargain at any price? Certainly, the vast majority of people would not be willing to risk their entire fortunes to play this game, which would result, in most cases, in a gain of several ducats, with a remote chance of gaining an enormous number of ducats. That is the paradox: (4.1) "suggests" that we should risk all, though, for most investors, that seems completely unreasonable.

Cramer (1728) and (Nikolaus's cousin, Daniel) Bernoulli (1738)[1] *postulated* that one should maximize the expected utility derived from wealth, where the utility from wealth is not linearly related to wealth, W, but increases at a decreasing rate. This is plausible. The utility function — which expresses the utility associated with a given level of wealth — should increase, since more ought to be preferred to less; the rate of increase of the utility function should decrease, since an incremental unit of wealth should provide more utility for a person with limited wealth than for a person of greater wealth.

If we think about how much we should pay to play this game in terms of expected utility, we are not led to the expected payoff of the game, but rather to the expected utility of playing the game assuming that it costs c ducats to play. Let N denote the (random) number of tosses that occur up to and including the first head; i.e., the first head occurs on the (random) N^{th} toss. For a player with initial wealth W_0 and utility function, $U(\cdot)$, the expected utility of playing the game, if it costs c ducats to play, is given by

$$E[U(W_0 + 2^{N-1} - c)] = \sum_n \frac{1}{2^n} U(W_0 + 2^{n-1} - c). \tag{4.2}$$

Such an investor will pay any amount that increases his expected utility. That is, he will pay any amount for which

$$E[U(W_0 + 2^{N-1} - c)] > U(W_0). \tag{4.3}$$

---

[1]Daniel Bernoulli presented, in an article to the Imperial Academy of Sciences in St. Petersburg (hence, the name), a statement and resolution of the paradox.

That is, the investor would pay any number of ducats, c, up to the solution of

$$0 = E[U(W_0 + 2^{N-1} - c)] - U(W_0) = \sum_n \frac{1}{2^n} U(W_0 + 2^{n-1} - c) - U(W_0) \quad (4.4)$$

to play the game.

For $U(W) = \log(W)$, the utility function used by Cramer and Daniel Bernoulli, the investor would pay any number of ducats up to the solution (solving for c) of

$$0 = E[\log(W_0 + 2^{n-1} - c)] - \log(W_0) = \sum_n \frac{1}{2^n} \log(W_0 + 2^{n-1} - c) - \log(W_0)$$
$$(4.5)$$

to play the game. An investor with a logarithmic utility function has a utility which approaches $-\infty$ as $W \to 0^+$ and grows "slowly" as $W \to \infty$; such an investor would therefore be extremely averse to risking all or nearly all of his wealth on a gamble that usually doesn't pay too much. So, it would not be surprising should this expected logarithmic utility approach resolve the paradox. To see that this approach does indeed resolve the paradox, note that it can be shown that the sum in (4.5) is finite (see Exercise 1) and monotone decreasing in c with a range of $(-\infty, \infty)$ (see Exercise 2), so there is always a finite solution, c, to (4.5). Thus, the St. Petersburg paradox is resolved: an expected utility maximizing investor with a logarithmic utility function and initial wealth, W_0, will pay only a finite amount to play this game.

As the Figure 4.1, produced by numerically solving (4.5), indicates, the amount that the logarithmic utility investor will pay is quite modest: even a ducat-trillionaire would pay at most 20.90 ducats to play this game. Investors with less initial wealth will pay less to play the game. For example, an investor with initial wealth 1 ducat will borrow .67 ducats and pay at most[2] 1.67 ducats to play the game. An investor with 2 ducats will pay at most 2 ducats to play the game, an investor with 1,000 ducats will pay at most 5.96 ducats to play, while an investor with 1,000,000 ducats will pay at most 10.93 ducats to play (see Exercise 3).

We note that there are other resolutions of the St. Petersburg paradox (for example, asserting that no real casino would offer a game with unbounded expected payoff) and related problem formulations that are not resolved for an expected logarithmic utility investor, for example, for the Super St. Petersburg paradox:

(i) a fair coin is to be tossed until a head appears, and,

(ii) if the first head appears on the nth toss, then the game ends and the payoff from playing the game is $e^{2^{n-1}}$ ducats.

---

[2] We consider only multiples of .01 ducats.

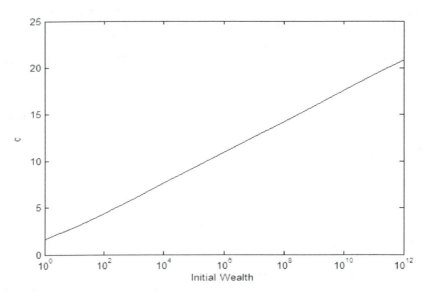

FIGURE 4.1: Maximum amount an expected logarithmic utility investor would pay to play the St. Petersburg game, as a function of initial wealth.

Our purpose was to describe the beginnings of utility theory, so we do not discuss these other resolutions of the St. Petersburg paradox, or other more subtle paradoxes. We refer the reader to Samuelson (1977) for further discussion along these lines.

Thus, the idea of maximizing expected utility was established in the 18th century; for a more general gamble, with discrete wealth outcomes in the set A, the expected utility is given by

$$E[U(W)] = \sum_{w \in A} prob\{W = w\}U(w). \tag{4.6}$$

This idea of an expected utility maximizing investor, originally *postulated* by Bernoulli and Cramer, is used throughout the remainder of this book. In the next section, we trace elements of the axiomatic approach of von Neumann and Morgenstern (1944) developed about 200 years later.

4.2 Axiomatic Approach

In this discussion, which follows that of J. Ingersoll, Jr. (1987), we do not derive results from the most primative assumptions, nor do we seek completeness or generality. Rather, we indicate how certain results of utility theory, for example, the idea of maximizing expected utility, may be derived from more primitive assumptions. Under the axiomatic approach, we define goods, preferences for different bundles of goods, and state fundamental axioms. As we shall see, in this context, the idea of an expected utility maximizing investor *follows as a logical consequence* of more fundamental axioms. The reader content to accept the idea of an expected utility maximizing investor as a *postulate* can skip this section.

We assume that there are goods:

Definition 4.1 *(Bundle of goods) Let x denote a bundle, with x_i units of good i, $x \in X$, where X is a closed, convex subset of R^n that denotes the set of all bundles of goods.*

The goods may represent tangible goods, such as cotton or steel.

We define two types of preference relationships:

Definition 4.2 *(Weakly preferred) The statement $x \succeq z$ is read "x is weakly preferred to z."*

Definition 4.3 *(Strictly preferred) $x \succ y$ (x is strictly preferred to y) if $x \succeq y$, but not $y \succeq x$.*

We define equivalence as follows:

Definition 4.4 *(Equivalence) $x \sim y$ (x is equivalent to y) if $x \succeq y$, and $y \succeq x$.*

Now we state three natural axioms:

Axiom 1 *(Completeness on bundles) For every pair of bundles, $x \in X$ and $y \in X$, either $x \succeq y$, or $y \succeq x$.*

Axiom 2 *(Reflexivity on bundles) For every bundle, $x \in X$, $x \succeq x$.*

Axiom 3 *(Transitivity on bundles) For every pair of bundles, $x \in X$ and $y \in X$, if $x \succeq y$, and $y \succeq z$, then $x \succeq z$.*

We now introduce the notion of an ordinal utility function — a function that encodes the preference of the ordering relation:

Definition 4.5 *(Ordinal utility function) $\Upsilon: X \rightarrow R$ is an ordinal utility function if and only if, for bundles x and $z \in X$,*

$$\Upsilon(x) > \Upsilon(z) \text{ if and only if } x \succ z, \text{ and}$$
$$\Upsilon(x) = \Upsilon(z) \text{ if and only if } x \sim z.$$

We note that ordinal utility functions encode only the ordering of preferences, and not the magnitude of the preferences.[3]

The axioms stated above are not sufficient to insure the existence of an ordinal utility function. In order to insure the existence of an ordinal utility function, it is necessary to introduce an additional axiom. It can be shown that the following axiom, the continuity axiom, guarantees the existence of an ordinal utility function, as well as the continuity of the utility function.

Axiom 4 *(Continuity on bundles) For every bundle $x \in X$, the two subsets:*

(i) all bundles that are strictly preferred to x, and

(ii) all bundles that are strictly worse than x

are open subsets of X.

That is, it can be shown[4] that:

Theorem 4.1 *(Existence of ordinal utility function) Under Axioms 1-4 above, there exists a continuous ordinal utility function Υ mapping X to the real line that satisfies*

$$\Upsilon(x) > \Upsilon(z) \text{ if and only if } x \succ z, \text{ and}$$
$$\Upsilon(x) = \Upsilon(z) \text{ if and only if } x \sim z.$$

Observe that if $\Upsilon(x)$ is an ordinal utility, then so is $\theta[\Upsilon(x)]$, where $\theta(\cdot)$ is a strictly increasing function. That is, ordinal utility functions are equivalent up to strictly increasing monotone transformations. Also note that we have not yet introduced the notion of risk into our axiomatic discussion of utility theory. We do so now.

Definition 4.6 *(Lottery) A lottery is a pair consisting of*

(i) a collection of bundles $(x^1, \ldots, x^m)$, with $x^i \in X$, and

(ii) the probability measure $(\pi^1, \ldots, \pi^m)$ on these payoffs.

We assume that there is a preference ordering on the *lotteries*, with strict preference and indifference defined as above, that satisfies the following axioms:

Axiom 5 *(Completeness on lotteries) For every pair of lotteries (L_1, L_2), either $L_1 \succeq L_2$, or $L_2 \succeq L_1$.*

Axiom 6 *(Reflexivity on lotteries) For every lottery, L, $L \succeq L$.*

---

[3]Below, we describe cardinal utility functions which encode more information about the magnitude of the preferences than ordinal utility functions.

[4]For a proof, see, for example, Luce and Raiffa (1989).

Axiom 7 *(Transitivity on lotteries) If $L_1 \succeq L_2$, and $L_2 \succeq L_3$, then $L_1 \succeq L_3$.*

It can be shown that these axioms are sufficient to guarantee that preferences are consistent with an ordinal utility function defined on lotteries. We now state additional axioms from which it follows that a decision maker who subscribes to the full set of axioms will maximize an expected utility function.

Axiom 8 *(Independence) Let L_1 be the lottery with*

(i) the collection of bundles $(x^1, \ldots, x^v, \ldots, x^m)$, and

(ii) probability measure $(\pi^1, \ldots, \pi^v, \ldots, \pi^m)$.

Let z denote a bundle of goods or another lottery. If z is a lottery, let it be the lottery with

(i) the collection of bundles $(y^1, \ldots, y^n)$, and

(ii) probability measure $(p^1, \ldots, p^n)$.

Let L_2 be the lottery with

(i) the collection of bundles $(x^1, \ldots, z, \ldots, x^m)$, and

(ii) probability measure $(\pi^1, \ldots, \pi^v, \ldots, \pi^m)$.

If $x^v \sim z$, then $L_1 \sim L_2$, whether z is a bundle or another lottery. If z is another lottery, then $L_1 \sim L_2 \sim L_3$, where L_3 denotes the lottery with

(i) the collection of bundles $(x^1, \ldots, x^{v-1}, y^1, \ldots, y^n, x^{v-1}, \ldots, x^m)$, and

(ii) probability measure $(\pi^1, \ldots, \pi^{v-1}, \pi^v p^1, \ldots, \pi^v p^n, \pi^{v-1}, \ldots, \pi^m)$.

Under this axiom, only the preferences on final payoffs and final probabilities matter — not the path taken (through a single lottery or compound lottery). In that sense, preferences are *independent* of the path taken, hence the name of the axiom.

Axiom 9 *(Continuity on lotteries) If $x^1 \succeq x^2 \succeq x^3$, then there exists a probability $\pi, 0 \leq \pi \leq 1$, such that $x^2 \sim L$, where L denotes the lottery with*

(i) the collection of bundles (x^1, x^3), and

(ii) probability measure $(\pi, 1 - \pi)$.

This probability is unique unless $x^1 \sim x^3$.

Axiom 10 *(Dominance) Let L_1 denote the lottery with*

(i) the collection of bundles (x^1, x^2), and

(ii) probability measure $(\pi, 1 - \pi)$

and let L_2 denote the lottery with

(i) *the collection of bundles (x^1, x^2), and*

(ii) *probability measure $(p, 1 - p)$.*

If $x^1 \succ x^2$, then $L_1 \succ L_2$ if and only if $\pi > p$.

We now state what is perhaps the main result from utility theory. It can be shown that[5]

Theorem 4.2 *(Expected utility maximization) A decision maker, who subscribes to Axioms 1-10, facing two or more lotteries, will choose the lottery with maximum expected cardinal utility, $\Psi(x)$.*

The cardinal utility function of this theorem is called a von Neumann-Morgenstern utility function. For a cardinal utility function, the numerical value of the utility has a precise meaning (up to a linear transformation). Thus, unlike an ordinal utility function (which, when composed with any monotone function, produces another ordinal utility function), a cardinal utility function encodes information on preferences beyond rank. Two different cardinal utility functions can be consistent with the same ordinal utility function. Thus, two consumers who make the same choices under certainty may choose different lotteries.

4.2.1 Utility of Wealth

So far, we have expressed outcomes in terms of bundles of goods. In much of the finance literature and in the remainder of this book, utility functions are typically described in terms of wealth. In the case where the bundles consist of a single good — wealth — in varying quantity, the previous discussion leads directly to utility of wealth as a special case. In general, given market prices for the goods, we can define a utility function as a function of wealth as follows:

$$U(W; \mathbf{p}) = max\{\Psi(x) | \mathbf{p}^T x = W\}, \quad (4.7)$$

where $\mathbf{p}$ is the price vector, i.e., $\mathbf{p}_i$ is the price of good i. Below, whenever we use a utility function, we mean the utility of wealth, as defined here.

4.3 Risk Aversion

Many investors are not thrill-seekers. They do not seek risk for its own sake; rather, they possess varying degrees of risk aversion. In this section, we give

---

[5]For a proof, see J. Ingersoll, Jr. (1987).

precise definitions of risk aversion and the certainty equivalent.

A decision maker is *risk averse* if he would avoid a gamble with expectation zero. Formally,

Definition 4.7 (*Risk aversion*) *A decision maker with utility function U is risk averse if*

$$U(W) > E[U(W + \epsilon)] \qquad (4.8)$$

for a gamble with $E[\epsilon] = 0$.

It can be shown that

Theorem 4.3 *A decision maker is risk averse if and only if his utility function is strictly concave.*

Usually, it is assumed that $U(W)$ is twice continuously differentiable (for tractability), increasing (more is preferred to less), and strictly concave (risk aversion). For tractability, in Section 4.6, we shall further restrict the class of utility functions that we consider in this book.

Given the notions of a utility function, the Expected Utility Maximization Theorem, and a risk averse investor, it is natural to ask the following questions:

(*i*) How much would we have to compensate an investor to take a fair gamble?

(*ii*) How much of an insurance premium would a risk averse investor pay to avoid a fair gamble?

Indeed, as we shall see in Chapter 8.6, such notions arise naturally in the context of monetary measures of model performance.

We now develop precise, implicit formulas to answer these questions. Let ϵ denote a fair gamble, i.e., we suppose that $E[\epsilon] = 0$. Then the amount, Π_C that we would have to compensate an investor to *take* a fair gamble is given by

$$E[U(W + \Pi_C + \epsilon)] = U(W) \qquad (4.9)$$

and the insurance premium, Π_i, that an investor would pay to *avoid* a fair gamble is given by

$$E[U(W + \epsilon)] = U(W - \Pi_i). \qquad (4.10)$$

If risk is small and utility function is smooth, $\Pi_C \approx \Pi_i$ The quantity $W - \Pi_i$ is known as the certainty equivalent of the gamble $W + \epsilon$.

Next, we work toward measures of risk aversion; the first is based on (4.10). In (4.10), the insurance premium, Π_i, is subtracted from wealth; that is, Π_i represents an *absolute* insurance payment. After some manipulation, we shall obtain a corresponding so-called absolute measure of risk aversion. Assuming that the risk is small and U is smooth, the Taylor expansion of (4.10) is given by:

$$E[U(W) + \epsilon U'(W) + \frac{1}{2}\epsilon^2 U''(W) + o(\epsilon^2)] = U(W) - \Pi_i U'(W) + o(\Pi_i) \quad (4.11)$$

so

$$\frac{1}{2} var(\epsilon) U''(W) \approx -\Pi_i U'(W).$$

Solving,

$$\Pi_i \approx -\frac{1}{2} \frac{U''(W)}{U'(W)} var(\epsilon).$$

Thus, Π_i, the insurance premium that a risk averse investor would pay to avoid a fair gamble, is, approximately, proportional to the *Arrow-Pratt absolute risk aversion coefficient*

$$A(W) = -\frac{U''(W)}{U'(W)}. \tag{4.12}$$

By assuming that the payment is a fraction, Π, of wealth (that is, the payment amount is relative to the level of wealth), we obtain the defining relation for a corresponding so-called relative risk aversion:

$$E[U(W(1 + \epsilon))] = U(W(1 - \Pi)). \tag{4.13}$$

For smooth U and small risk, one can deduce (see Exercise 4.14) the relative risk aversion measure

$$-W \frac{U''(W)}{U'(W)}. \tag{4.14}$$

4.4 Some Popular Utility Functions

In this section, we provide examples of several popular utility functions. All are special (sometimes limiting) cases of the HARA (hyperbolic absolute risk aversion) utility function:

$$U(W) = \frac{1 - \gamma}{\gamma} \left(\frac{aW}{1 - \gamma} + b \right)^{\gamma}. \tag{4.15}$$

The utility functions in the following examples are normalized so that $U(1) = 0$ and $U'(1) = 1$.

Example 4.1 (*Linear utility function*)

$$U(W) = W - 1. \tag{4.16}$$

This utility function is, in effect, the utility function used in risk neutral pricing (see, for example, Duffie (1996)).

Example 4.2 (*Exponential utility function, with risk aversion parameter, κ*)

$$\frac{1 - e^{-\kappa(w-1)}}{\kappa}. \tag{4.17}$$

This utility function can be particularly tractable for certain portfolio allocation problems, due to its separability when $W = \sum_i b_i Y_i$, where i is the index of the instrument in the portfolio, Y_i is the price relative for the ith instrument, and b_i is the allocation to the ith instrument (see, for example, Madan (2006)).

Example 4.3 (*Power utility function, with constant relative risk aversion κ*)

$$U(W) = \frac{W^{1-\kappa} - 1}{1 - \kappa}. \tag{4.18}$$

This utility function is commonly used (see, for example, Morningstar (2002)), has constant relative risk aversion, and possesses striking optimality properties:

(i) an investor who wants to maximize the probability that the growth rate of invested wealth will exceed a targeted growth rate has a power utility function (see Stutzer (2003)), and

(ii) (from Luenberger (1998), p.427, under certain assumptions):

> **Growth efficiency proposition** An investor who considers only long-term performance will evaluate a portfolio on the basis of its logarithm of single-period return, using only the expected value and the variance of this quantity.

Luenberger then notes that the power utility function is approximately a weighted sum of the expected logarithm of return and the variance of that logarithm. That is, the power utility function suits the long-term investor for whom the Growth efficiency proposition holds.

Example 4.4 (*Logarithmic utility function*)

$$U(W) = log(W). \tag{4.19}$$

This is, in effect, the utility function that is maximized by the Kelly investor of Section 3.3; it follows that investors with this utility function inherit the optimality properties of Kelly investors.

Example 4.5 (*Generalized logarithmic utility function*)

$$U(W) = (1 - \gamma) \ln \left(\frac{w - \gamma}{1 - \gamma} \right). \tag{4.20}$$

Some believe that there are compelling financial reasons for all investors to adopt precisely this particular utility function (see, for example, Rubinstein (1976) and Wilcox (2003)). As we shall see, this utility function arises naturally in subsequent chapters and will play a prominent role in this book, though it can appear with different context-appropriate parameterizations. We note (see Exercise 8) that this utility function has a wealth dependent relative risk aversion of $\frac{W}{W-\gamma} \in (0,1)$.

4.5   Field Studies

Is utility theory a reasonable approximation of behavior? In some empirical and field studies (see, for example, Schoemaker (1980)), an attempt is made to approximate an investor's utility function by asking him a series of questions that reveal his preferences with regard to simple lotteries. For example, if we assume that $U(1) = 0$, and $U(100) = 1$, to find $U(50)$, a subject is asked to consider the choice:

(*i*) Obtain a wealth of 50 with certainty, or

(*ii*) Obtain wealth of 100 with probability p, or wealth of 1 with probability $1 - p$.

The subject is then asked which p value makes him indifferent, i.e., for which value of p is

$$U(50) = (1 - p)U(1) + pU(100) ? \tag{4.21}$$

In this manner, it is possible to calibrate and interpolate utility functions.

The results of such experiments are mixed. According to Schoemaker (1980), "People are poor intuitive statisticians." Thus, it might not be easy to identify an investor's utility function, and, of course, allocation decisions (and their consequences) based on expected utility maximization may be sensitive to assumptions on the utility function.

4.6   Our Assumptions

We make the following assumptions about our investor throughout this book.

Assumption 4.1 *Our investor subscribes to the axioms of utility theory.*[6] *The investor has a utility function, U, that*

(i) *is strictly concave,*

(ii) *is twice differentiable,*

(iii) *is strictly monotone increasing, and*

(iv) *has the property* $(0, \infty) \subseteq range(U')$.

Unless otherwise noted, these assumptions will hold for the remainder of this book.

4.6.1 Blowup and Saturation

We note that condition (*iv*) of Assumption 4.1 is a technical condition that we shall use to streamline the exposition, and that some of the results in this book could be developed in more general settings. According to this last condition, there exists a level of wealth, $W_b \in [-\infty, \infty]$, such that

$$\lim_{W \to W_b^+} U'(W) = \infty \qquad (4.22)$$

and there exists a level of wealth, $W_s \in [-\infty, \infty]$, such that

$$\lim_{W \to W_s^-} U'(W) = 0. \qquad (4.23)$$

We formalize this discussion with the definitions

Definition 4.8 *(Blowup) The utility function, U, blows up at the wealth level,* W_b, *if*

$$\lim_{W \to W_b^+} U'(W) = \infty. \qquad (4.24)$$

Definition 4.9 *(Saturation) The utility function, U, is saturated at the wealth level,* W_s, *if*

$$\lim_{W \to W_s^-} U'(W) = 0. \qquad (4.25)$$

In Chapter 5, we shall revisit the notion of a horse race market and make additional assumptions on the blowup point, W_b, and the saturation point, W_s, and relate them to the payoffs in the market that we consider.

We note that many popular utility functions are consistent with the above conditions.

---

[6]As we have seen in Section 4.2, Theorem 4.2, such an investor maximizes his expected utility with respect to the probability measure that he believes.

Example 4.6 (*Blowup and saturation: power utility*) The power utility blows up at wealth zero and saturates at infinite wealth.

Example 4.7 (*Blowup and saturation: logarithmic utility*)
The generalized logarithmic utility

$$U(W) = \alpha \log(W - \gamma) + \beta \tag{4.26}$$

blows up at $W = \gamma$ and saturates at $W = \infty$.

Example 4.8 (*Blowup and saturation: exponential utility*)
The exponential utility

$$U(W) = 1 - e^{-W} \tag{4.27}$$

blows up at $W = -\infty$ and saturates at $W = \infty$.

Example 4.9 (*Blowup and saturation: quadratic utility*)
The quadratic utility

$$U(W) = (W - 1) - \frac{1}{2}(W - 1)^2 \tag{4.28}$$

blows up at $W = -\infty$ and saturates at $W = 2$.

However, not all utility functions are consistent with the blowup and saturation conditions listed above.

Example 4.10 (*Blowup and saturation: a utility function that never saturates*)
The utility function
$$U(W) = W - e^{-W} \tag{4.29}$$

does satisfy the first three conditions of Assumption 4.1 and does blow up at $W = -\infty$, but never saturates.

4.7   Exercises

1. Show that the sum in (4.5) converges.

2. Show that the sum in (4.5) is monotone decreasing in c with a range of $(-\infty, \infty)$.

3. Show numerically that for the St. Petersburg paradox game, a logarithmic utility investor with initial wealth 1 ducat will borrow .67 ducats and pay at most (assuming multiples of .01 ducats) 1.67 ducats to play the game, an investor with 2 ducats will pay at most 2 ducats to play the game, an investor with 1,000 ducats will pay at most 5.96 ducats to play, while an investor with 1,000,000 ducats will pay at most 10.93 ducats to play.

4. Which utility functions have constant Arrow-Pratt absolute risk aversion? (Solve the appropriate ordinary differential equation.)

5. For small risk and smooth utility functions, derive the approximation for the measure of relative risk aversion (4.14). (Hint: use Taylor expansions.)

6. Which utility functions have constant relative risk aversion? (Solve the appropriate ordinary differential equation.)

7. Justify the term HARA for the HARA utility function

$$U(W) = \frac{1-\gamma}{\gamma} \left(\frac{aW}{1-\gamma} + b \right)^{\gamma}. \tag{4.30}$$

Show how to set the parameters for the special cases described in Examples 4.1 to 4.5 (you may need additional additive constants).

8. Show that the generalized logarithmic utility function

$$U(W) = (1-\gamma) \ln \left(\frac{W-\gamma}{1-\gamma} \right) \tag{4.31}$$

has a wealth dependent relative risk aversion given by

$$\frac{W}{W-\gamma} \in (0,1). \tag{4.32}$$

9. Confirm the blowup and saturation properties listed in Examples 4.6 to 4.10.

10. Confirm the constant relative risk aversion for the power utility function of Example 4.3.

Chapter 5

The Horse Race and Utility

In Chapter 3, we considered the actions of a Kelly investor in the horse race environment — the simplest possible probabilistic investment setting, where each state in the probabilistic model has its own payoff. In this chapter, we consider the actions of a (more general) expected utility maximizing investor who operates in horse race environments.

We explore four settings: the unconditional discrete horse race, the conditional discrete horse race, an unconditional "continuous," and, finally, a conditional horse race where the probabilities of the various states can be described by a mixture of density functions and point masses. The material in this chapter is invoked when we consider model performance and model estimation issues.

5.1   The Discrete Unconditional Horse Races

In this section, we revisit the ideas of the discrete unconditional horse race of Chapter 3, but we consider investors with more general utility functions. In this setting, we discuss the compatibility of the horse race and the investor's utility function (as we shall see, not all horse races and utility functions are compatible, in a certain sense), optimal allocation for general (compatible) utility functions, horse races with homogeneous returns, the Kelly investor (revisited), generalized logarithmic utility functions (which, as we shall see, play a prominent role in this book), and the power utility.

5.1.1   Compatibility

We begin with an example.

Example 5.1 *(An incompatible horse race and utility.)* An investor has the utility

$$U(W) = log(W - 1). \tag{5.1}$$

Suppose that this investor is to operate in a two-state horse race environment with payoffs $\mathcal{O}_1 = 2$ and $\mathcal{O}_2 = 2$ that occur with probabilities p_1 and p_2, respectively. Allocating b_1 to horse 1 and $1 - b_1$ to horse 2, his expected utility would be

$$p_1 U(b_1 \mathcal{O}_1) + p_2 U((1 - b_1)\mathcal{O}_2) = p_1 log(2b_1 - 1) + p_2 log(2(1 - b_1) - 1). \quad (5.2)$$

This utility is equal to $-\infty$ for any value of b_1, since, for any allocation to the two states, at least one of the terms in (5.2) will blow up.

Thus, not all horse races and utility functions are compatible (we give a formal definition for compatibility below). There is nothing inherently wrong with either the utility function or the horse race of Example 5.1. The problem is that the utility function blows up at a (too large) value of wealth that is not attainable with certainty in this particular horse race. No allocation will be free of states where the payoff is less than the blowup point, resulting in an expected utility that blows up (is $-\infty$) for all allocations.

We see from the following example that in a more favorable horse race (with higher odds ratios), the optimal allocation problem is well defined.

Example 5.2 *(A compatible horse race and utility.)* An investor has the utility

$$U(W) = log(W - 1). \quad (5.3)$$

Suppose that this investor is to operate in a two-state horse race environment with payoffs $\mathcal{O}_1 = 3$ and $\mathcal{O}_2 = 3$ that occur with probabilities p_1 and p_2, respectively. Allocating b_1 to horse 1 and $1 - b_1$ to horse 2, his expected utility would be

$$p_1 U(b_1 \mathcal{O}_1) + p_2 U((1 - b_1)\mathcal{O}_2) = p_1 log(3b_1 - 1) + p_2 log(3(1 - b_1) - 1) \quad (5.4)$$

Note that this utility is a well-defined concave function of b_1, suitable for maximization, for $b_1 \in \left(\frac{1}{3}, \frac{2}{3}\right)$.

Fortunately, it is possible to identify conditions that guarantee that the horse race and the investor's utility are jointly tractable. We now seek such conditions.

An investor operating in the discrete unconditional horse race setting with the allocation b will derive utility

$$U(b_y \mathcal{O}_y) \quad (5.5)$$

should state y occur. Suppose that the investor's utility function blows up at the wealth level W_b (possibly, $W_b = -\infty$). In order for the utility of the allocation b to be defined, his allocation must satisfy the constraint

$$b_y \mathcal{O}_y > W_b, \quad (5.6)$$

for all y. We refer to allocations that do not cause the utility to blow up as *admissible allocations*.

Definition 5.1 *An allocation is admissible if, for each horse race outcome, the utility is finite.*

It follows from the wealth constraint, (5.6), and the definition of the bank account, B, that if there exists an admissible allocation b, we must have

$$1 = \sum_y b_y > \sum_y \frac{W_b}{\mathcal{O}_y} = \frac{W_b}{B}; \tag{5.7}$$

so, we obtain the natural compatibility condition

$$W_b < B. \tag{5.8}$$

That is, for the utility function U and the odds ratios $\mathcal{O}_y$ to admit an admissible allocation, we must have $W_b < B$.

Next, using the representation of B, (3.1), we note that

$$B = \frac{1}{\sum_y \frac{1}{\mathcal{O}_y}} < \frac{1}{\frac{1}{\mathcal{O}'_y}} = \mathcal{O}_{y'}, \forall y' \in \mathcal{Y}, \tag{5.9}$$

so

$$B < \min_y \mathcal{O}_y. \tag{5.10}$$

If we want our utility function to be increasing (not yet saturated) for each of the horse race payoffs, we must have

$$W_s > \max_y \mathcal{O}_y. \tag{5.11}$$

From these two inequalities, we see that we must have

$$W_s > B. \tag{5.12}$$

In this book, we confine our attention to utility functions and horse races that are compatible in the sense that there always exists an admissible allocation and the utility function is, in fact, strictly increasing for each of the horse race payoffs. This discussion leads us to the following definition.

Definition 5.2 *(Compatibility) The utility function, U, and a horse race with bank account, B, are compatible if*

(i) U *blows up at the value W_b, with $W_b < B$ (equivalently, $(U')^{-1}(\infty) < B$), and*

(ii) U *saturates at a value W_s, with $W_s > B$ (equivalently, $(U')^{-1}(0) > B$),*

where B denotes the bank account derived from the odds ratios, given by (3.1).

We shall see below (Lemma 5.1) that if a utility function and a horse race are compatible, then there indeed exists an admissible allocation.

We note that in the discrete market setting, whether these conditions are satisfied or not depends on the investor's utility function and the horse race payoffs (the market). In the discrete market setting, these conditions have nothing whatsoever to do with the probabilities associated with the horse race.[1]

Example 5.3 (*Compatibility: power utility*) The power utility is compatible with all horse races.

Example 5.4 (*Compatibility: generalized logarithmic utility*) The generalized logarithmic utility

$$U(W) = \alpha \log(W - \gamma B) + \beta \tag{5.13}$$

is compatible with all horse races provided that $\gamma < 1$. Thus, the usual logarithmic utility, $U(W) = log(W)$, is compatible with all horse races.

Example 5.5 (*Compatibility: exponential utility*) The exponential utility

$$U(W) = 1 - e^{-W} \tag{5.14}$$

is compatible with all horse races.

Example 5.6 (*Compatibility: quadratic utility*) The quadratic utility

$$U(W) = (W - 1) - \frac{1}{2}(W - 1)^2 \tag{5.15}$$

is compatible with all horse races for which $B < 2$.

As we shall see in Section 5.1.2, the compatibility conditions given in Definition 5.2 render the allocation problem tractable.

5.1.2   Allocation

We assume that our investor believes the model q and therefore allocates his \$1 so as to maximize his expected utility under q. It follows then from that fact the investor's wealth after the bet is $b_y \mathcal{O}_y$, if horse y won the race, that our investor chooses the following allocation.

Definition 5.3 *The optimal allocation for an investor who believes the model q is*

$$b^*(q) = \arg \max_{\{b:\sum_{y\in\mathcal{Y}} b_y = 1\}} \sum_{y\in\mathcal{Y}} q_y U(b_y \mathcal{O}_y) . \tag{5.16}$$

---

[1]When we go beyond the discrete market setting, the probabilities will play a role in the notion of compatibility.

The following lemma gives an explicit expression for the optimal allocation.

Lemma 5.1 *If the utility function U and the odds ratios $\mathcal{O}$ are compatible, then*

$$\sum_{y \in \mathcal{Y}} \frac{1}{\mathcal{O}_y} (U')^{-1} \left(\frac{\lambda}{q_y \mathcal{O}_y} \right) \tag{5.17}$$

is a strictly monotone decreasing function of λ, there exists a unique solution, λ, to the equation

$$\sum_{y \in \mathcal{Y}} \frac{1}{\mathcal{O}_y} (U')^{-1} \left(\frac{\lambda}{q_y \mathcal{O}_y} \right) = 1, \tag{5.18}$$

and

$$b_y^*(q) = \frac{1}{\mathcal{O}_y} (U')^{-1} \left(\frac{\lambda}{q_y \mathcal{O}_y} \right). \tag{5.19}$$

We note that it is straightforward to find the optimal allocation by numerical methods: by the monotonicity of (5.17) in λ, one can solve (5.18) for λ numerically by root search and obtain the optimal allocation by substituting λ into (5.19).

Proof: The Lagrangian for the convex optimization problem posed by (5.16) is

$$\mathcal{L} = -\sum_{y \in \mathcal{Y}} q_y U(b_y \mathcal{O}_y) + \lambda \left(\sum_{y \in \mathcal{Y}} b_y - 1 \right). \tag{5.20}$$

In order to find the optimal allocation, we have to solve

$$\frac{\partial \mathcal{L}}{\partial b_y} \bigg|_{b=b^*} = 0, \tag{5.21}$$

i.e.,

$$-q_y \mathcal{O}_y U'(b_y^* \mathcal{O}_y) + \lambda = 0 \tag{5.22}$$

for b^*. So we must have

$$U'(b_y^* \mathcal{O}_y) = \frac{\lambda}{q_y \mathcal{O}_y}, \tag{5.23}$$

i.e.,

$$b_y^* = \frac{1}{\mathcal{O}_y} (U')^{-1} \left(\frac{\lambda}{q_y \mathcal{O}_y} \right). \tag{5.24}$$

From

$$\sum_{y \in \mathcal{Y}} b_y^* = 1 \tag{5.25}$$

it follows that λ must satisfy (5.18), i.e.,

$$\sum_{y \in \mathcal{Y}} \frac{1}{\mathcal{O}_y} (U')^{-1} \left(\frac{\lambda}{q_y \mathcal{O}_y} \right) = 1 . \tag{5.26}$$

We note that by the strict concavity of $U(W)$, $U'(W)$ is a strictly decreasing function of W, so $(U')^{-1}$ is a strictly decreasing function. It follows that the left hand side of (5.26) is a strictly decreasing function of λ.

Since U and B are compatible, we have

$$(U')^{-1}(0) = W_s > B \tag{5.27}$$

and

$$(U')^{-1}(\infty) = W_b < B. \tag{5.28}$$

Since the left hand side of (5.26) is a strictly decreasing function of λ, the maximum (minimum) value for the left hand side of (5.26) can be obtained by allowing λ to approach 0 (∞). Under the conditions (5.27) and (5.28), we see that the maximum (minimum) value is greater (less) than 1. Since (5.26) depends continuously and monotonically on λ, by the Intermediate Value Theorem, there exists a λ that satisfies (5.26). □

Corollary 5.1 b_y *is bounded above and below, as follows:*

(i)

$$b_y^*(q) \geq \frac{W_b}{\mathcal{O}_y} \tag{5.29}$$

and

$$b_y^*(q_y) = \frac{W_b}{\mathcal{O}_y} \text{ if and only if } q_y = 0. \tag{5.30}$$

(ii)

$$b_y^*(q) \leq 1 - W_b \sum_{y \neq y'} \frac{1}{\mathcal{O}_y} \tag{5.31}$$

and

$$b_y^*(q_y) = 1 - W_b \sum_{y \neq y'} \frac{1}{\mathcal{O}_y} \text{ if and only if } q_y = 1. \tag{5.32}$$

Proof:

(i) $(U')^{-1}$ is a decreasing function with

$$\lim_{x \to \infty} (U')^{-1}(x) = W_b. \tag{5.33}$$

From (5.19), since $(U')^{-1}$ is a decreasing function, $(U')^{-1}$ must be greater than W_b for positive values of q_y, with (considering the extended value $x = \infty$) $(U')^{-1}(\infty) = W_b$. So, by (5.19), we have

$$b_y^*(q) \geq \frac{W_b}{\mathcal{O}_y}, \tag{5.34}$$

and

$$b_y^*(q_y) = \frac{W_b}{\mathcal{O}_y} \text{ if and only if } q_y = 0. \tag{5.35}$$

(*ii*) By the definition of b, we have

$$b_{y'}^* = 1 - \sum_{y \neq y'} b_y \tag{5.36}$$

$$\leq 1 - \sum_{y \neq y'} \frac{W_b}{\mathcal{O}_y} \text{ (by (}i\text{))} \tag{5.37}$$

$$= 1 - W_b \sum_{y \neq y'} \frac{1}{\mathcal{O}_y} \tag{5.38}$$

and

$$b_y^*(q_y) = 1 - W_b \sum_{y \neq y'} \frac{1}{\mathcal{O}_y} \text{ if and only if } q_y = 1. \tag{5.39}$$

□

Corollary 5.2 *If U blows up at a nonnegative value, then $b_y^*(q) > 0, \forall q > 0$.*

Proof: If $W_b \geq 0$, then

$$\lim_{x \to \infty} (U')^{-1}(x) = W_b \geq 0. \tag{5.40}$$

Since $(U')^{-1}$, is a decreasing function, $(U')^{-1}$ must be positive, and so, by (5.19), must b_y^*. □

We now consider the excess allocation over the bank account allocation

$$\tilde{b}_y \equiv b_y - \frac{B}{\mathcal{O}_y}. \tag{5.41}$$

Summing both sides, we see that

$$\sum_y \tilde{b}_y = 0. \tag{5.42}$$

Rewriting, we have

$$b_y = \frac{B}{\mathcal{O}_y} + \tilde{b}_y. \tag{5.43}$$

That is, the allocation b_y can be thought of as the sum of the bank account allocation and a "market neutral" allocation in the sense of (5.42). Equations (5.43) and (5.42) collectively reveal that, in a horse race, an investor effectively bets so that \$1 is allocated to the bank account and a net allocation of \$0 is distributed among the horses. It follows from Lemma 5.1 and (5.41) that the optimal "market neutral" excess allocation is given by

$$\tilde{b}_y^*(q) = \frac{1}{\mathcal{O}_y}(U')^{-1}\left(\frac{\lambda}{q_y \mathcal{O}_y}\right) - \frac{B}{\mathcal{O}_y}. \tag{5.44}$$

In the following sections, we examine allocation in four important contexts that will recur throughout this book.

5.1.3   Horse Races with Homogeneous Returns

As we shall soon see, horse races with homogeneous returns are particularly tractable. Moreover, as we shall see when we discuss model performance measurement and model estimation, it is possible to make certain approximations for horse races with nearly homogeneous returns.

Consider the homogeneous expected return measure from Definition 3.3, in Chapter 3, which is given by

$$p^{(h)} = \left\{ p_y^{(h)} = \frac{B}{\mathcal{O}_y}, \ y \in \mathcal{Y} \right\}, \tag{5.45}$$

where B is the bank account payoff. Recall that, under this measure, the expected payoff for a bet placed on a single horse, y, is always B, independent of y, i.e., that

$$p_y^{(h)} \mathcal{O}_y + \left(1 - p_y^{(h)}\right) 0 = B, \ \forall y \in \mathcal{Y}. \tag{5.46}$$

In order to compute the optimal allocation for our investor, we substitute (5.46) into Lemma 5.1, (5.19), and obtain

$$b_y^*\left(p^{(h)}\right) = \frac{1}{\mathcal{O}_y}(U')^{-1}\left(\frac{\lambda}{B}\right) = \frac{p_y^{(h)}}{B}(U')^{-1}\left(\frac{\lambda}{B}\right).$$

Summing and using (5.18), we see that

$$1 = \frac{1}{B}(U')^{-1}\left(\frac{\lambda}{B}\right), \tag{5.47}$$

so

$$\lambda = BU'(B), \text{ and} \tag{5.48}$$

$$b_y^*\left(p^{(h)}\right) = p_y^{(h)}. \tag{5.49}$$

That is, under the homogeneous expected return measure, we allocate according to the bank account allocation.

5.1.4   The Kelly Investor Revisited

In Section 3.3, we discussed the Kelly investor who allocates his assets so as to maximize his expected wealth growth rate (see Definition 3.6, in Chapter 3). As we shall soon see, the Kelly investor can be viewed as an expected utility maximizing investor with a particular utility function.

According to Definition 3.5, the expected wealth growth rate corresponding to a probability measure q and a betting strategy b is given by

$$W(b,q) = E_q\left[\log\left(b,\mathcal{O}\right)\right] = \sum_{y\in\mathcal{Y}} q_y \log(b_y \mathcal{O}_y) . \tag{5.50}$$

Comparing this to the expected utility, $E_q\left[U\left(b,\mathcal{O}\right)\right]$, we see that the Kelly investor is, in fact, an expected utility maximizing investor with the utility function

$$U(W) = \log W . \tag{5.51}$$

It follows from Lemma 5.1, (5.18), that $\lambda = 1$ and from (5.19) that

$$b_y^*(q) = q_y , \tag{5.52}$$

which is consistent with Theorem 3.1.

The logarithmic utility function is important both for its tractability and the following optimality properties:[2]

 (*i*) the probability that the ratio of the wealth of a Kelly investor to the wealth of a non-Kelly investor after n trials will exceed any constant can be made as close to 1 as we please, for n sufficiently large, and

 (*ii*) the expected time to double the wealth is smaller for the Kelly investor than for any other investor.

However, in spite of these optimality properties, the logarithmic utility may not be appropriate for many investors. For a criticism of the logarithmic utility function, see Samuelson (1971) and Samuelson (1979); for a justification of nearly, but not quite, logarithmic utility functions, see Luenberger (1998) pp. 426-427. Browne and Whitt (1996) and Janeček (2002) discuss drawdown consequences of the logarithmic and power utilities and show that the logarithmic utility function can lead to investment strategies that are quite aggressive.

---

[2]For proofs of these statements, see, for example, Cover and Thomas (1991), Chapter 6, or Breiman (1961). Some of these properties also hold for Kelly investors who invest in continuous-time markets (see Merton (1971), Karatzas et al. (1991), Jamishidian (1992), and Browne and Whitt (1996)).

5.1.5 Generalized Logarithmic Utility Function

In this section, we discuss the generalized logarithmic utility function, which will play a crucial role in the remainder of this book. We motivate this utility function by first considering two extremes with respect to conservatism in allocation. Recall that the bank account allocation

$$b_y = \frac{B}{\mathcal{O}_y} \tag{5.53}$$

results in the payoff B, no matter which horse wins the race. This is the ultimate conservative strategy — there is no uncertainty in the outcome — no risk. It is possible to identify another extreme: the growth-optimal and doubling-time-optimal Kelly allocation strategy

$$b_y = q_y, \tag{5.54}$$

which is too aggressive for many investors (there are riskier strategies, but none that have the optimality properties of the Kelly strategy, so, at least from the perspective of these optimality properties, there is nothing to be gained by considering them, since they provide more risk and suboptimal growth).

It is natural to consider the weighted allocation strategy

$$b_y = (1 - \gamma)q_y + \gamma\frac{B}{\mathcal{O}_y} \tag{5.55}$$

and ask the question: for what utility function (if any) is the weighted strategy optimal for an expected utility maximizing investor, for general odds ratios $\mathcal{O}$ and probability measures, q? To answer this question, we solve

$$b_y^*(q) = (1 - \gamma)q_y + \gamma\frac{B}{\mathcal{O}_y}. \tag{5.56}$$

From (5.19), we must have

$$\frac{1}{\mathcal{O}_y}(U')^{-1}\left(\frac{\lambda}{q_y\mathcal{O}_y}\right) = (1 - \gamma)q_y + \gamma\frac{B}{\mathcal{O}_y}. \tag{5.57}$$

It follows that

$$U'\left((1 - \gamma)q_y\mathcal{O}_y + \gamma B\right) = \frac{\lambda}{q_y\mathcal{O}_y}. \tag{5.58}$$

Integrating with respect to $q_y\mathcal{O}_y$, we obtain

$$\frac{1}{1 - \gamma}U\left((1 - \gamma)q_y\mathcal{O}_y + \gamma B\right) = \lambda\log(q_y\mathcal{O}_y) + const. \tag{5.59}$$

Putting

$$W = (1 - \gamma)q_y\mathcal{O}_y + \gamma B, \tag{5.60}$$

we obtain

$$U(W) = (1-\gamma)\lambda \log\left(\frac{W-\gamma B}{1-\gamma}\right) + const = \alpha \log(W - \gamma B) + \beta, \quad (5.61)$$

where

$$\alpha = (1-\gamma)\lambda \quad (5.62)$$

and β is a constant incorporating an integration constant. Since the blowup point must be less than B, for this function to be compatible with the horse race, we must have $\gamma < 1$.

Thus, by considering weighted combinations of the two extreme allocations optimal for

(*i*) the rather aggressive Kelly investor, and

(*ii*) the conservative investor who tolerates no risk,

we are naturally led to the generalized logarithmic family of utility functions, defined as follows.

Definition 5.4 *The generalized logarithmic utility function is defined by*

$$U(W) = \alpha \log(W - \gamma B) + \beta \,, \ \alpha > 0\,. \quad (5.63)$$

Here γB represents the blowup wealth level; as we have seen, for this utility to be compatible with all horse races, we must have $\gamma < 1$. As we have seen, the special case of this utility function with $\gamma = 0$ was suggested by Bernoulli (1738) in his solution to the St. Petersburg paradox. The more general form (5.63) has been advocated by Rubinstein (1976) in a widely cited article entitled *The Strong Case for the Generalized Logarithmic Utility as the Premier Model of Financial Markets* and by Wilcox (2003). We note that neither Rubinstein nor Wilcox focuses on statistical learning problems and their motivations are quite different from ours. Rubinstein is more interested in utility functions that can be used to make effective financial models. Wilcox is more interested in investment tools that can accommodate conservative investors who wish to avoid shortfalls.

One can show directly, as an exercise, that a generalized logarithmic utility investor allocates according to

$$b_y^*(q) = (1-\gamma)\, q_y + \gamma \frac{B}{\mathcal{O}_y}\,. \quad (5.64)$$

That is, the allocation is the weighted sum of the allocation of the Kelly investor and the bank account allocation, where the weights are given by $1-\gamma$ and γ, respectively.

This investor will never experience a wealth less than γB. To see this, note that, from (5.64),

$$b_y^*(q)\mathcal{O}_y = (1-\gamma)\, q_y \mathcal{O}_y + \gamma B > \gamma B, \text{ for } \gamma < 1. \quad (5.65)$$

It follows from (5.65) that

$$b_y^*(q) > \gamma \frac{B}{\mathcal{O}_y} > 0, \text{ for } \gamma > 0. \tag{5.66}$$

Thus, we have shown that

Theorem 5.1 *An expected utility maximizing investor allocates according to the weighted average of the Kelly allocation and the bank account allocation*

$$(1 - \gamma)q_y + \gamma \frac{B}{\mathcal{O}_y} \tag{5.67}$$

if and only if his utility function is a generalized logarithmic utility function,

$$U(W) = \alpha \log(W - \gamma B) + \beta \ , \ \alpha > 0, \gamma < 1. \tag{5.68}$$

An investor with this utility function is guaranteed to have wealth exceeding γB.

As we shall see, the generalized logarithmic utility will play an important role in this book in the context of model performance measurement and model estimation.

An investor who allocates according to such a generalized logarithmic utility acts exactly as an investor who first allocates the fraction γ of his wealth to a bank account, and then allocates the remaining fraction $1 - \gamma$ of his wealth as would a Kelly investor with wealth $1 - \gamma$.

Among strategies that "insure" a payoff not less than γB, this strategy inherits the optimality properties of the Kelly investor on the remaining fraction, $1 - \gamma$, of initial wealth (see Exercises 5 and 6):

(*i*) the probability that the ratio of the wealth of a generalized logarithmic utility expected utility maximizing investor to the wealth of an alternative investor, who "banks" γB, after n trials will exceed any constant can be made as close to 1 as we please, for n sufficiently large, and,

(*ii*) the expected time to double the generalized logarithmic utility investor's wealth is smaller than for any alternative investor.

5.1.6   The Power Utility

As an additional optimal allocation example, we consider an investor with a power utility function, which is given by

$$U_\kappa(W) = \frac{W^{1-\kappa} - 1}{1 - \kappa} \to log(W) \text{ as } \kappa \to 1, \tag{5.69}$$

where $\kappa \geq 0$. The power utility family is compelling because it is the most general family with constant relative risk aversion. We see from (5.18) that

$$\lambda = \left(\sum_{y'} \frac{1}{\mathcal{O}_{y'}} (q_{y'} \mathcal{O}_{y'})^{\frac{1}{\kappa}} \right)^{\kappa} ; \qquad (5.70)$$

from (5.19), we see that

$$b_y^*(q) = \frac{\frac{1}{\mathcal{O}_y} (q_y \mathcal{O}_y)^{\frac{1}{\kappa}}}{\sum_{y'} \frac{1}{\mathcal{O}_{y'}} (q_{y'} \mathcal{O}_{y'})^{\frac{1}{\kappa}}}. \qquad (5.71)$$

It follows that

$$U_\kappa(b_y^*(q)\mathcal{O}_y) = \frac{1}{1-\kappa} \left[\left(\frac{(q_y \mathcal{O}_y)^{\frac{1}{\kappa}}}{\sum_{y'} \frac{1}{\mathcal{O}_{y'}} (q_{y'} \mathcal{O}_{y'})^{\frac{1}{\kappa}}} \right)^{1-\kappa} - 1 \right]. \qquad (5.72)$$

5.2 Discrete Conditional Horse Races

We shall see that most of the definitions and results from Section 5.1 can be easily generalized to the discrete conditional horse race setting.

We consider probabilities of a random variable Y with state space $\mathcal{Y}$ given values of the random variable X with state space $\mathcal{X}$, and denote marginal and conditional probability measures simply by q_x and $q_{y|x}$, respectively, rather than q_x^X and $q_{y|x}^{Y|X}$. We take the point of view of a conditional investor similar to the one defined in Definition 3.11 (only now, we consider investors with more general utility functions), in a conditional horse race, as defined in Definition 3.7. A number of probabilistic learning modeling problems can be viewed as taking place in such a discrete conditional horse race setting, for example, the corporate default probability model discussed in Section 12.1.1, or the text classification model discussed in Section 12.3.

5.2.1 Compatibility

Before discussing the question of allocating to maximize expected utility, we discuss conditions which must hold for the allocation problem to be well-defined.

Given an observation, x, an investor operating in the discrete unconditional horse race setting with the allocation b will derive the utility

$$U(b_{y|x} \mathcal{O}_{y|x}) \qquad (5.73)$$

should state y occur. Suppose that the investor's utility function blows up at the wealth level W_b (possibly, $W_b = -\infty$). In order for the utility of the allocation b to be defined for any x, his allocation must satisfy the constraint

$$b_{y|x}\mathcal{O}_{y|x} > W_b; \tag{5.74}$$

for all x and y.

It follows from the wealth constraint, (5.74), and the definition of B, that if there exists an admissible allocation for each x, then we must have

$$1 = \sum_y b_{y|x} > \sum_y \frac{W_b}{\mathcal{O}_{y|x}} = \frac{W_b}{B_x}; \tag{5.75}$$

so, we obtain the natural compatibility condition

$$W_b < B_x, \forall x. \tag{5.76}$$

That is, for the utility function U and the odds ratios $\mathcal{O}_{y|x}$ to admit an allocation with utility greater than negative infinity for all x, we must have $W_b < B_x, \forall x$.

Next, we note that

$$B_x = \frac{1}{\sum_y \frac{1}{\mathcal{O}_{y|x}}} < \frac{1}{\frac{1}{\mathcal{O}_{y'|x}}} = \mathcal{O}_{y'|x}, \forall y' \in \mathcal{Y}, \tag{5.77}$$

so

$$B_x < \mathcal{O}_{y|x}, \forall y \in \mathcal{Y}. \tag{5.78}$$

If we want our utility function to be increasing (not yet saturated), for all x, for at least one of the horse race payoffs, we must have

$$W_s > \min_y \mathcal{O}_{y|x} > B_x, \forall x. \tag{5.79}$$

This discussion leads us to

Definition 5.5 *(Compatibility, conditional discrete horse race) The utility function, U, and a conditional horse race with bank accounts, B_x, are compatible if*

(i) *U blows up at the value W_b, with $W_b < B_x, \forall x \in \mathcal{X}$ (equivalently, $(U')^{-1}(\infty) < B_x, \forall x \in \mathcal{X}$), and*

(ii) *U saturates at a value W_s, with $W_s > B_x, \forall x \in \mathcal{X}$ (equivalently, $(U')^{-1}(0) > B_x, \forall x \in \mathcal{X}$),*

where B_x denotes the conditional bank account derived from the odds ratios, $\mathcal{O}_{y|x}$ (see Definition 3.8).

With, at most, slight modification, the conditions under which the utility functions listed in Examples 5.3 to 5.6 are essentially identical to those that render the utilities compatible in this discrete conditional horse race setting.

Example 5.7 (*Power utility*) The power utility is compatible with all horse races.

Example 5.8 (*Generalized logarithmic utility*) The generalized logarithmic utility

$$U(W) = \alpha \log(W - \gamma B) + \beta \tag{5.80}$$

is compatible with all horse races if $\gamma < 1$, where B denotes the worst conditional bank account defined in Definition 3.9.

Example 5.9 (*Exponential utility*) The exponential utility

$$U(W) = 1 - e^{-W} \tag{5.81}$$

is compatible with all horse races.

Example 5.10 (*Quadratic utility*) The quadratic utility

$$U(W) = (W - 1) - \frac{1}{2}(W - 1)^2 \tag{5.82}$$

is compatible with all horse races for which $B_x < 2, \forall x \in \mathcal{X}$.

5.2.2 Allocation

We generalize Definition 5.3 as follows.

Definition 5.6 *The optimal allocation for a conditional investor who believes the (conditional probabilistic) model q is*

$$b^*(q) = \arg \max_{\{b: \sum_{y \in \mathcal{Y}} b_{y|x}=1\}} \sum_{y \in \mathcal{Y}} q_{y|x} U(b_{y|x} \mathcal{O}_{y|x}) . \tag{5.83}$$

Other definitions are possible; in particular, one could define the optimal allocation via

$$b^*(q) = \arg \max_{\{b: \sum_{y \in \mathcal{Y}} b_{y|x}=1\}} \sum_{x \in \mathcal{X}} q_x \sum_{y \in \mathcal{Y}} q_{y|x} U(b_{y|x} \mathcal{O}_{y|x}) . \tag{5.84}$$

This definition leads to exactly the same allocation as Definition 5.6. The reason for this equivalence is that the alternative definition prescribes a maximization of a linear combination of the terms that are maximized under Definition 5.6. Since the coefficients of the linear combination are all positive, each term can be maximized independently, as per Definition 5.6.

The following lemma, which generalizes Lemma 5.1, gives an explicit expression for the optimal allocation.

Lemma 5.2 *If the utility function U and the odds ratios $\mathcal{O}_{y|x}$ are compatible, then*

$$\sum_{y \in \mathcal{Y}} \frac{1}{\mathcal{O}_{y|x}} (U')^{-1} \left(\frac{\lambda_x}{q_{y|x}\mathcal{O}_y} \right) \tag{5.85}$$

is a strictly monotone decreasing function of λ_x, there exists a unique solution, in λ_x, to the equation

$$\sum_{y \in \mathcal{Y}} \frac{1}{\mathcal{O}_{y|x}} (U')^{-1} \left(\frac{\lambda_x}{q_{y|x}\mathcal{O}_y} \right) = 1, \tag{5.86}$$

and

$$b^*_{y|x}(q) = \frac{1}{\mathcal{O}_{y|x}} (U')^{-1} \left(\frac{\lambda_x}{q_{y|x}\mathcal{O}_{y|x}} \right). \tag{5.87}$$

Proof: The proof is essentially the same as the one for Lemma 5.1 in Section 5.1.2.

5.2.3 Generalized Logarithmic Utility Function

It follows from Lemma 5.2 that for a utility function of the form,

$$U(W) = \alpha \log(W - \gamma B) + \beta \tag{5.88}$$

with $\alpha > 0$ and $\gamma B < \min_{x \in \mathcal{X}}(B_x)$, the optimal betting weights are given by

$$b^*_{y|x}(q) = q_{y|x} \left[1 - \frac{\gamma B}{B_x} \right] + \frac{\gamma B}{\mathcal{O}_{y|x}}. \tag{5.89}$$

5.3 Continuous Unconditional Horse Races

Until now we have discussed horse races on a random variable Y with a discrete (finite) state space. We now consider horse races where payoffs depend on random variables that can take any value in a volume element by taking the small-size limit of horse races associated with a series of discretizations of the support of Y.

This section contains no new ideas beyond a technical discussion allowing us to pass from the sums of the discrete setting of the previous sections in this chapter to a continuous setting which makes use of integrals. Readers who are not interested in such technical details may skip this section.

5.3.1 The Discretization and the Limiting Expected Utility

We denote the support of the continuous, possibly vector valued, random variable Y by $\mathcal{Y}$ and assume that $\mathcal{Y} \subset \mathbf{R}^d$. For the sake of convenience, we have chosen the same notation ($\mathcal{Y}$) to represent the support of the random variable considered in this section as the notation for the finite state spaces in the previous section; it will always be clear from the context which type of support we are referring to. We denote by q a bounded probability density on $\mathcal{Y}$, which has the property

$$\int_{\mathcal{Y}} q(y)dy = 1 \ . \tag{5.90}$$

In order to define horse races for probability densities, we discretize $\mathcal{Y}$. To this end, we define a set of partitions,

$$\{\mathcal{Y}_k^{(n)}\}_{k=1}^n, \tag{5.91}$$

of $\mathcal{Y}$, indexed by n, with

$$\bigcup_{k=1}^n \mathcal{Y}_k^{(n)} = \mathcal{Y}, \tag{5.92}$$

$$\mathcal{Y}_k^{(n)} \cap \mathcal{Y}_j^{(n)} = \emptyset \text{ for } j \neq k \ . \tag{5.93}$$

We assume that this sequence of partitions has the following property.

$$\max_k \Delta_k^{(n)} \to 0, \text{ as } n \to \infty \ , \tag{5.94}$$

$$\text{where } \Delta_k^{(n)} = \int_{\mathcal{Y}_k^{(n)}} dy \ . \tag{5.95}$$

Let $y_k^{(n)} \in \mathcal{Y}_k^{(n)}, k = 1 \ldots, n$. The probability that $Y \in \mathcal{Y}_k^{(n)}$ is then

$$q_k^{(n)} = \int_{\mathcal{Y}_k^{(n)}} q(y)dy = q(y_k^{(n)})\Delta_k^{(n)} + o(\Delta_k^{(n)}) \ . \tag{5.96}$$

We now introduce the "odds ratio density" $\mathcal{O}(y)$ and assume that the odds ratio for a bet on $Y \in \mathcal{Y}_k^{(n)}$, $\mathcal{O}_k^{(n)}$, is related to this "odds ratio density" via

$$\mathcal{O}_k^{(n)} = \frac{\mathcal{O}(y_k^{(n)})}{\Delta_k^{(n)}} + \frac{o(\Delta_k^{(n)})}{\Delta_k^{(n)}}. \tag{5.97}$$

This crucial assumption is plausible; as the partition elements, $\mathcal{Y}_k^{(n)}$, become smaller, more numerous, and less probable, the market maker must pay more to attract an investor. We can think of the odds ratio density, together with a particular discretization, as generating a discrete horse race. Together with (5.96), (5.97) implies that

$$E_q(\text{payoff on a \$1 bet on } Y \in \mathcal{Y}_k^{(n)}) = q_k^{(n)}\mathcal{O}_k^{(n)} + 0(1 - q_k^{(n)})$$

$$= q(y_k^{(n)})\mathcal{O}(y_k^{(n)}) + \frac{o(\Delta_k^{(n)})}{\Delta_k^{(n)}}. \tag{5.98}$$

Suppose that our investor allocates

$$b_k^{(n)} = \int_{\mathcal{Y}_k^{(n)}} b(y)dy = b(y_k^{(n)})\Delta_k^{(n)} + o(\Delta_k^{(n)}) \tag{5.99}$$

to a bet that $Y \in \mathcal{Y}_k^{(n)}$, where $b(y)$ denotes a bounded "allocation density." Under this betting scheme, we have expected utility (under q)

$$\bar{u}^{(n)} = \sum_{k=1}^{n} q_y^{(n)} U(b_k^{(n)} \mathcal{O}_k^{(n)})$$

$$= \sum_{k=1}^{(n)} [q(y_k^{(n)})\Delta_k^{(n)} + o(\Delta_k^{(n)})]U\left([b(y_k^{(n)})\Delta_k^{(n)} + o(\Delta_k^{(n)})]\left[\frac{\mathcal{O}(y_k^{(n)})}{\Delta_k^{(n)}} + \frac{o(\Delta_k^{(n)})}{\Delta_k^{(n)}}\right]\right)$$

(by (5.96), (5.97) and (5.99))

$$= \sum_{k=1}^{n} [q(y_k^{(n)})\Delta_k^{(n)} + o(\Delta_k^{(n)})]U\left([b(y_k^{(n)})\mathcal{O}(y) + o(\Delta_k^{(n)})\frac{\mathcal{O}(y_k^{(n)})}{\Delta_k^{(n)}} + o(\Delta_k^{(n)})]\right).$$

Expanding U to leading order, we have

$$\bar{u}^{(n)} = \sum_{k=1}^{n} [q(y_k^{(n)})\Delta_k^{(n)} + o(\Delta_k^{(n)})]\left(U(b(y_k^{(n)})\mathcal{O}(y)) + \frac{o(\Delta_k^{(n)})\mathcal{O}(y)}{\Delta_k^n}U'(b(y_k^n)\mathcal{O}(y_k^n))]\right)$$

$$= \sum_{k=1}^{n} [q(y_k^n)U(b(y_k^n)\mathcal{O}(y_k^n))\Delta_k^n + o(\Delta_k^n)].$$

Under the assumption of equation (5.94), we see that

$$\lim_{n\to\infty} \bar{u}^n = \int_{\mathcal{Y}} q(y)U(b(y)\mathcal{O}(y))dy , \tag{5.100}$$

i.e., that the limit of the expected utilities associated with the sequence of discrete horse races defined by the method of this section, will be given by the right hand side of (5.100). We note that, informally, the right hand side of this expression can be derived from its discrete counterpart by replacing a summation by an integral.

5.3.2 Compatibility

Before discussing the question of allocating to maximize expected utility, we discuss conditions which must hold for the allocation problem to be well-defined.

Assuming that

$$\frac{1}{\mathcal{O}(y)} \tag{5.101}$$

is an integrable function, we note that the allocation

$$b(y) = \frac{\frac{1}{\mathcal{O}(y)}}{\int \frac{1}{\mathcal{O}(y)} dy} \tag{5.102}$$

"pays"

$$b(y)\mathcal{O}(y) = \frac{1}{\int \frac{1}{\mathcal{O}(y)} dy} \tag{5.103}$$

in each state; that is, the allocation in (5.102) has a fixed, deterministic payoff density, no matter which state occurs. From the point of view of the sequence of discrete horse races discussed in the previous section, note that we have observed that

$$b_k^{(n)} \mathcal{O}_k^{(n)} = b(y_k^{(n)})\mathcal{O}(y) + o(\Delta_k^{(n)}); \tag{5.104}$$

so under the discrete allocations derived from (5.102), the discrete horse races will have the payoff

$$\frac{1}{\int \frac{1}{\mathcal{O}(y)} dy} + o(\Delta_k^{(n)}). \tag{5.105}$$

This suggests the following definition for the bank account in this continuous horse race setting:

$$B = \frac{1}{\int \frac{1}{\mathcal{O}(y)} dy}. \tag{5.106}$$

We define compatibility in this context in a manner similar to that used to define compatibility in Sections 5.1 and 5.2 (but, here, the probability measure enters the picture, and we require two additional technical conditions).

Definition 5.7 *(Compatibility, continuous unconditional horse race) The utility function, U, continuous horse race with odds ratio density $\mathcal{O}(y)$, probability measure $q(y)$, and bank account, B, are compatible if*

(i) *U blows up at the value W_b, with $W_b < B$ (equivalently, $(U')^{-1}(\infty) < B$),*

(ii) *U saturates at a value W_s, with $W_s > B$ (equivalently, $(U')^{-1}(0) > B$),*

(iii) *the function*

$$\frac{1}{\mathcal{O}(y)} \tag{5.107}$$

is integrable, over $\mathcal{Y}$, and

(iv) *the quantity $q(y)\mathcal{O}(y)$ is bounded above,*

where B denotes the bank account, defined in (5.106), derived from the odds ratio density.

Condition (*iii*) is necessary to guarantee that the bank account exists. However, conditions (*i*) to (*iii*) are not sufficient to guarantee the existence of a maximum expected utility allocation. By adding condition (*iv*), as we shall see, we can always explicitly calculate the maximum expected utility allocation when U, $\mathcal{O}(y)$, and $q(y)$ are compatible.

Example 5.11 (*Power utility*) The power utility is compatible with all continuous horse races with odds ratio densities $\mathcal{O}(y)$, and all probability measures $q(y)$ for which conditions (*iii*) and (*iv*) in Definition 5.7 hold.

Example 5.12 (*Generalized logarithmic utility*) The generalized logarithmic utility

$$U(W) = \alpha \log(W - \gamma B) + \beta \tag{5.108}$$

is compatible with all continuous horse races with odds ratio densities $\mathcal{O}(y)$, and all probability measures $q(y)$ for which conditions (*iii*) and (*iv*) in Definition 5.7 hold provided that $\gamma < 1$.

Example 5.13 (*Exponential utility*) The exponential utility

$$U(W) = 1 - e^{-W} \tag{5.109}$$

is compatible with all continuous horse races with odds ratio densities $\mathcal{O}(y)$, and all probability measures $q(y)$ for which conditions (*iii*) and (*iv*) in Definition 5.7 hold.

Example 5.14 (*Quadratic utility*) The quadratic utility

$$U(W) = (W - 1) - \frac{1}{2}(W - 1)^2 \tag{5.110}$$

is compatible with all continuous horse races with odds ratio densities $\mathcal{O}(y)$, and all probability measures $q(y)$ for which conditions (*iii*) and (*iv*) in Definition 5.7 hold, provided that $B < 2$.

5.3.3 Allocation

In light of the limit expressed in (5.100), an expected utility maximizing investor in this setting will allocate according to

$$b^*[q](y) = \arg \max_{\{b : \int_y b(y)dy = 1\}} \int_y q(y) U(b(y)\mathcal{O}(y))dy. \tag{5.111}$$

The following lemma gives an explicit expression for the optimal allocation.

Lemma 5.3 *If the utility function U, the odds ratios $\mathcal{O}(y)$, and the probability measure $q(y)$ are compatible, then*

$$\int_y \frac{1}{\mathcal{O}(y)} (U')^{-1} \left(\frac{\lambda}{q(y)\mathcal{O}(y)} \right) dy \tag{5.112}$$

is a strictly monotone decreasing function of λ, there exists a unique solution, λ, to the equation

$$\int_{\mathcal{Y}} \frac{1}{\mathcal{O}(y)} (U')^{-1} \left(\frac{\lambda}{q(y)\mathcal{O}(y)} \right) dy = 1, \tag{5.113}$$

and

$$b^*[q](y) = \frac{1}{\mathcal{O}(y)} (U')^{-1} \left(\frac{\lambda}{q(y)\mathcal{O}(y)} \right), \tag{5.114}$$

where λ is the solution of (5.113).

Proof: The proof is similar to that given for Lemma 5.1, but in this continuous setting, we make use of the first variation of the Lagrangian.

The Lagrangian for the convex optimization problem posed by (5.111) is

$$\mathcal{L} = - \int_{y \in \mathcal{Y}} q(y)U(b(y)\mathcal{O}(y))dy + \lambda \left(\int_{y \in \mathcal{Y}} b(y)dy - 1 \right). \tag{5.115}$$

Taking the first variation of the Lagrangian resulting from a variation in $b(y)$, $\delta b(y)$, we obtain

$$\delta \mathcal{L} = - \int_{y \in \mathcal{Y}} q(y)U'(b(y)\mathcal{O}(y))\mathcal{O}(y)\delta b(y)dy + \lambda \int_{y \in \mathcal{Y}} \delta b(y)dy. \tag{5.116}$$

If $\delta \mathcal{L} = 0$ for all variations, $\delta b(y)$, we must have

$$-q(y)U'(b(y)\mathcal{O}(y))\mathcal{O}(y)\delta b(y) + \lambda \delta b(y) = 0, \tag{5.117}$$

which implies that for $b^*[q]$ we must have

$$U'(b^*[q](y)\mathcal{O}(y)) = \frac{\lambda}{q(y)\mathcal{O}(y)}, \tag{5.118}$$

i.e.,

$$b^*[q](y) = \frac{1}{\mathcal{O}(y)} (U')^{-1} \left(\frac{\lambda}{q(y)\mathcal{O}(y)} \right). \tag{5.119}$$

Next, we show that under the assumption that $q(y)\mathcal{O}(y)$ is bounded above (condition (iv) of Definition 5.7), the integral

$$\int_{y \in \mathcal{Y}} \frac{1}{\mathcal{O}(y)} (U')^{-1} \left(\frac{\lambda}{q(y)\mathcal{O}(y)} \right) dy \tag{5.120}$$

exists for all positive values of λ. Suppose that $q(y)\mathcal{O}(y)$ is bounded above by M. Then

$$\frac{\lambda}{q(y)\mathcal{O}(y)} > \frac{\lambda}{M}. \tag{5.121}$$

We note that by the strict concavity of $U(W)$, $U'(W)$ is a strictly decreasing function of W, so $(U')^{-1}$ is a strictly decreasing function; so, applying (5.121), we see that

$$\int_{y \in \mathcal{Y}} \frac{1}{\mathcal{O}(y)} (U')^{-1} \left(\frac{\lambda}{q(y)\mathcal{O}(y)} \right) dy \leq \int_{y \in \mathcal{Y}} \frac{1}{\mathcal{O}(y)} (U')^{-1} \left(\frac{\lambda}{M} \right) dy, \quad (5.122)$$

which is integrable, by condition (*iii*) of Definition 5.7.

From

$$\int_{y \in \mathcal{Y}} b^*[q](y) dy = 1 \qquad (5.123)$$

it follows that λ must satisfy (5.113), i.e.,

$$\int_{y \in \mathcal{Y}} \frac{1}{\mathcal{O}(y)} (U')^{-1} \left(\frac{\lambda}{q(y)\mathcal{O}(y)} \right) dy = 1 . \qquad (5.124)$$

We also see (from the fact that $(U')^{-1}$ is a strictly decreasing function) that the left hand side of (5.124) is a strictly decreasing function of λ.

Since U and B are compatible, we have

$$(U')^{-1}(0) = W_s > B \qquad (5.125)$$

and

$$(U')^{-1}(\infty) = W_b < B. \qquad (5.126)$$

Since the left hand side of (5.124) is a strictly decreasing function of λ, the maximum (minimum) value for the left hand side of (5.124) can be obtained by allowing λ to approach 0 (∞). Under the conditions (5.125) and (5.126), we see that the maximum (minimum) value is greater (less) than 1. Since (5.124) depends continuously and monotonically on λ, by the Intermediate Value Theorem, there exists a λ that satisfies (5.124). $\square$

5.3.4   Connection with Discrete Random Variables

Suppose that there is a sequence of bounded probability density functions

$$q_m(y) \to \sum_i q_{y_i} \delta(y - y_i) \text{ as } m \to \infty , \qquad (5.127)$$

where the y_i are a set of points in $\mathcal{Y}$.

Each q_m is a bounded density on $\mathcal{Y}$, so we may apply the expression (5.100) for the expected utility. We see that each q_m has associated expected utility

$$\bar{u}_m = \int_{\mathcal{Y}} q_m(y) U(b(y)\mathcal{O}(y)) dy. \qquad (5.128)$$

So we have

$$\lim_{m \to \infty} \bar{u}_m = \int_Y \sum_i q_{y_i} \delta(y - y_i) U(b(y)\mathcal{O}(y)) dy = \sum_i q_{y_i} U(b(y_i)\mathcal{O}(y_i)) ,$$

(5.129)

i.e., we have recovered the expected utility for a discrete setting.

Thus, this scheme applied to horse race type betting on continuous densities leads to betting consistent with the discrete case.

5.4 Continuous Conditional Horse Races

In this section, we describe a still more general horse race, where the probability of various outcomes can be described by a conditional density model which may include point masses. We suppose that certain individual states can occur with finite probability and that others are best described with a probability density function. That is, the vector valued random variable Y has the continuous conditional probability density $q(y|x)$ on the set $\mathcal{Y} \subset R^n$ and the finite conditional point probabilities $q_{\rho|x}$ on the set of points $\{y_\rho \in R^n, \rho = 1, 2, ..., m\}$, where x denotes a value of the vector X of explanatory variables which can take any of the values $x_1, ..., x_M, x_i \in \mathbf{R}^d$. This setting can be associated with interesting applications such as the modeling of recovery values of defaulted debt that we will discuss in Section 12.1.2 (from historical defaulted debt data, it is evident that discounted defaultable debt indeed has point masses that occur at 0 and 100% of the amount borrowed, with a continuous distribution over other values).

This section contains no new ideas beyond a technical discussion allowing us to pass from the sums of the discrete conditional setting to a continuous conditional setting which makes use of integrals. Readers who are not interested in such technical details may skip this section.

We can think of this setting as the limit of a series of discrete horse races, with horses of two types:

(i) horses associated with the partition elements of Section 5.3.1, and

(ii) horses associated with the discrete points $\{y_\rho \in R^n, \rho = 1, 2, ..., m\}$.

5.4.1 Compatibility

Before discussing the question of allocating to maximize expected utility, we discuss conditions which must hold for the allocation problem to be well-defined.

Assuming that, given x,

$$\frac{1}{\mathcal{O}(y|x)} \tag{5.130}$$

is an integrable function, we note that, given x, the allocation

$$b(y|x) = \frac{\frac{1}{\mathcal{O}(y|x)}}{\int \frac{1}{\mathcal{O}(y|x)} dy + \sum_\rho \frac{1}{\mathcal{O}_{\rho|x}}}, \quad b_{\rho|x} = \frac{\frac{1}{\mathcal{O}_{\rho|x}}}{\int \frac{1}{\mathcal{O}(y|x)} dy + \sum_\rho \frac{1}{\mathcal{O}_{\rho|x}}} \tag{5.131}$$

"pays"

$$b(y|x)\mathcal{O}(y|x) = \frac{1}{\int \frac{1}{\mathcal{O}(y|x)} dy + \sum_\rho \frac{1}{\mathcal{O}_{\rho|x}}} \quad \text{or} \quad b_{\rho|x}\mathcal{O}_{\rho|x} = \frac{1}{\int \frac{1}{\mathcal{O}(y|x)} dy + \sum_\rho \frac{1}{\mathcal{O}_{\rho|x}}} \tag{5.132}$$

in each state; that is, the allocation in (5.131) has a fixed, deterministic payoff density, no matter which state occurs. From the point of view of the sequence of discrete horse races discussed in the previous section, note that for the discretization of the continuous horse race, we have already observed that

$$b_k^{(n)}\mathcal{O}_k^{(n)} = b(y_k^{(n)})\mathcal{O}(y) + o(\Delta_k^{(n)}); \tag{5.133}$$

so under the discrete allocations derived from (5.131), the discrete horse races will have the payoff

$$\frac{1}{\int \frac{1}{\mathcal{O}(y|x)} dy + \sum_\rho \frac{1}{\mathcal{O}_{\rho|x}}} + o(\Delta_k^{(n)}). \tag{5.134}$$

This suggests the following definition for the bank account in this setting:

$$B_x = \frac{1}{\int \frac{1}{\mathcal{O}(y|x)} dy + \sum_\rho \frac{1}{\mathcal{O}_{\rho|x}}}. \tag{5.135}$$

We define compatability in this context.

Definition 5.8 *(Compatibility, continuous conditional horse race) The utility function, U, the continuous conditional horse race with odds ratio density $\mathcal{O}(y|x)$ and discrete conditional odds ratios $\mathcal{O}_{\rho|x}$, bank accounts, B_x, and probability measure with conditional density $q(y|x)$ and point probabilities $q_{\rho|x}$ are compatible if*

(i) *U blows up at the value W_b, with $W_b < \inf_x B_x$ (equivalently, $(U')^{-1}(\infty) < \inf_x B_x$),*

(ii) *U saturates at a value W_s, with $W_s > \sup_x B_x$ (equivalently, $(U')^{-1}(0) > \sup_x B_x$),*

(iii) *the function*

$$\frac{1}{\mathcal{O}(y|x)} \tag{5.136}$$

for all x is integrable, over $\mathcal{Y}$, and

(iv) the quantity $q(y|x)\mathcal{O}(y|x)$ is bounded above, for all x

where B_x denotes the bank account, defined in (5.135), derived from the odds ratio density.

5.4.2 Allocation

We assume that our investor allocates $b(y|x)$ to the event[3] $Y = y$ and $b_{\rho|x}$ to the event $Y = y_\rho$, if $X = x$ was observed, where

$$1 = \int_{\mathcal{Y}} b(y|x)dy + \sum_{\rho=1}^{m} b_{\rho|x} . \tag{5.137}$$

This means that an investor who believes the model q allocates according to

$$b^*[q] = \arg\max_{\{b \in \mathcal{B}\}} \left[\int_{\mathcal{Y}} q(y|x)U(b(y|x)\mathcal{O}(y|x))dy + \sum_{y} q_{\rho|x}U(b_{\rho|x}\mathcal{O}_{x,\rho}) \right] ,$$

where

$$\mathcal{B} = \{(b(y|x), b_{\rho|x}) : \int_{\mathcal{Y}} b(y|x)dy + \sum_{\rho=1}^{m} b_{\rho|x} = 1\}$$

denotes the set of betting weights consistent with (5.137). The following lemma gives an explicit expression for the optimal allocation.

Lemma 5.4 *If the utility function U, the horse race with odds ratios $\mathcal{O}(y|x)$ and $\mathcal{O}_{\rho|x}$, and the probability measure with density $q(y|x)$ and point probabilities $q_{\rho|x}$ are compatible for every x, then*

$$\sum_{\rho} \frac{1}{\mathcal{O}_{\rho|x}}(U')^{-1}\left(\frac{\lambda}{q_{\rho|x}\mathcal{O}_{\rho|x}}\right) + \int_{\mathcal{Y}} \frac{1}{\mathcal{O}(y|x)}(U')^{-1}\left(\frac{\lambda}{q(y|x)\mathcal{O}(y|x)}\right)dy \tag{5.138}$$

is a strictly monotone decreasing function of λ, there exists a unique solution, λ, to the equation

$$\sum_{\rho} \frac{1}{\mathcal{O}_{\rho|x}}(U')^{-1}\left(\frac{\lambda}{q_{\rho|x}\mathcal{O}_{\rho|x}}\right) + \int_{\mathcal{Y}} \frac{1}{\mathcal{O}(y|x)}(U')^{-1}\left(\frac{\lambda}{q(y|x)\mathcal{O}(y|x)}\right)dy = 1,$$
$$\tag{5.139}$$

and

$$b^*[q](y|x) = \frac{1}{\mathcal{O}(y|x)}(U')^{-1}\left(\frac{\lambda}{q(y|x)\mathcal{O}(y|x)}\right) \quad and$$
$$b^*_{\rho|x} = \frac{1}{\mathcal{O}_{\rho|x}}(U')^{-1}\left(\frac{\lambda}{q_{\rho|x}\mathcal{O}_{\rho|x}}\right) ,$$

---

[3]To be precise, we consider a sequence of betting schemes over finite partitions of $\mathcal{Y}$ as described in Section 5.3.1.

where λ is the solution of (5.139).

Proof: The proof is similar to those of Lemmas 5.1, 5.2, and 5.3. The Lagrangian for the convex optimization problem posed by (5.16) is

$$\mathcal{L} = -\int_{y \in \mathcal{Y}} q(y|x) U(b(y|x)\mathcal{O}(y|x)) dy + \sum_{\rho} q_{\rho|x} U(b_{\rho|x}\mathcal{O}_{\rho|x})$$

$$+ \lambda \left(\int_{y \in \mathcal{Y}} b(y|x) dy + \sum_{\rho} b_{\rho|x} - 1 \right).$$

Taking the first variation of the Lagrangian with respect to $b(y|x)$, $\delta b(y|x)$, we obtain

$$\delta \mathcal{L} = -\int_{y \in \mathcal{Y}} q(y|x) U'(b(y|x)\mathcal{O}(y|x))\mathcal{O}(y|x)\delta b(y|x) dy + \lambda \int_{y \in \mathcal{Y}} \delta b(y|x) dy.$$

$$(5.140)$$

If $\delta \mathcal{L} = 0$ for all variations, $\delta b(y|x)$, we must have

$$-q(y|x) U'(b(y|x)\mathcal{O}(y|x))\mathcal{O}(y|x)\delta b(y|x) + \lambda \delta b(y|x) = 0, \qquad (5.141)$$

which implies that for $b^*[q]$ we must have

$$U'(b^*[q](y|x)\mathcal{O}(y|x)) = \frac{\lambda}{q(y|x)\mathcal{O}(y|x)}, \qquad (5.142)$$

i.e.,

$$b^*[q](y|x) = \frac{1}{\mathcal{O}(y|x)} (U')^{-1} \left(\frac{\lambda}{q(y|x)\mathcal{O}(y|x)} \right). \qquad (5.143)$$

Taking the partial derivative of the Lagrangian with respect to $b_{\rho|x}$ and setting to zero, we must have

$$\left. \frac{\partial \mathcal{L}}{\partial b_{\rho|x}} \right|_{b=b^*} = 0, \qquad (5.144)$$

i.e.,

$$-q_{\rho|x}\mathcal{O}_{\rho|x} U'(b^*_{\rho|x}\mathcal{O}_{\rho|x}) + \lambda = 0 \qquad (5.145)$$

for b^*. So we must have

$$U'(b^*_{\rho|x}\mathcal{O}_{\rho|x}) = \frac{\lambda}{q_{\rho|x}\mathcal{O}_{\rho|x}}, \qquad (5.146)$$

i.e.,

$$b^*_{\rho|x} = \frac{1}{\mathcal{O}_{\rho|x}} (U')^{-1} \left(\frac{\lambda}{q_{\rho|x}\mathcal{O}_{\rho|x}} \right). \qquad (5.147)$$

From

$$\int_{\mathcal{Y}} b(y|x) dy + \sum_{\rho=1}^{m} b_{\rho|x} = 1 \qquad (5.148)$$

it follows that λ must satisfy the following equation:

$$\sum_\rho \frac{1}{\mathcal{O}_{\rho|x}} (U')^{-1} \left(\frac{\lambda}{q_{\rho|x} \mathcal{O}_{\rho|x}} \right) + \int_\mathcal{Y} \frac{1}{\mathcal{O}(y|x)} (U')^{-1} \left(\frac{\lambda}{q(y|x)\mathcal{O}(y|x)} \right) dy = 1 ,$$

(5.149)

where we know that the integral on the left hand side exists by reasoning similar to that in the proof of Lemma 5.3.

We note that by the strict concavity of $U(W)$, $U'(W)$ is a strictly decreasing function of W; so $(U')^{-1}$ is a strictly decreasing function. It follows that the left hand side of (5.149) is a strictly decreasing function of λ.

Since U and this horse race are compatible, we have

$$(U')^{-1}(0) = W_s > \sup_x B_x \tag{5.150}$$

and

$$(U')^{-1}(\infty) = W_b < \inf_x B_x. \tag{5.151}$$

Since the left hand side of (5.149) is a strictly decreasing function of λ, the maximum (minimum) value for the left hand side of (5.149) can be obtained by allowing λ to approach 0 (∞). Under the conditions (5.150) and (5.151), we see that the maximum (minimum) value is greater (less) than 1. Since (5.149) depends continuously and monotonically on λ, by the Intermediate Value Theorem, there exists a λ that satisfies (5.149). $\square$

5.4.3 Generalized Logarithmic Utility Function

One can show (see Exercise 7) that for a utility function of the form

$$U(W) = \alpha \log(W - \gamma B) + \beta, \tag{5.152}$$

with $\alpha > 0$ and $\gamma B < \min_{x \in \mathcal{X}}(B_x)$, the optimal betting weights are given by

$$b^*_{\rho|x}(q) = q_{\rho|x} \left[1 - \frac{\gamma B}{B_x} \right] + \frac{\gamma B}{\mathcal{O}_{\rho|x}} \tag{5.153}$$

and

$$b^*[q](y|x) = q(y|x) \left[1 - \frac{\gamma B}{B_x} \right] + \frac{\gamma B}{\mathcal{O}(y|x)} . \tag{5.154}$$

5.5 Exercises

1. Starting with the generalized logarithmic utility, use Lemma 5.1, (5.18), and Definition 3.2 to calculate λ and show directly that for this utility

function, the allocation is the weighted sum of the allocation of the Kelly investor and the bank account allocation, where the weights are given by $1 - \gamma$ and γ, respectively.

2. Derive the optimal allocation for a discrete horse race for an investor with a HARA utility function.

3. Prove the statements, regarding compatibility, in Examples 5.3 to 5.6.

4. Show directly that if $W_b \geq B$, then the expected utility in the discrete horse race setting, under any probability measure, is $-\infty$.

5. Show that in the discrete horse race setting, the probability that the ratio of the wealth of an expected generalized logarithmic utility maximizing investor to the wealth of an alternative investor, who "banks" γB, after n trials will exceed any constant can be made as close to 1 as we please, for n sufficiently large. You may rely on the analogous statement for the $U(W) = log(W)$ investor.

6. Show that in the discrete horse race setting, the expected time to double the generalized logarithmic utility investor's wealth is smaller than for any alternative investor. You may rely on the analogous statement for the $U(W) = log(W)$ investor.

7. Show that for a utility function of the form

$$U(W) = \alpha \log(W - \gamma B) + \beta, \qquad (5.155)$$

with $\alpha > 0$ and $\gamma B < \min_{x \in \mathcal{X}}(B_x)$, the optimal betting weights in a continuous conditional horse race are given by

$$b^*_{\rho|x}(q) = q_{\rho|x}\left[1 - \frac{\gamma B}{B_x}\right] + \frac{\gamma B}{\mathcal{O}_{\rho|x}} \qquad (5.156)$$

and

$$b^*[q](y|x) = q(y|x)\left[1 - \frac{\gamma B}{B_x}\right] + \frac{\gamma B}{\mathcal{O}(y|x)} \quad . \qquad (5.157)$$

8. Derive an expression for the allocation b when the investor has a linear utility.

Chapter 6

Select Methods for Measuring Model Performance

In this chapter, we introduce some methods for measuring the performance of a probabilistic model. It is not our aim to give a comprehensive overview over this topic. We discuss the rather popular rank-based approaches (not related to our utility-based approach) for two-state problems in Section 6.1. We then discuss approaches more closely related to our utility-based approach: the likelihood in Section 6.2 and performance measurement via a loss function in Section 6.3. We discuss model performance measurement in our utility-based framework in detail in Chapter 8.

6.1 Rank-Based Methods for Two-State Models

There are a great number of practical problems that involve classification or conditional probability estimation on a two-state outcome. For example:

- Given a woman's age, age at menarche, number of previous breast biopsies, number of first degree relatives who have developed breast cancer, and age at first live birth, what is the probability that she will develop breast cancer?

- Given financial statement data for a particular firm (which may default over the next year, or not), what is the probability that the firm will default?

- Given the text of an article, is it a sports article or a business article?

These and other applications will be discussed more fully in Chapter 12. Special methods, appropriate for two-state classification and conditional probability estimation, have been developed to measure model performance for such problems. Perhaps the most popular concept is the ROC (Receiver Operator Characteristic) curve. There are a number of closely related ideas including the accuracy ratio, Power Curves (also known as Gini Curves), Cumulative Accuracy Profiles, Kolmogorov Smirnov methods, and the U statistic (see Hosmer and Lemeshow (2000)) and the references therein for further discussion). In this section, we shall confine our discussion to ROC methods.

ROC methods were first used in the 1940's by the U.S. military to better understand why the U.S. RADAR "receiver operators" failed to identify the Japanese attack on Pearl Harbor (see Wikipedia (2010)). ROC methods arise naturally from basic concepts in statistical hypothesis testing. Consider the two hypotheses:

H_0: the signal does not represent an enemy aircraft (null hypothesis — a straw man hypothesis that we may reject), and

H_1: the signal represents an enemy aircraft (the alternative hypothesis that we want to test).

There are four possible outcomes of a statistical test on these hypotheses. These outcomes can be represented in the "confusion matrix" displayed in Table 6.1. There are two ways to make an error: rejecting the null hypothesis

TABLE 6.1: Confusion Matrix.

Signal\ Test Outcome	H_0 Not Rejected	H_0 Rejected
Not Enemy Aircraft	$prob\{$accept $H_0\|H_0$ is true$\}$ Specificity $=$1-False Positive Rate	$prob\{$reject $H_0\|H_0$ is true$\}$ Type I error
Enemy Aircraft	$prob\{$accept $H_0\|H_0$ is false$\}$ Type II error	$prob\{$reject $H_0\|H_0$ is false$\}$ Sensitivity $=$true positive rate

when it is true (Type I error) and accepting the null hypothesis when it is false (Type II error). It is desirable to have a statistical test with

- high specificity (we are likely to accept H_0 when it is true — for the example above, we identify a nonenemy aircraft as a nonenemy aircraft), and

- high sensitivity (we are likely to reject H_0 when it is false — for the example above, we identify an enemy aircraft as an enemy aircraft).

Equivalently, we would prefer statistical tests that generate low false positive rates and high true positive rates. ROC plots display information on the false positive rates and true positive rates for a collection of tests generated by a model. To gain insight into how this plot is constructed, we can first imagine two overlapping populations of signals, as depicted in Figure 6.1.

Consider the test criterion: reject H_0 if the signal exceeds the cutoff level, c. As the cutoff level is varied, the false positive rate and true positive rates of the

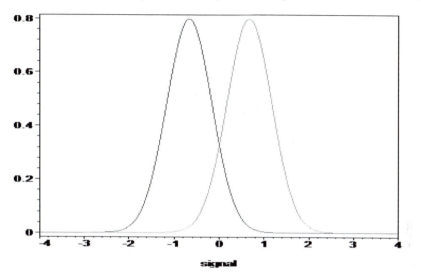

FIGURE 6.1: Probability densities for signals from the two populations. Here, the curve to the left denotes the enemy aircraft probability density and the curve to the right denotes the nonenemy aircraft probability density.

test vary. For extremely high cutoffs, there is almost no chance of rejecting H_0; so, for high cutoffs, the false positive rate will be low. However, for extremely high cutoffs, since there is almost no chance of rejecting H_0, the true positive rate will be low. For extremely low cutoffs, the false positive rate will be high and the true positive rate will be high. As the cutoff is varied, the results of the test trace out points in the (false positive)-(true positive plane). The ROC curve generated by the populations in Figure 6.1 is displayed in Figure 6.2.

In the case of a two-state classification or conditional probability model, it is possible to construct ROC curves by considering a collection of hypothesis tests of the form: reject H_0 if $p(Y = 1|x) > c$, and construct ROC curves by plotting the false positive rates and true positive rates associated with various levels of the cutoff, c. Likewise, one can construct an ROC curve for a classification model, where classification is based on the output of the classifier relative to the cutoff, c.

It is interesting to note that, for a conditional probability model, any measure of model performance derived from the ROC curve will depend only on the ranks of the probabilities and not on the actual values of the probabilities.[1] To see this, let $FP(c, q)$ denote the percentage of observations classified

---

[1]We assume in our discussion that the ROC based performance measure is constructed from the ROC curve, a collection of points $\{(FP_i, TP_i), i = 1, \ldots, n\}$. If the ROC curve is

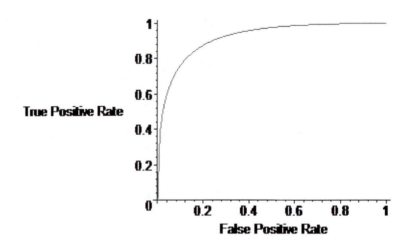

FIGURE 6.2: ROC Curve. For our example: True Positive Rate — enemy aircraft identified as enemy aircraft. False Positive Rate — nonenemy aircraft identified as enemy aircraft

as positive ($y_i = 1$) that are actually negative ($y_i = 0$), given the cutoff value c and the conditional probability model q. Let $TP(c,q)$ denote the percentage of observations classified as positive that are actually positive. We have

$$FP(c,q) = \frac{\sum_{\{i:y_i=0\}} 1_{q_{Y=1|x_i} \geq c}}{\sum_{\{i:y_i=0\}} 1} \tag{6.1}$$

and

$$TP(c,q) = \frac{\sum_{\{i:y_i=1\}} 1_{q_{Y=1|x_i} \geq c}}{\sum_{\{i:y_i=1\}} 1}. \tag{6.2}$$

Let

$$t(q_{Y=1|x_i}), \tag{6.3}$$

where $t : [0,1] \rightarrow [0,1]$ denotes a monotone increasing function. From our monotonicity assumption, it follows that the condition

$$q_{Y=1|x_i} \geq c \text{ is equivalent to } t(q_{Y=1|x_i}) \geq t(c). \tag{6.4}$$

---

supplemented with additional information related to the relation of the index i with the cutoff c, it may be possible to construct a more general performance measure; in this case, of course, the performance measure depends on more than just the ROC curve.

It therefore follows that

$$FP(c,q) = FP(t(c), t(q)) \text{ and } TP(c,q) = TP(t(c), t(q)). \quad (6.5)$$

We see that the ROC curve is preserved under the transformation t. Since t is a generic rank preserving transformation, it follows that any performance measure derived from the ROC curve is based only on the ranks of the probabilities, not their actual values. Thus, there is some question as to whether ROC analysis could lead to definitive conclusions for a model user whose actions would reflect the *values* of the conditional probabilities, rather than merely the *ranks* of the probabilities.

Model builders often wish to compare or benchmark models in the hope of choosing a good model. Given two models, each model would generate its own ROC curve. Under the assumption that a good model is one that leads to good statistical tests, in the sense of high specificity and sensitivity, one ROC curve is viewed as dominating another if for every level of the false positive rate, the true positive rate is higher. Equivalently, one ROC curve is viewed as dominating another if for every level of the true positive rate, the false positive rate is lower. That is, if the ROC curve generated by model A lies to the "northwest" of the ROC curve generated by model B, then model A can be viewed as dominating model B. This is illustrated in Figure 6.3.

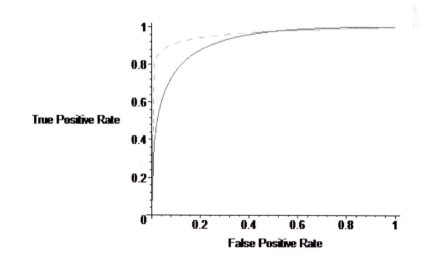

FIGURE 6.3: The model that generated the dashed ROC curve dominates the model that generated the solid ROC curve.

In practice, ROC curves often cross each other. In order to generate a single summary statistic to determine which curve is "better" in this circumstance, people often consider the area under the curve (AUC), which is equivalent to the Mann-Whitney U-statistic. It is known (see, for example, Hosmer and Lemeshow (2000)) that the AUC can be interpreted as the percentage of occurrences where $prob(Y_i = 1|x_i) \geq prob(Y_j = 1|x_j)$, given that $y_i = 1$ and $y_j = 0$. Given two AUC values corresponding to the ROC curves generated by two models, there are statistical tests on the hypothesis that the higher AUC value is due to chance.[2]

ROC curves are often compared in these ways. However, there are a number of questions that arise. The area under the curve is strongly affected by cutoff values that generate tests with large false positive rates. It is not clear whether the AUC is perhaps unduly influenced by these cutoff values (few would actually use such values for classification or statistical testing purposes).

6.2   Likelihood

The likelihood of a model is the probability of observing a given set of data under the assumption that the model is correct. This concept, which was first introduced by Fisher (1922), plays a central role in classical statistics; it is widely used for measuring the performance of probabilistic models and for building such models. In this section, we discuss the likelihood and some of the reasons why it is useful. For a more detailed review, we refer the reader to textbooks such as the ones by Jaynes (2003), Robert (1994), Berger (1985), Davidson and MacKinnon (1993), or Bernardo and Smith (2000).

There are a number of reasons for measuring model performance in terms of the likelihood. One of these reason is that Bayesian logic implies that the probability of a model given observed data is proportional to the likelihood; another one is that the likelihood provides an optimal model selection criterion in the sense of the Neyman-Pearson lemma.

Measuring model performance by means of the likelihood is closely related to the likelihood principle, which states that the information provided be an observation about a model is entirely contained in the likelihood of the model. This principle is equivalent to the conjunction of the sufficiency and the conditionality principle.

There are also decision-theoretic arguments that underpin model performance measurement via likelihood. This type of argument is, of course, the main focus of this book. Therefore, we will explore the decision-theoretic approach to likelihood at various places. In this section, we will do so in the

---

[2]See DeLong et al. (1988).

setting of a horse race gambler who maximizes his expected wealth growth rate. decision makers with more general risk preferences will be considered in later chapters.

First, we will introduce and discuss the likelihood in the context of discrete unconditional probabilities, and subsequently we will generalize to conditional probabilities and modify it for the case of probability densities.

6.2.1   Definition of Likelihood

We consider a random variable Y that can take values in a finite set $\mathcal{Y}$ and a probability measure p for this random variable.

Definition 6.1 *(Likelihood for unconditional probabilities) The likelihood of a probability measure p, given the set $\mathcal{D} = \{y_1, ..., y_N\}$ of N independent observations for Y, is given by*

$$L_{\mathcal{D}}(p) = \prod_{i=1}^{N} p_{y_i} \, . \tag{6.6}$$

It is straightforward to generalize the likelihood to observations that are not necessarily independent. However, the above definition is general enough for the methods we shall discuss in this book.

Definition 6.2 *(Log-likelihood ratio for unconditional probabilities) The log-likelihood ratio of the measures $p^{(1)}$ and $p^{(2)}$, given the set $\mathcal{D} = \{y_1, ..., y_N\}$ of N independent observations for Y corresponding to the empirical probability measure (relative frequencies) $\tilde{p}$, is given by*

$$l\left(p^{(1)}, p^{(2)}\right) = \log\left(\frac{L_{\mathcal{D}}\left(p^{(2)}\right)}{L_{\mathcal{D}}\left(p^{(1)}\right)}\right)^{\frac{1}{N}} = \sum_{y \in \mathcal{Y}} \tilde{p}_y \log\left(\frac{p_y^{(2)}}{p_y^{(1)}}\right) . \tag{6.7}$$

For the sake of convenience, we have not indicated the dependence on the observed data in our notation; it will always be clear from the context which data we refer to. It would perhaps be more precise to call l the sample-averaged log-likelihood ratio of an observation, or to define l with a prefactor N so that it can be interpreted as the logarithm of the likelihood ratio on the set of N observations. None of these modifications, however, would lead to any substantial change; so we use the above definition, which is more convenient.

6.2.2   Likelihood Principle

Let us consider a family of parametric models, p^θ, that is parameterized by the vector θ. The likelihood principle (see Fisher (1922), Fisher (1959), Barnard (1949), Birnbaum (1962), or Robert (1994) for a review) gives us guidance as to how to compare models that belong to the same family and infer their parameters from data. It can be formulated as follows.

Likelihood principle: The information about the parameter vector θ contained in the data $\mathcal{D}$ is entirely contained in the likelihood function $L_{\mathcal{D}}\left(p^{\theta}\right)$. Moreover, if for two observed datasets, $\mathcal{D}_1$ and $\mathcal{D}_2$, $L_{\mathcal{D}_1}\left(p^{\theta}\right) = cL_{\mathcal{D}_2}\left(p^{\theta}\right)$, $\forall\theta$ with some constant c, then the two datasets contain the same information about θ.

The most commonly used implementation of the likelihood principle is the maximum-likelihood method, which we shall discuss below, in Section 9.1.3. In this method, one usually conjectures a family of parametric models, p^{θ}, and finds the parameter vector θ such that it maximizes the likelihood function on a set of observations. This means that one implicitly assumes that models are ranked according to their likelihood function, i.e., that one chooses the likelihood as a model performance measure, at least for ranking models. We note that this argument is restricted to parametric families of models, although the usage of the likelihood as a model performance measure is not.

Birnbaum's motivation of the likelihood principle

The likelihood principle itself might seem somewhat ad hoc; however, it can be related to the following more basic principles.

Sufficiency principle: Two observations, y_1 and y_2, that lead to the same value of a sufficient statistic, T, i.e., with $T(y_1) = T(y_2)$, lead to the same inference on the parameter vector θ. Here, a sufficient statistic is a function of Y such that, if Y has the probability distribution p^{θ}, the probabilities of Y given $T(Y)$ are independent of θ. Intuitively, a sufficient statistic, if it exists, contains all the information about the parameter vector θ that can be extracted from the data.

Conditionality principle: If two experiments, $\mathcal{E}_1$ and $\mathcal{E}_2$, are available and one of these experiments is selected randomly, the resulting inference on the parameter vector θ should only depend on the actually selected experiment.

These two principles are plausible and generally accepted, and they lead to the likelihood principle, as the following theorem states.

Theorem 6.1 *(Equivalence result from Birnbaum (1962)) The likelihood principle is equivalent to the conjunction of the sufficiency and the conditionality principles.*

Proof: (We follow the logic of the proof of Theorem 1.3.8 from Robert (1994) here.) We first prove that the conjunction of the sufficiency and the conditionality principle implies the likelihood principle. To this end, let us denote by $\mathcal{E}^*$ the sequential experiment of first picking j with probability π_j and then executing $\mathcal{E}_j$. Furthermore, we denote by $e(\mathcal{E}_j, y_j)$ the evidence provided by the outcome y_j of the experiment $\mathcal{E}_j$ and by $e(\mathcal{E}^*, (j, y_j))$ the evidence provided by the outcome (j, y_j) of the experiment $\mathcal{E}^*$. In this notation, the conditionality principle can be formulated as

$$e(\mathcal{E}^*, (j, y_j)) = e(\mathcal{E}_j, y_j) \,, \ j = 1, 2 \,. \tag{6.8}$$

Next, we consider two outcomes y_1^0 and y_2^0 of the experiments $\mathcal{E}_1$ and $\mathcal{E}_2$, respectively, with the property

$$L_{y_1^0}\left(p^\theta\right) = cL_{y_2^0}\left(p^\theta\right) \,, \ \forall \theta \,, \tag{6.9}$$

and we define the function

$$T(j, y_j) = \begin{cases} (1, y_1^0) \text{ if } j = 2 \text{ and } y_2 = y_2^0 \\ (j, y_j) \text{ otherwise.} \end{cases} \tag{6.10}$$

This function defines a sufficient statistic for the experiment $\mathcal{E}^*$. In order to see this, we denote by Y^* the outcome of $\mathcal{E}^*$, and compute $P(Y^* = (j, y_j) | T(Y^*) = t)$. We obtain the following

(i) If $t \neq (1, y_1^0)$,

$$P_{p^\theta}(Y^* = (j, y_j) | T(Y^*) = t) = \begin{cases} 1 \text{ if } t = (j, y_j) \\ 0 \text{ otherwise} \end{cases} \text{ (by (6.10))} \,,$$

and,

(ii) if $t = (1, y_1^0)$, under the probabilities p^θ for each experiment,

$$P_{p^\theta}(Y^* = (j, y_j) | T(Y^*) = t) = \begin{cases} \frac{c\pi_1}{c\pi_1 + \pi_2} \text{ if } (j, y_j) = (1, y_1^0) \\ \frac{\pi_2}{c\pi_1 + \pi_2} \text{ if } (j, y_j) = (2, y_2^0) \\ 0 \text{ otherwise} \end{cases}$$

(by (6.9) and (6.10)),

which is independent of θ. Hence, T is a sufficient statistic for the experiment $\mathcal{E}^*$.

The sufficiency principle, in conjunction with (6.10), then implies that

$$e(\mathcal{E}^*, (1, y_1^0)) = e(\mathcal{E}^*, (2, y_2^0)) \,. \tag{6.11}$$

Combining this equation with (6.8), we obtain

$$e(\mathcal{E}_1, y_1^0) = e(\mathcal{E}_2, y_2^0) \,. \tag{6.12}$$

Hence, any outcomes y_1^0 and y_2^0 of the experiments $\mathcal{E}_1$ and $\mathcal{E}_2$, for which (6.9) holds, provide the same evidence about θ. This is the likelihood principle, i.e., the conjunction of the sufficiency and the conditionality principle implies the likelihood principle. So we have proved that part of theorem that is the most important for our ends, which is to motivate the likelihood as a model performance measure.

That the likelihood principle implies the conditionality principle follows from the fact that the likelihood functions corresponding to (i, y_j) and y_j are proportional to each other. That the likelihood principle implies the sufficiency principle follows from the factorization theorem (for more details, see for example, Robert (1994), Section 1.3). $\square$

Bayesian motivation of the likelihood principle

The Bayesian framework provides an alternative to the classical approach to estimating parameters of a probabilistic model. We will discuss this approach in Section 9.3, and refer the reader to that section or to the textbooks by Bernardo and Smith (2000), Robert (1994), Jaynes (2003), or Gelman et al. (2000) for a more comprehensive review. Here, we only state the main ideas of the Bayesian approach and shall see how it leads to the likelihood as a model performance measure.

In the Bayesian framework, we assume that the parameter vector, θ, of a probability measure, is a random variable, the probability distribution of which needs to be estimated. Before we have observed the data we would like to base our inference on, we perceive the so-called prior probability measure, $P(\theta)$, for θ. This probability measure reflects the knowledge about θ we have prior to analyzing the data. After having seen the data, we can derive, using Bayes' law, the so-called posterior distribution, which gives the probability $P(\theta|\mathcal{D})$ of observing the parameter value θ given the observed data $\mathcal{D}$ (if θ is continuous, $P(\theta|\mathcal{D})$ is a conditional probability density). By Bayes' rule, we obtain

$$P(\theta|\mathcal{D}) = \frac{P(\theta)P(\mathcal{D}|\theta)}{P(\mathcal{D})}$$
$$\propto P(\theta)P(\mathcal{D}|\theta) ,$$

where by '$\propto$' we mean that equality holds up to a θ-independent proportionality factor. If we assume that the data are sampled from a member of the family of parametric model, p^θ, we have

$$P(\theta|\mathcal{D}) \propto P(\theta)L_\mathcal{D}\left(p^\theta\right) ,$$

where $L_\mathcal{D}$ is given by Definition 6.1.

If we assume furthermore that the prior probability measure is uniform, then the probability of a parameter-vector value θ is proportional to the likelihood. Hence, if we compare models within the family of models of the form p^θ, we measure their relative probabilities by their likelihood ratio. So, if we identify a model's performance with the probability of the model, the likelihood ratio is implicitly used as a relative model performance measure. In the case of a nonuniform prior measure, this statement is not true anymore; however, the likelihood of a model still affects its probability.

6.2.3   Likelihood Ratio and Neyman-Pearson Lemma

The likelihood ratio also plays an important role in statistical testing. This role derives from the Neyman-Pearson lemma (see Neyman and Pearson (1933)), which states that the likelihood ratio provides an optimal decision criterion for model selection tests. In what follows, we explain this result in more detail. Our exposition is close to the one from Cover and Thomas (1991), Section 12.7.

Suppose we have data, $\mathcal{D}$, which are generated by the probability measure p, and we have two candidate models, $p^{(1)}$ and $p^{(2)}$, that might explain the data. So we consider the following two hypotheses.

$$H_1 : p = p^{(1)} \text{ , and}$$
$$H_2 : p = p^{(2)} \text{ .}$$

When we design a statistical test, we are interested in the probabilities of the two possible errors (type I and type II):

$$\alpha = P(H_2 \; accepted | H_1 \; true) \text{ , and}$$
$$\beta = P(H_1 \; accepted | H_2 \; true) \text{ .}$$

Ideally, we would like to find a test that minimizes the probabilities of both errors. However, there is a trade-off: decreasing one of the error probabilities usually increases the other one. All we can hope for is a test that minimizes one of the error probabilities under a given upper bound of the other error probability. As the following lemma states, the likelihood ratio test has this property.

Theorem 6.2 *(Neyman-Pearson lemma) Let $\mathcal{D} = \{y_1, ..., y_N\}$ be independent observations of a random variable $Y \in \mathcal{Y}$ with probability measure p. Let R denote a decision region for the following statistical test: accept the hypothesis $H_1 : p = p^{(1)}$ if $\mathcal{D} \in R$ and accept the hypothesis $H_2 : p = p^{(2)}$ if $\mathcal{D} \notin R$; and let*

$$\alpha_R = P(H_2 \; accepted | H_1 \; true) = P^{(1)} (\mathcal{D} \notin R) \text{ , and} \qquad (6.13)$$
$$\beta_R = P(H_1 \; accepted | H_2 \; true) = P^{(2)} (\mathcal{D} \in R) \text{ ,} \qquad (6.14)$$

where $P^{(i)}(...)$ denotes the probability of an event under the assumption that $p = p^{(i)}$, denote the error probabilities associated with this test. For $t \geq 0$, we define the decision region

$$A(t) = \left\{ \mathcal{D} : \frac{L_\mathcal{D} \left(p^{(1)} \right)}{L_\mathcal{D} \left(p^{(2)} \right)} > t \right\} , \qquad (6.15)$$

in $\mathcal{Y}^{\times N}$. Let $B \neq A(t)$ be another decision region in $\mathcal{Y}^{\times N}$. Then, $\alpha_B \leq \alpha_{A(t)}$ implies $\beta_B \geq \beta_{A(t)}$, and, $\beta_B \leq \beta_{A(t)}$ implies $\alpha_B \geq \alpha_{A(t)}$.

Proof: (We follow the logic of the proof of Theorem 12.7.1 from Cover and Thomas (1991) here.) For all $\mathcal{D} \in \mathcal{Y}^{\times N}$, we have

$$\left[I_{A(t)}(\mathcal{D}) - I_B(\mathcal{D})\right] \left[L_{\mathcal{D}}\left(p^{(1)}\right) - tL_{\mathcal{D}}\left(p^{(2)}\right)\right] \geq 0 \,,$$

which can be seen by separately considering the cases $\mathcal{D} \in A(t)$ and $\mathcal{D} \notin A(t)$ and using (6.15). Inserting $L_{\mathcal{D}}\left(p^{(i)}\right) = P^{(i)}(\mathcal{D})$, multiplying out and summing over $\mathcal{D}$, we obtain

$$\begin{aligned}
0 &\leq \sum_{\mathcal{D} \in \mathcal{Y}^{\times N}} I_{A(t)}(\mathcal{D}) \left[P^{(1)}(\mathcal{D}) - tP^{(2)}(\mathcal{D})\right] \\
&\quad - \sum_{\mathcal{D} \in \mathcal{Y}^{\times N}} I_B(\mathcal{D}) \left[P^{(1)}(\mathcal{D}) - tP^{(2)}(\mathcal{D})\right] \\
&\leq \left[P^{(1)}(A(t)) - tP^{(2)}(A(t))\right] - \left[P^{(1)}(B) - tP^{(2)}(B)\right] \\
&\leq \left[(1 - \alpha_{A(t)}) - t\beta_{A(t)}\right] - \left[(1 - \alpha_B) - t\beta_B\right] \text{ (by (6.13) and (6.14))} \\
&\leq t(\beta_B - \beta_{A(t)}) - (\alpha_{A(t)} - \alpha_B) \,.
\end{aligned}$$

The theorem follows from this equation and $t \geq 0$. $\square$

Theorem 6.2 states that we cannot simultaneously lower the probabilities of both types of error below those of the likelihood ratio test, defined by the decision region $A(t)$. Therefore, it is optimal in some sense to choose the likelihood ratio as a decision criterion when we have to decide between two models, i.e., rank model performance according to the likelihood ratio.

6.2.4 Likelihood and Horse Race

In order to give a first decision-theoretic motivation for using the likelihood as a model performance measure, we consider a Kelly investor in a horse race with odds ratios $\mathcal{O}$. We follow closely the logic from Cover and Thomas (1991), Chapter 6. Later in this book, we shall revisit the horse race gambler, and consider more general risk preferences than the ones of the Kelly investor.

We have seen in Theorem 3.1 in Section 3.3 that a Kelly investor who believes the model p chooses the (expected-wealth-growth-rate optimal) allocation

$$b_y^*(p) = p_y \,. \tag{6.16}$$

Let us evaluate the model p in terms of the test set-averaged wealth growth rate resulting from our investor's optimal strategy, which, according to Definition 3.5, is given by

$$\begin{aligned}
W(b^*(p), \tilde{p}) &= \sum_{y \in \mathcal{Y}} \tilde{p}_y \log(b_y^*(p)\mathcal{O}_y) \\
&= \sum_{y \in \mathcal{Y}} \tilde{p}_y \log(p_y\mathcal{O}_y) \text{ (by (6.16))} \,, \tag{6.17}
\end{aligned}$$

where $\tilde{p}$ denotes the empirical probabilities (relative frequencies) on the test set. Consistent with the above performance measure, we evaluate the relative performance of the two models, $p^{(1)}$ and $p^{(2)}$, by means of

$$W\left(b^*\left(p^{(2)}\right),\tilde{p}\right) - W\left(b^*\left(p^{(1)}\right),\tilde{p}\right) = \sum_{y\in\mathcal{Y}}\tilde{p}_y[\log(p_y^{(2)}\mathcal{O}_y)] \qquad (6.18)$$

$$-\sum_{y\in\mathcal{Y}}\tilde{p}_y[\log(p_y^{(1)}\mathcal{O}_y)]$$

$$= \sum_{y\in\mathcal{Y}}\tilde{p}_y\log\left(\frac{p_y^{(2)}}{p_y^{(1)}}\right)$$

$$= l\left(p^{(1)},p^{(2)}\right) \qquad (6.19)$$

(by Definition 6.2) ,

i.e., by means of the log-likelihood ratio between the two models. So (6.20) gives a decision-theoretic motivation for the log-likelihood ratio as a model performance measure. We will revisit and generalize the above logic in Chapter 8.

6.2.5 Likelihood for Conditional Probabilities and Probability Densities

So far we have discussed the likelihood for unconditional probabilities of a discrete random variable Y. In this section, we generalize this concept to conditional probabilities and adapt it to probability densities.

Conditional probabilities

We consider the random variables X and Y with state spaces $\mathcal{X}$ and $\mathcal{Y}$, respectively, and a conditional probability measure $p = \{p_{y|x} , x \in \mathcal{X} , x \in \mathcal{X}\}$.

Definition 6.3 *(Likelihood for conditional probabilities) The likelihood of a conditional probability measure p, given the set $\mathcal{D} = \{((x_1,y_1), ..., (x_N,y_N)\}$ of N independent observations for (X,Y), is given by*

$$L_{\mathcal{D}}(p) = \prod_{i=1}^{N} p_{y_i|x_i} . \qquad (6.20)$$

Definition 6.4 *(Log-likelihood ratio for conditional probabilities) The log-likelihood ratio of the conditional probability measures $p^{(1)}$ and $p^{(2)}$, given the set $\mathcal{D} = \{((x_1,y_1), ..., (x_N,y_N)\}$ of N independent observations for (X,Y) corresponding to the empirical joint probability measure (relative frequencies)*

$\tilde{p}$, *is given by*

$$l\left(p^{(1)}, p^{(2)}\right) = \log\left(\frac{L_{\mathcal{D}}\left(p^{(2)}\right)}{L_{\mathcal{D}}\left(p^{(1)}\right)}\right)^{\frac{1}{N}} = \sum_{y\in\mathcal{Y}}\sum_{x\in\mathcal{X}}\tilde{p}_{x,y}\log\left(\frac{p^{(2)}_{y|x}}{p^{(1)}_{y|x}}\right). \quad (6.21)$$

We will give a decision-theoretic interpretation of the log-likelihood ratio for conditional probabilities in Section 8.3.

Probability densities

In the case of a continuous random variable Y with an unconditional (conditional) probability density p, we define the likelihood by the same equation as in Definition 6.1 (Definition 6.3), and the log-likelihood ratio by the same equation as in Definition 6.2 (Definition 6.4). We will give a decision-theoretic interpretation of the log-likelihood ratio for probability densities in Section 8.4.

6.3 Performance Measurement via Loss Function

In this section, we briefly review the main ideas behind measuring model performance by means of a loss function. For an in-depth discussion of this topic, we refer the reader to Robert (1994) or Berger (1985).

The concept of a loss function is a useful tool that can be used to evaluate the performance of a parametric model p^{θ}. In the context of model evaluation, a loss function is usually defined as a function Λ that assigns a loss, i.e., a positive number, to a pair $(\theta, \hat{\theta})$, where θ is a parameter (vector) and $\hat{\theta}$ is a parameter estimate. The value $\Lambda(\theta, \hat{\theta})$ can be viewed as the loss incurred by a model builder who estimates the model parameter by $\hat{\theta}$, i.e., who believes the model $p^{\hat{\theta}}$, in a world that can be described by p^{θ}.

In practical applications, it is often not obvious what the appropriate loss function should be. For this reason, many practitioners resort to generic loss function such as

(*i*) the quadratic loss

$$\Lambda(\theta, \hat{\theta}) = \|\theta - \hat{\theta}\|^2,$$

where $\|...\|$ denotes the quadratic norm of a vector,

(*ii*) the absolute error loss (for one-dimensional θ)

$$\Lambda(\theta, \hat{\theta}) = |\theta - \hat{\theta}|,$$

(*iii*) the relative entropy

$$\Lambda(\theta, \hat{\theta}) = D\left(p^\theta \| p^{\hat{\theta}}\right)$$

(see Definition 2.7 for D), or

(*iv*) the Hellinger distance

$$\Lambda(\theta, \hat{\theta}) = \frac{1}{2} E_{p^\theta}\left[\left(\sqrt{\frac{p^{\hat{\theta}}}{p^\theta}} - 1\right)^2\right].$$

From a Bayesian point of view, an appropriate performance measure for the model $p^{\hat{\theta}}$ is the posterior expected loss,

$$E_{post}[L|\mathcal{D}, \hat{\theta}] = \int_\theta \Lambda(\theta, \hat{\theta}) P(\theta|\mathcal{D}) d\theta. \qquad (6.22)$$

Here $\mathcal{D}$ is the observed dataset and $P(\theta|\mathcal{D}) = \frac{L_\mathcal{D}(p^\theta)P(\theta)}{P(\mathcal{D})}$ is the posterior probability measure of θ, where $L_\mathcal{D}$ is the likelihood from Definition 6.1.

Since the choice of a parameter estimate $\hat{\theta}$ can be viewed as a decision by the model builder, the above approach is decision-theoretic in nature, and the loss function can be identified with the negative of a utility function. We shall not discuss the above loss-function approach further, but rather focus on a somewhat different, but related, decision-theoretic approach. The latter approach, which we shall introduce in Chapter 8, differs from the above approach in the following respects.

(*i*) The performance measure from Chapter 8 differs from the posterior expected loss, (6.22), in that it is not restricted to parametric models.

(*ii*) The performance measure from Chapter 8 is an average over a test dataset as opposed to an average over a theoretical distribution. This seems to be the most natural choice in a framework that is not restricted to parametric models.

(*iii*) In Chapter 8, we specify the loss function as the utility loss experienced by someone who uses the model to invest in a horse race.

6.4 Exercises

1. Show that if

    (*i*) nonenemy aircraft signals are drawn from a population that is distributed normally with mean 0 and variance 1, and

    (*ii*) enemy aircraft signals are drawn from a population that is distributed normally with mean μ and variance σ^2,

then the ROC curve is given by

$$ROC(x) = N\left(\mu + \frac{N^{-1}(x)}{\sigma}\right), \text{ for } 0 < x < 1, \qquad (6.23)$$

where $N(\cdot)$ is the cumulative standard normal distribution function.

2. Let p^θ be the family of normal distributions for the one-dimensional random variable Y. Compute the posterior expected loss for the estimator $\hat{\theta}$, assuming a Gaussian prior for θ and a single observation y'.

3. Let us define the Bayes estimator as the estimator $\hat{\theta}$ that minimizes the posterior expected loss for a given observed dataset $\mathcal{D}$. Prove that, for a quadratic loss function, the Bayes estimator is the posterior mean.

4. Derive the maximum likelihood estimate for a normal probability density function.

5. Construct an example where the Bayesian logic gives higher posterior probabilities to models with a lower likelihood than to models with a higher likelihood.

Chapter 7

A Utility-Based Approach to Information Theory

Information theory provides powerful tools that have been successfully applied in a great variety of diverse fields, including statistical learning theory, physics, communication theory, probability theory, statistics, economics, finance, and computer science (see, for example, Cover and Thomas (1991)). As we have seen in Chapter 3, the fundamental quantities of information theory, such as entropy and Kullback-Leibler relative entropy, can be interpreted in terms of the expected wealth growth rate for a Kelly investor who operates in a complete market. Alternatively, as we shall see below, one can describe these information theoretic quantities in terms of expected utilities for an investor with a logarithmic utility function.

In this chapter, we extend these interpretations and explore decision-theoretic generalizations of the fundamental quantities of information theory along the lines of Friedman and Sandow (2003b) and Friedman et al. (2007). Our discussion takes place in the discrete setting. Slomczyński and Zastawniak (2004) and Harańczyk et al. (September, 2007) discuss related ideas in a more general setting.

We shall see that some of the quantities and results of classical information theory, discussed in Section 2.3, have more general analogs. U-entropy and relative U-entropy, generalizations of entropy and Kullback-Leibler relative entropy, respectively, are particularly important because they share a great number of properties with entropy and Kullback-Leibler relative entropy (for example, a form of the Second Law of Thermodynamics). Moreover, U-entropy and relative U-entropy, as well as the more general $(U, \mathcal{O})$-entropy and relative $(U, \mathcal{O})$-entropy, lend themselves to framing statistical learning problems that are robust, in a sense to be made precise later (Chapter 10). Later in this book, we shall also use ideas from this chapter to describe model performance measures (Chapter 8).

In Sections 7.1 to 7.3, we motivate and define decision-theoretic generalizations of entropy and Kullback-Leibler relative entropy, the $(U, \mathcal{O})$-entropy and relative $(U, \mathcal{O})$-entropy, respectively, and discuss various properties possessed by these quantities. In Section 7.4, we define U-entropy and relative U-entropy and state a number of properties that generalize well-known results from classical information theory.

7.1 Interpreting Entropy and Relative Entropy in the Discrete Horse Race Context

Theorems 3.2 and 3.3 from Section 3.4, restated below for convenience, suggest that we can interpret entropy and Kullback-Leibler relative entropy, respectively, in terms of the expected wealth growth rate for a certain type of an investor who operates in the horse race setting.

Theorem 3.2 (from Section 3.4) *A Kelly investor who knows that the horses in a horse race win with the probabilities given by the measure p has the expected wealth growth rate*

$$W_p^*(p) = W_p^{**} - H(p) , \qquad (7.1)$$

where

$$W_p^{**} = E_p [\log \mathcal{O}] \qquad (7.2)$$

is the wealth growth rate of a clairvoyant investor, i.e., of an investor who wins every bet.

Interpretation: From (7.1), we see that entropy, $H(p)$, is the discrepancy between

(*i*) the expected wealth growth rate of the clairvoyant investor, who attains the growth rate W_p^{**}, and

(*ii*) the expected wealth growth rate of the expected wealth growth rate maximizing investor, who attains expected wealth growth rate $W_p^*(p)$.

Thus, entropy is the gap, in expected wealth growth rate, between a clairvoyant and an expected wealth growth rate maximizing investor.

Theorem 3.3 (from Section 3.4) *In a horse race where horses win with the probabilities given by the measure p, the difference in expected wealth growth rates between a Kelly investor who knows the probability measure p and a Kelly investor who believes the (misspecified) probability measure q is given by*

$$W_p^*(p) - W_p^*(q) = D(p\|q) . \qquad (7.3)$$

Interpretation: From (7.3), we see that Kullback-Leibler relative entropy, $D(p\|q)$, is the discrepancy between

(*i*) the expected wealth growth rate for an expected wealth growth rate maximizing investor who believes and allocates according to the correctly specified probability measure, p, and (*ii*) the expected wealth growth rate for an expected wealth growth rate maximizing investor who believes and allocates according to the misspecified probability measure, q.

Thus, Kullback-Leibler relative entropy is the gap, in expected wealth growth rate, between the investor who allocates according to a correctly specified probability measure and an investor who allocates according to a misspecified measure.

7.2 $(U, \mathcal{O})$-Entropy and Relative $(U, \mathcal{O})$-Entropy for Discrete Unconditional Probabilities

Before generalizing entropy and relative entropy, we recall the definition of the expected wealth growth rate in the discrete unconditional horse race setting, restated here for convenience:

Definition 3.5 (from Section 3.4) *The expected wealth growth rate corresponding to a probability measure p and a betting strategy b is given by*

$$W(b, p) = E_p\left[\log\left(b, \mathcal{O}\right)\right] = \sum_{y \in \mathcal{Y}} p_y \log(b_y \mathcal{O}_y) . \tag{7.4}$$

Note that the expected wealth growth rate can be interpreted as the expected utility, for an investor with a logarithmic utility function. Thus the interpretations of Section 7.1 can be expressed in terms of expected utility for an investor with a logarithmic utility function.

Motivated by this interpretation, we shall define the $(U, \mathcal{O})$-entropy, $H_{U,\mathcal{O}}(p)$, to be the difference between

(*i*) the expected utility of the clairvoyant investor, who has expected utility $E_p[U(\mathcal{O})] = \sum_y p_y U(\mathcal{O}_y)$, and

(*ii*) the expected utility of the expected utility maximizing investor, who has expected utility $E_p[U(b^*(p), O)] = \sum_y p_y U(b_y^*(p)\mathcal{O}_y)$.

Formally, we have:

Definition 7.1 *($(U, \mathcal{O})$-entropy) Given a utility function, U, a compatible system of market prices, $\mathcal{O}$, and a probability measure, p, the $(U, \mathcal{O})$-Entropy is given by:*

$$H_{U,\mathcal{O}}(p) = E_p[U(\mathcal{O})] - E_p[U(b^*(p), O)] \tag{7.5}$$

$$= \sum_y p_y U(\mathcal{O}_y) - \sum_y p_y U(b_y^*(p)\mathcal{O}_y), \tag{7.6}$$

where the optimal allocation $b^*(p)$ is with respect to the utility function, U.

Thus, $(U, \mathcal{O})$-entropy is a measure of uncertainty, expressing the gap in expected utility between an expected utility maximizing investor and a clairvoyant. Entropy is a special case, where the investor has logarithmic utility.

Similarly, generalizing the interpretation below (7.3), we shall define the relative $(U, \mathcal{O})$-entropy, $D_{U,\mathcal{O}}(p\|q)$, to be the difference between

(*i*) the expected utility of the expected utility maximizing investor, betting under the correctly specified probability measure, p, who has expected utility $E_p[U(b^*(p), O)] = \sum_y p_y U(b_y^*(p)\mathcal{O}_y)$, and

(*ii*) the expected utility of the expected utility maximizing investor, betting under the misspecified probability measure, q, who has expected utiltiy $E_p[U(b^*(q), O)] = \sum_y p_y U(b_y^*(q)\mathcal{O}_y)$.

Formally, we have:

Definition 7.2 *(Relative $(U, \mathcal{O})$-entropy) Given a utility function, U, a compatible system of market prices, $\mathcal{O}$, and probability measures, p and q, the relative $(U, \mathcal{O})$-entropy is given by:*

$$D_{U,\mathcal{O}}(p\|q) = E_p[U(b^*(p), O)] - E_p[U(b^*(q), O)] \qquad (7.7)$$

$$= \sum_y p_y U(b_y^*(p)\mathcal{O}_y) - \sum_y p_y U(b_y^*(q)\mathcal{O}_y), \qquad (7.8)$$

where the optimal allocations $b^*(p)$ and $b^*(q)$ are with respect to the utility function, U.

Thus, relative $(U, \mathcal{O})$-entropy is a measure of the discrepancy between probability measures, expressing the gap in expected utility between two investors, one of whom allocates according to the correctly specified probability measure, with the other allocating according to a misspecified probability measure. In the special case, where the investor has logarithmic utility, we obtain the Kullback-Leibler relative entropy.

Throughout this section, and in the remainder of the book, whenever we refer to a quantity involving the pair $(U, \mathcal{O})$, we shall assume that the market, the utility function, and associated probability measures are compatible in the sense of Definitions 5.2, 5.5, 5.7, or 5.8 as appropriate.

7.2.1   Connection with Kullback-Leibler Relative Entropy

It so happens that relative $(U, \mathcal{O})$-entropy essentially reduces to Kullback-Leibler relative entropy for a logarithmic family of utilities. In this case, remarkably, the odds ratios drop out of the relative entropy. Moreover, this is the most general family of utility functions for which this is so. We summarize this in the following theorem.

Theorem 7.1 *The relative $(U, \mathcal{O})$-entropy, $D_{U,\mathcal{O}}(p||q)$, is independent of the odds ratios, $\mathcal{O}$, for any candidate model p and prior measure, q, if and only if the utility function, U, is a member of the logarithmic family*

$$U(W) = \gamma_1 \log(W - \gamma) + \gamma_2 \ , \tag{7.9}$$

where $\gamma_1 > 0$, γ_2 and $\gamma < B$ are constants. In this case,

$$D_{U,\mathcal{O}}(p||q) = \gamma_1 E_p \left[\log \left(\frac{p}{q} \right) \right] = \gamma_1 D(p||q) \ . \tag{7.10}$$

Proof: It follows from Lemma 5.1, (5.18), and Definition 3.2 that

$$\lambda = \frac{\gamma_1}{1 - \frac{\gamma}{B}} \ , \tag{7.11}$$

and from Lemma 5.1, (5.19), that the optimal betting weights are given by

$$b_y^*(q) = \frac{\gamma_1}{\lambda} q_y + \frac{\gamma}{\mathcal{O}_y} \ . \tag{7.12}$$

Inserting (7.11) into (7.12), we obtain

$$b_y^*(q) = q_y \left[1 - \frac{\gamma}{B} \right] + \frac{\gamma}{\mathcal{O}_y} \ . \tag{7.13}$$

From (7.13), we have

$$U(b_y^*(q)\mathcal{O}_y) = \gamma_1 \log q_y + \gamma_1 \log \left(\mathcal{O}_y \left[1 - \frac{\gamma}{B} \right] \right) + \gamma_2 \ . \tag{7.14}$$

By inserting this expression into Definition 7.2 one can see that the relative $(U, \mathcal{O})$-entropy reduces to (7.10).

For the converse, see Friedman and Sandow (2003a), Theorem 3. $\square$

It follows that relative $(U, \mathcal{O})$-entropy reduces, up to a multiplicative constant, to Kullback-Leibler relative entropy, if and only if the utility is a member of the logarithmic family (7.9). We note that the odds ratios, $\mathcal{O}$, drop out of both $(U, \mathcal{O})$-entropy and relative $(U, \mathcal{O})$-entropy in the special case of logarithmic family utility functions. The odds ratios do not drop out for more general utility functions. This will have important implications. In cases where the odds ratios are not known with certainty, it may make sense for the model builder or model performance assessor to approximate his utility function with a utility from the logarithmic family. We shall elaborate on these points in Chapters 8 and 10.

7.2.2 Properties of $(U, \mathcal{O})$-Entropy and Relative $(U, \mathcal{O})$-Entropy

The Kullback-Leibler relative entropy satisfies the information inequality, that is, the relative entropy between two probability measures is nonnegative and is zero if and only if the measures are the same. We shall now show that $(U, \mathcal{O})$-relative entropy also has this property. Note that since

$$b^*(p) = \arg \max_{\{b : \sum_y b_y = 1\}} \sum_y p_y U(b_y \mathcal{O}_y), \tag{7.15}$$

we must have

$$\sum_y p_y U(b_y \mathcal{O}_y) \leq \sum_y p_y U(b_y^*(p) \mathcal{O}_y) \tag{7.16}$$

for general b, so, in particular,

$$\sum_y p_y U(b_y^*(q) \mathcal{O}_y) \leq \sum_y p_y U(b_y^*(p) \mathcal{O}_y), \tag{7.17}$$

with equality for $q = p$, so

$$\sum_y p_y U(b_y^*(p) \mathcal{O}_y) - \sum_y p_y U(b_y^*(q) \mathcal{O}_y) \geq 0, \text{ with equality for } q = p. \tag{7.18}$$

Note that $0 \leq H_{U,\mathcal{O}}(p) \leq \sum_y p_y \left(U(\mathcal{O}_y) - U\left(\frac{\mathcal{O}_y}{m}\right) \right)$, where $m = |\mathcal{Y}|$. The first inequality follows from the fact that $b^*(p) \leq 1$ and U is monotone increasing. The first inequality is tight. To see this, note that in the deterministic case $m = 1$, $b^*(p) = 1$ and the right hand side of (7.6) is zero. To prove the second inequality, note that from (7.16), with $q = \frac{1}{m}$, we have

$$\sum_y p_y U\left(\frac{\mathcal{O}_y}{m}\right) \leq \sum_y p_y U(b_y^*(p) \mathcal{O}_y). \tag{7.19}$$

Substituting into (7.6), we obtain the second inequality. The second inequality is not tight. Note that for $U(W) = log(W)$, we recover the entropy and Kullback-Leibler relative entropy.

$H_{U,\mathcal{O}}(p)$ and $D_{U,\mathcal{O}}(p\|q)$ have additional important properties summarized in the following theorem.

Theorem 7.2 *The relative $(U, \mathcal{O})$-entropy, $D_{U,\mathcal{O}}(p\|q)$, and the $(U, \mathcal{O})$-entropy, $H_{U,\mathcal{O}}(p)$, have the following properties*

(i) $D_{U,\mathcal{O}}(p\|q)$ is a strictly convex function of p,

(ii) $D_{U,\mathcal{O}}(p\|q) \geq 0$ with equality if and only if $p = q$ (information inequality),

(iii) $H_{U,\mathcal{O}}(p) \geq 0$, and

(iv) $H_{U,\mathcal{O}}(p)$ is a strictly concave function of p.

Proof: For property (i), we first show convexity.[1] Note that by Definition 7.2,

$$D_{U,\mathcal{O}}(p\|q) = \sup_b \left(E_p[U(b,\mathcal{O})] - E_p[U(b^*(q),\mathcal{O})] \right). \quad (7.20)$$

That is, $D_{U,\mathcal{O}}(p\|q)$ is the supremum over a set of functions that are linear in p. By Lemma 2.2, $D_{U,\mathcal{O}}(p\|q)$ is convex in p.

Suppose that $D_{U,\mathcal{O}}(p\|q)$ is not strictly convex in p. Then there exists a pair $p_1 \neq p_2$ and a $\mu \in (0,1)$ such that

$$D_{U,\mathcal{O}}(\mu p_1 + (1-\mu)p_2\|q) = \mu D_{U,\mathcal{O}}(p_1\|q) + (1-\mu)D_{U,\mathcal{O}}(p_2\|q). \quad (7.21)$$

Equivalently,

$$\sum_y (\mu p_1 + (1-\mu)p_2)_y U(b_y^*(\mu p_1 + (1-\mu)p_2)\mathcal{O}_y)$$

$$- \sum_y (\mu p_1 + (1-\mu)p_2)_y U(b_y^*(q)\mathcal{O}_y)$$

$$= \mu \left(\sum_y p_{1y} U(b_y^*(p_1)\mathcal{O}_y) - \sum_y p_{1y} U(b_y^*(q)\mathcal{O}_y)) \right) \quad (7.22)$$

$$+ (1-\mu) \left(\sum_y p_{2y} U(b_y^*(p_2)\mathcal{O}_y) - \sum_y p_{2y} U(b_y^*(q)\mathcal{O}_y)) \right).$$

Canceling terms involving q, we obtain

$$\sum_y (\mu p_1 + (1-\mu)p_2)_y U(b_y^*(\mu p_1 + (1-\mu)p_2)\mathcal{O}_y)$$

$$= \mu \sum_y p_{1y} U(b_y^*(p_1)\mathcal{O}_y) + (1-\mu) \sum_y p_{2y} U(b_y^*(p_2)\mathcal{O}_y).$$

By definition of b^* we have

$$\mu \sum_y p_{1y} U(b_y^*(p_1)\mathcal{O}_y) \geq \sum_y \mu p_{1y} U(b_y^*(\mu p_1 + (1-\mu)p_2)\mathcal{O}_y) \quad (7.23)$$

and

$$(1-\mu) \sum_y p_{2y} U(b_y^*(p_2)\mathcal{O}_y) \geq \sum_y (1-\mu)p_{2y} U(b_y^*(\mu p_1 + (1-\mu)p_2)\mathcal{O}_y). \quad (7.24)$$

For equality to hold in (7.22), we must have

$$\sum_y p_{1y} U(b_y^*(p_1)\mathcal{O}_y) = \sum_y p_{1y} U(b_y^*(\mu p_1 + (1-\mu)p_2)\mathcal{O}_y) \quad (7.25)$$

---

[1] Key ideas for the following proof were provided by Huang (2003).

and

$$\sum_y p_{2_y} U(b_y^*(p_2)\mathcal{O}_y) = \sum_y p_{2_y} U(b_y^*(\mu p_1 + (1-\mu)p_2)\mathcal{O}_y). \tag{7.26}$$

The problem

$$\sup_{\{b:\sum_y b_y = 1\}} \sum_y p_{1_y} U(b_y(p_1)\mathcal{O}_y) \tag{7.27}$$

is a strictly concave problem that has a unique solution, by Theorem 2.8. Given the equality in (7.25) and the uniqueness of the solution to the optimization problem (7.27), which implies that there can be only one argument that maximizes both sides, we must have

$$b_y^*(p_1)\mathcal{O}_y = b_y^*(\mu p_1 + (1-\mu)p_2)\mathcal{O}_y, \forall y, \tag{7.28}$$

which, by Lemma 5.1, implies that

$$(U')^{-1}\left(\frac{\lambda_1}{p_{1_y}\mathcal{O}_y}\right) = (U')^{-1}\left(\frac{\lambda'}{(\mu p_1 + (1-\mu)p_2)_y\mathcal{O}_y}\right), \forall y, \tag{7.29}$$

where λ' denotes the Lagrange multiplier for the optimal allocation under the probability measure $\mu p_1 + (1-\mu)p_2$. The strict concavity of U implies that $(U')^{-1}$ is a strictly monotone function, so we must have

$$p_{1_y} = \frac{\lambda_1}{\lambda'}(\mu p_1 + (1-\mu)p_2)_y, \forall y. \tag{7.30}$$

Summing over y, we see that $\lambda' = \lambda_1$ and $\mu = 1$, contradicting the assumption that $\mu \in (0,1)$. Thus, we have shown that $D_{U,\mathcal{O}}(p\|q)$ is strictly convex in p.

For property (*ii*), see (7.18), which indicates that $D_{U,\mathcal{O}}(p\|q)$ is nonnegative and that $D_{U,\mathcal{O}}(p\|q) = 0$, if $p = q$. The fact that $D_{U,\mathcal{O}}(p\|q) = 0$, only if $p = q$ follows from property (*i*) and Theorem 2.8. We showed that property (*iii*) is true in the discussion following Definition 7.1. Property (*iv*) follows from property (*ii*) and the fact that the sum of $H_{U,\mathcal{O}}(p)$ and $D_{U,\mathcal{O}}(p\|q)$ is linear in p. $\square$

Thus, from Theorem 7.2, we see that the relative $(U,\mathcal{O})$-entropy and $(U,\mathcal{O})$-entropy, which include relative entropy and entropy, respectively, as special cases, preserve a number of important properties of relative entropy and entropy.

7.2.3 Characterization of Expected Utility under Model Misspecification

Suppose that an investor operates in a horse race environment described by the probability measure p, but allocates so as to maximize his expected utility under the misspecified probability measure, q. It is easy to show (see Exercise 2) that

$$E_p[U(b^*(q), O)] = E_p[U(O)] - H_{U,\mathcal{O}}(p) - D_{U,\mathcal{O}}(p\|q); \qquad (7.31)$$

that is, the expected utility that can be attained by such an investor can be expressed as the expected utility that can be attained by a clairvoyant, minus the expected utility loss related to the uncertainty (expressed by entropy), minus the expected utility loss from allocating according to a misspecified measure.

7.2.4 A Useful Information-Theoretic Quantity

In Chapter 10, when we shall formulate optimization problems to estimate probabilistic models, we shall see that such problems can be elegantly expressed in terms of the following information-theoretic quantity, the gain in expected utility under the probability measure p from allocating according to measure p^2, rather than p^1.

Definition 7.3

$$G_{U,\mathcal{O}}(p^2, p^1; p) = \sum_y p_y[U(b_y^*(p^2)\mathcal{O}_y) - U(b_y^*(p^1)\mathcal{O}_y)]. \qquad (7.32)$$

We note that it is possible to express $D_{U,\mathcal{O}}(p\|q)$ in terms of G:

$$D_{U,\mathcal{O}}(p\|q) = \sum_y p_y U(b_y^*(p)\mathcal{O}_y) - \sum_y p_y U(b_y^*(q)\mathcal{O}_y) \qquad (7.33)$$

$$= G_{U,\mathcal{O}}(p, q; p). \qquad (7.34)$$

7.3 Conditional $(U, \mathcal{O})$-Entropy and Conditional Relative $(U, \mathcal{O})$-Entropy for Discrete Probabilities

In this subsection, we (briefly) generalize the notions of conditional entropy and conditional Kullback-Leibler relative entropy.

Using the same ideas and interpretations that we used to define the $(U, \mathcal{O})$-entropy and relative $(U, \mathcal{O})$-entropy for discrete unconditional probabilities in Section 7.2, we define these quantities for discrete conditional probabilities.

Definition 7.4 *(Conditional $(U, \mathcal{O})$-entropy for discrete probabilities) Given a utility function, U, and a system of market prices, $\mathcal{O}$, the conditional $(U, \mathcal{O})$-entropy is given by:*

$$H_{U,\mathcal{O}}(p_{y|x}) = \sum_x p_x \sum_y p_{y|x} U(\mathcal{O}_{y|x}) - \sum_x p_x \sum_y p_{y|x} U(b_{y|x}^*(p_{y|x})\mathcal{O}_{y|x}). \qquad (7.35)$$

That is, the conditional $(U, \mathcal{O})$-entropy is the difference between the expected utility that a clairvoyant could attain in a conditional horse race market governed by the conditional probability measure $p_{y|x}$, and the expected utility that could be attained by an expected utility maximizing investor allocating under the conditional probability measure $p_{y|x}$.

Definition 7.5 *(Conditional relative $(U, \mathcal{O})$-entropy) Given a utility function, U and a system of market prices, $\mathcal{O}$, the conditional relative $(U, \mathcal{O})$-entropy is given by:*

$$D_{U,\mathcal{O}}(p_{y|x}\|q_{y|x}) = \sum_x p_x \sum_y p_{y|x} U(b^*(p_{y|x})\mathcal{O}_{y|x})$$
$$- \sum_x p_x \sum_y p_{y|x} U(b^*_{y|x}(q_{y|x})\mathcal{O}_{y|x}).$$

That is, the conditional relative $(U, \mathcal{O})$-entropy is the difference between

(i) the expected utility that could be attained in a conditional horse race market governed by the conditional probability measure $p_{y|x}$ when allocation is computed according to the conditional probability measure $p_{y|x}$, and

(ii) the expected utility that could be attained by an expected utility maximizing investor allocating under the misspecified conditional probability measure $q_{y|x}$.

$H_{U,\mathcal{O}}(p_{y|x})$ and $D_{U,\mathcal{O}}(p_{y|x}\|q_{y|x})$ have important properties summarized in the following theorem, which we state without proof.[2]

Theorem 7.3 *The conditional relative $(U, \mathcal{O})$-entropy, $D_{U,\mathcal{O}}(p_{y|x}\|q_{y|x})$, and the conditional $(U, \mathcal{O})$-entropy, $H_{U,\mathcal{O}}(p_{y|x})$, have the following properties*

(i) $D_{U,\mathcal{O}}(p_{y|x}\|q_{y|x}) \geq 0$ *with equality if and only if $p = q$,*

(ii) $D_{U,\mathcal{O}}(p_{y|x}\|q_{y|x})$ *is a strictly convex function of p,*

(iii) $H_{U,\mathcal{O}}(p_{y|x}) \geq 0$, *and*

(iv) $H_{U,\mathcal{O}}(p_{y|x})$ *is a strictly concave function of p.*

---

[2]The proof is analogous to the proof of Theorem 7.2.

7.4 *U*-Entropy for Discrete Unconditional Probabilities

We have seen that $D_{U,\mathcal{O}}(p\|q) \geq 0$ represents the expected gain in utility from allocating under the "true" measure p, rather than allocating according to the misspecified measure q under market odds $\mathcal{O}$ for an investor with utility U. One of the goals of this chapter is to explore what happens when we set q_y equal to the homogeneous expected return measure of Definition 3.3,

$$q_y = \frac{B}{\mathcal{O}_y}, \tag{7.36}$$

where B represents the value of the bank account of Definition 3.1. Much of the material in this section can be found in Friedman et al. (2007), which generalizes some results from Cover and Thomas (1991).

Readers familiar with finance will recognize q_y as the risk neutral pricing measure[3] generated by the odds ratios. In finance, a risk neutral pricing measure is a measure under which the price of any contingent claim is equal to the discounted (by the bank account) expectation of the payoff of the contingent claim. In the horse race setting, there is only one such measure, given by (7.36). We shall see in Chapter 10 that there are compelling reasons to consider this case — statistical learning problems formulated under this specialization are robust in a well-defined sense.

In this case, with q_y equal to the homogeneous expected return measure of Definition 3.3,

$$\mathcal{O}_y = \frac{B}{q_y}, \tag{7.37}$$

and

$$D_{U,\frac{B}{q}}(p\|q) = \sum_y p_y U\left(B\frac{b_y^*(p)}{q_y}\right) - \sum_y p_y U(B),$$

so

$$D_{U,\frac{B}{q}}(p\|q) = \sum_y p_y U\left(B\frac{b_y^*(p)}{q_y}\right) - U(B). \tag{7.38}$$

Thus, in this special case, $D_{U,\frac{B}{q}}(p\|q)$ can be interpreted as the excess performance from allocating according to the true measure p over the (risk-neutral) measure q derived from the odds ratios.

---

[3]See, for example, Duffie (1996). We note that the risk neutral pricing measure generated by the odds ratios need not coincide with any "real world" measure.

Likewise, under (7.37), we have

$$H_{U,\frac{B}{q}}(p) = \sum_y p_y U\left(\frac{B}{q_y}\right) - \sum_y p_y U\left(B\frac{b_y^*(p)}{q_y}\right)$$

$$= \sum_y p_y U\left(\frac{B}{q_y}\right) - D_{U,\frac{B}{q}}(p\|q) - U(B).$$

In the special case where q is the uniform distribution, $q_y = \frac{1}{|\mathcal{Y}|}$, which we denote by $\frac{1}{|\mathcal{Y}|}$, we obtain

$$H_{U,|\mathcal{Y}|B}(p) = \sum_y p_y U\left(|\mathcal{Y}|B\right) - D_{U,B|\mathcal{Y}|}\left(p\|\frac{B}{|\mathcal{Y}|}\right) - U(B). \qquad (7.39)$$

For simplicity, we assume that $B = 1$ and $U(B) = 0$ from now on.[4]

7.4.1 Definitions of U-Entropy and Relative U-Entropy

Let

$$\mathcal{B}_y = \{b : \sum_{y\in\mathcal{Y}} b_y = 1\}. \qquad (7.40)$$

Motivated by (7.38) and the fact that

$$b^* = \arg\max_b E_p[U(b,\mathcal{O})], \qquad (7.41)$$

we make the following definition:

Definition 7.6 *The relative U-entropy from the probability measure p to the probability measure q is given by*

$$D_U(p\|q) = \sup_{b\in\mathcal{B}_y} \sum_y p_y U\left(\frac{b_y}{q_y}\right). \qquad (7.42)$$

Motivated by (7.39), we define the U-entropy:[5]

Definition 7.7 *The U-entropy of the probability measure p is given by*

$$H_U(p) = U(|\mathcal{Y}|) - D_U\left(p\|\frac{1}{|\mathcal{Y}|}\right). \qquad (7.43)$$

---

[4]It is straightforward to develop the material below under more general assumptions.

[5]Harańczyk et al. (September, 2007) compare this definition of U-entropy in the discrete setting with the definition of u-entropy in Slomczyński and Zastawniak (2004), which is made in a more general setting; they note that U-entropy and u-entropy in the discrete setting are related by monotone transformations.

Of course, these specializations of $(U, \mathcal{O})$-entropy and relative $(U, \mathcal{O})$-entropy inherit all of the properties stated in Section 7.2.2. It is easy to show that for logarithmic utilities, they reduce to entropy and Kullback-Leibler relative entropy, respectively; so these quantities are generalizations of entropy and Kullback-Leibler relative entropy, respectively.

We see that the information-theoretic quantities that we have just defined fall midway between the classical information-theoretic quantities and the more general $(U, \mathcal{O})$ quantities in terms of generality; we shall see below, in Section 7.4.2 of this chapter, that they also fall midway between the classical information-theoretic quantities and the more general $(U, \mathcal{O})$ quantities in terms of their information-theoretic properties, since they share certain "thermodynamic" properties with the entropy and relative entropy.

Before exploring these properties, we extend the above definitions to conditional probability measures. To keep our notation simple, we use p_x and p_y to represent the probability distributions for the random variables X and Y, and $p_{y|x}$ to denote the conditional distribution for $Y = y$, given $X = x$. We use $H_U(p)$ to represent the U-entropy for the random variable Y which has probability measure p.

Definition 7.8 *The conditional relative U-entropy from the probability measure p to the probability measure q is given by*

$$D_U(p_{y|x} \| q_{y|x}) = \sum_x p_x D_U(p_{y|x} \| q_{y|x})$$

$$= \sum_x p_x \sup_{b(\cdot|x) \in \mathcal{B}_{\mathcal{Y}}} \sum_y p_{y|x} U\left(\frac{b_{y|x}}{q_{y|x}}\right).$$

Definition 7.9 *The conditional U-entropy from the probability measure p to the probability measure q is given by*

$$H_U(Y|X) = \sum_{x \in \mathcal{X}} p_x H_U(Y|X = x)$$

$$= U(|\mathcal{Y}|) - \sum_{x \in \mathcal{X}} p_x \sup_{\{b_{y|x} : \sum_y b_{y|x} = 1, b_{y|x} > 0, \forall y\}} \sum_y p_{y|x} U(|\mathcal{Y}| b_{y|x}).$$

We note that we could have stated the preceding definitions as special cases of conditional $(U, \mathcal{O})$-entropy and conditional relative $(U, \mathcal{O})$-entropy of Sections 7.3.

The following definition generalizes mutual information.

Definition 7.10 *The mutual U-information between the probability measures p and q is defined as*

$$I_U(X; Y) = D_U(p_{x,y} \| p_x p_y). \tag{7.44}$$

Thus, the mutual information between X and Y is symmetric in X and Y and captures the discrepancy between the joint distribution of X and Y and the distribution that has marginal distributions, X and Y, where X and Y are independent. When the X and Y are independent, the mutual information is zero. When the joint distribution is far from independent in X and Y, the mutual information is large. Thus, the greater the mutual information between X and Y, the greater the amount of information that X contains about Y and the greater the amount of information that Y contains about X.

7.4.2   Properties of U-Entropy and Relative U-Entropy

As a special case of Theorem 7.2, we have

Corollary 7.1 *The relative $(U, \mathcal{O})$- entropy, $D_U(p\|q)$, and the $(U, \mathcal{O})$-entropy, $H_U(p)$, have the following properties*

(i) $D_U(p\|q) \geq 0$ *with equality if and only if $p = q$,*

(ii) $D_U(p\|q)$ *is a strictly convex function of p,*

(iii) $H_U(p) \geq 0$*, and*

(iv) $H_U(p)$ *is a strictly concave function of p.*

We now establish that many properties that hold for the classical quantities of information theory (but need not hold for $(U, \mathcal{O})$-entropy and relative $(U, \mathcal{O})$-entropy) also hold in this more general setting.

Theorem 7.4 $H_U(Z|X, Y) \leq H_U(Z|X)$ *(extra conditioning information reduces entropy).*

Proof: By definition,

$$H_U(Z|X, Y) = \sum_{x,y} p_{x,y} H_U(Z|X = x, Y = y) \tag{7.45}$$

$$= \sum_{x,y} p_{x,y} \left(U(|\mathcal{Z}|) - D_U \left(p_{z|x,y} \| \frac{1}{|\mathcal{Z}|} \right) \right)$$

$$= U(|\mathcal{Z}|) - \sum_x p_x \sum_y p_{y|x} D_U \left(p_{z|x,y} \| \frac{1}{|\mathcal{Z}|} \right) \tag{7.46}$$

$$\leq U(|\mathcal{Z}|) - \sum_x p_x D_U \left(\sum_y p_{y|x} p_{z|x,y} \| \frac{1}{|\mathcal{Z}|} \right) \tag{7.47}$$

$$\text{(by convexity of } D_U(\cdot \| \tfrac{1}{|\mathcal{Z}|})) \tag{7.48}$$

$$= U(|\mathcal{Z}|) - \sum_x p_x D_U \left(p_{z|x} \| \frac{1}{|\mathcal{Z}|} \right) \tag{7.49}$$

$$= H_U(Z|X). \quad \square \tag{7.50}$$

That is, knowing the outcome of an additional random variable (Y, in addition to X) can only reduce the uncertainty (entropy) of the random variable of interest (Z). This idea is most simply stated in the following corollary.

Corollary 7.2 $H_U(Y|X) \leq H_U(Y)$ *(Conditioning reduces entropy)*

Proof: The proof follows as a direct consequence of Theorem 7.4. □

The next corollary shows that if we "further randomize" a random variable X, by applying a random permutation T to X, the resulting random variable TX has a U-entropy that is at least as large as the U-entropy of X.

Corollary 7.3 *(Shuffles increase entropy). If T is a random shuffle (permutation) of a deck of cards and X is the initial position of the cards in the deck and if the choice of shuffle T is independent of X, then*

$$H_U(X) \leq H_U(TX) \tag{7.51}$$

where TX is the permutation of the deck induced by the shuffle T.

Proof: We follow the proof sketched on page 48 of Cover and Thomas (1991). First, notice that for any fixed permutation $T = t$, from the definition of H_U, we have

$$H_U(tX) = H_U(X) \tag{7.52}$$

(since we have only reordered the states, but we have not changed the probabilities associated with the states). So

$$
\begin{aligned}
H_U(TX) &\geq H_U(TX|T) \\
&\quad \text{(by Corollary 7.2)} \\
&= \sum_t p_t H_U(TX|T = t) \\
&= \sum_t p_t H_U(tX) \\
&\quad \text{(by (7.52))} \\
&= H_U(X). \quad \square
\end{aligned}
$$

There are a number of important information theoretic results on Markov chains and Markov processes, such as the second law of thermodynamics and the data processing inequality (see Cover and Thomas (1991)); in the remainder of this section, we discuss some U-entropy generalizations of these results. First, we define a Markov chain.

Definition 7.11 *(Markov chain) The random variables X, Y, and Z form a Markov chain in that order, denoted by $X \rightarrow Y \rightarrow Z$, if the conditional distribution of Z depends only on Y and is independent of X.*

It then follows (see Exercise 6) from the definition of a Markov chain that for the Markov chain $X \to Y \to Z$, the joint probability function of X, Y, and Z satisfies

$$p(x, y, z) = p(x)p(y|x)p(z|y). \qquad (7.53)$$

From Theorem 7.4, we know that, in general, extra conditioning reduces conditional U-entropy, i.e., that $H_U(Z|Y, X) \leq H_U(Z|Y)$. As we now show, this inequality is saturated if the random variables X, Y, Z form a Markov chain $X \to Y \to Z$.

Theorem 7.5 *If the random variables X, Y, Z form a Markov chain $X \to Y \to Z$, then $H_U(Z|X, Y) = H_U(Z|Y)$.*

Proof:

$$H_U(Z|X, Y) = \sum_{x,y} p_{x,y} \left(U(|\mathcal{Z}|) - D_U \left(p_{z|x,y} \| \frac{1}{|\mathcal{Z}|} \right) \right)$$

$$= U(|\mathcal{Z}|) - \sum_{x,y} p_{x,y} D_U \left(p(z|y) \| \frac{1}{|\mathcal{Z}|} \right)$$

(by the Markov property)

$$= U(|\mathcal{Z}|) - \sum_{y} p_y D_U \left(p(z|y) \| \frac{1}{|\mathcal{Z}|} \right)$$

$$= H_U(Z|Y). \quad \square$$

That is, for a Markov chain $X \to Y \to Z$, additional conditioning of Z on X (in addition to Y) does not reduce the conditional U-entropy.

The next result shows that for a Markov chain $X \to Y \to Z$, the conditional U-entropy that results from conditioning Z on X exceeds that obtained by conditioning Z on Y.

Corollary 7.4 *If the random variables X, Y, Z form a Markov chain $X \to Y \to Z$, then $H_U(Z|X) \geq H_U(Z|Y)$ and $H_U(X|Z) \geq H_U(X|Y)$.*

Proof: We prove the first inequality by noting that by Corollary 7.2,

$$H_U(Z|X) \geq H_U(Z|X, Y), \qquad (7.54)$$

and by Theorem 7.5,

$$H_U(Z|X, Y) = H_U(Z|Y). \qquad (7.55)$$

The second inequality follows from the fact that $X \to Y \to Z$ is equivalent to $Z \to Y \to X$. $\square$

Our next results pertain to interpretations of the second law of thermodynamics. The idea here is that in an isolated system, the uncertainty, in some sense, increases over time. Before stating such results precisely, we define a discrete time stochastic process, a stationary process, and a Markov process.

Definition 7.12 *(Discrete stochastic process) A stochastic process is an indexed sequence of random variables.*

Example 7.1 *Discrete time approximation to geometric Brownian motion.* We consider the stochastic process, $S_0, S_1, \ldots$, for the end-of-day closing stock prices for a particular stock, where

$$S_{n+1} = S_n(1 + \mu\Delta T + \sigma\sqrt{\Delta T}\epsilon_n), \qquad (7.56)$$

where $\epsilon_n \sim N(0,1)$, S_0 is given, the drift and volatility parameters μ and σ are known, and ΔT represents one trading day (in years) — usually taken to be $\frac{1}{252}$, based on a year of 252 trading days.

Definition 7.13 *(Stationary process) A stochastic process is stationary if for any subset of the sequence of random variables, the joint distribution function is invariant with respect to shifts of the index of the random variables, i.e., if*

$$prob(X_1 = x_1, \ldots, X_n = x_n) = prob(X_{k+1} = x_1, \ldots, X_{k+n} = x_n), \quad (7.57)$$

for every shift k and all $x_1, \ldots, x_n$.

Example 7.2 For the discrete time geometric Brownian motion approximation of Example 7.1, the stochastic process S_n is not stationary, but the stochastic process $\frac{S_{n+1}}{S_n}$ is stationary (see Exercise 8).

Definition 7.14 *(Markov process) A stochastic process $X_1, \ldots, X_n, \ldots$ is a Markov process if*

$$prob(X_{n+1} = x_{n+1} \mid X_n = x_n, X_{n-1}, x_{n-1} \ldots, X_1 = x_1)$$
$$= prob(X_{n+1} = x_{n+1} | X_n = x_n),$$

for all $x_1, \ldots, x_n, x_{n+1}$.

If X_n is a stationary Markov process, then the U-entropy, $H_U(X_n)$, is constant for all n. However, the conditional U-entropy, for a time horizon, n, increases with n, for a stationary Markov process. This is a generalization of one of the interpretations of the second law of thermodynamics stated in Cover and Thomas (1991).

Corollary 7.5 *(Second law of thermodynamics, I.) The conditional entropy $H_U(X_n|X_1)$ increases with n for a stationary Markov process.*

Proof: The proof is essentially the same as that given on page 36 of Cover and Thomas (1991). See Exercise 9. □

We now address a different interpretation of the second law of thermodynamics, formulated in terms of relative U-entropy.

Theorem 7.6 (*Second law of thermodynamics, II.*) *Let μ_n and μ'_n be two probability distributions on the state space of a Markov process at time n, and let μ_{n+1} and μ'_{n+1} be the corresponding distributions at time $n+1$. Then $D_U(\mu_n\|\mu'_n)$ decreases with n.*

Proof: Let the corresponding joint probability function be denoted by $p(x_n, x_{n+1})$ and $q(x_n, x_{n+1})$ and let $r(\cdot|\cdot)$ denote the probability transition function for the Markov process. Then $p(x_n, x_{n+1}) = p(x_n)r(x_{n+1}|x_n)$ and $q(x_n, x_{n+1}) = q(x_n)r(x_{n+1}|x_n)$.

$$D_U(p(x_n, x_{n+1}) \| q(x_n, x_{n+1}))$$

$$= \sup_{b\in\mathcal{B}_{\mathcal{X}\times\mathcal{X}}} \sum_{x_n}\sum_{x_{n+1}} p(x_n, x_{n+1})U\left(\frac{b(x_n, x_{n+1})}{q(x_n, x_{n+1})}\right)$$

$$= \sup_{b\in\mathcal{B}_{\mathcal{X}\times\mathcal{X}}} \sum_{x_n} p(x_n)\sum_{x_{n+1}} r(x_{n+1}|x_n)U\left(\frac{b(x_n, x_{n+1})}{q(x_n, x_{n+1})}\right)$$

$$\leq \sup_{b\in\mathcal{B}_{\mathcal{X}\times\mathcal{X}}} \sum_{x_n} p(x_n)U\left(\sum_{x_{n+1}} r(x_{n+1}|x_n)\frac{b(x_n, x_{n+1})}{q(x_n)r(x_{n+1}|x_n)}\right)$$

(Here, we have used Jensen's inequality
and the fact that U is concave.)

$$= \sup_{b\in\mathcal{B}_{\mathcal{X}\times\mathcal{X}}} \sum_{x_n} p(x_n)U\left(\frac{\sum_{x_{n+1}} b(x_n, x_{n+1})}{q(x_n)}\right)$$

$$\leq \sup_{b\in\mathcal{B}_{\mathcal{X}}} \sum_{x_n} p(x_n)U\left(\frac{b(x_n)}{q(x_n)}\right)$$

(since $b(x_n) = \sum_{x_{n+1}} b(x_n, x_{n+1}) \in \mathcal{B}_{\mathcal{X}}$)

$$= D_U(p(x_n)\|q(x_n)).$$

By Theorem 7.7,

$$D_U(p(x_n, x_{n+1})\|q(x_n, x_{n+1})) \geq D_U(p(x_{n+1})\|q(x_{n+1})), \qquad (7.58)$$

hence,

$$D_U(p(x_n)\|q(x_n)) \geq D_U(p(x_n, x_{n+1})\|q(x_n, x_{n+1})) \geq D_U(p(x_{n+1})\|q(x_{n+1})).$$

$\square$

We express another interpretation of the second law of thermodynamics, formulated in terms of the stationary distribution of a Markov process.

Corollary 7.6 (*Second law of thermodynamics, III.*) *The relative U-entropy $D_U(\mu_n\|\mu)$ between a distribution μ_n on the state space of a Markov process at* *time n and a stationary distribution μ of the same Markov process decreases with n.*

Proof: This is a special case of Theorem 7.6, where $\mu'_n = \mu$ for any n. $\square$

Specializing to the case where the uniform distribution is a stationary distribution on the state space of a Markov process, we obtain another interpretation of the second law of thermodynamics.

Corollary 7.7 *(Second law of thermodynamics, IV.) If the uniform distribution $\frac{1}{|\mathcal{X}|}$ is a stationary distribution on the state space of a Markov process, then the entropy $H_U(\mu_n)$ increases with n.*

Not all of the results that hold for relative entropy hold for relative U-entropy. In general, the chain rule for relative U-entropy does not hold, i.e.,

$$D_U(p_{x,y}\|q_{x,y}) \neq D_U(p_x\|q_x) + D_U(p_{y|x}\|q_{y|x}). \tag{7.59}$$

However, we do have the following two inequalities.

Theorem 7.7

(i) $D_U(p_x\|q_x) \leq D_U(p_{x,y}\|q_{x,y})$, and

(ii) $D_U(p_{y|x}\|q_{y|x}) \leq D_U(p_{x,y}\|q_{x,y})$.

Proof: By definition,

$$D_U(p_{x,y}\|q_{x,y}) = \sup_{b \in \mathcal{B}_{\mathcal{X} \times \mathcal{Y}}} \sum_{x,y} p_{x,y} U\left(\frac{b_{x,y}}{q_{x,y}}\right) \tag{7.60}$$

$$\geq \sup_{b \in \mathcal{B}_{\mathcal{X}}} \sum_{x,y} p_{x,y} U\left(\frac{b_x q_{y|x}}{q_x q_{y|x}}\right) \tag{7.61}$$

$$(\text{since } b_x q_{y|x} \in \mathcal{B}_{\mathcal{X} \times \mathcal{Y}}) \tag{7.62}$$

$$= \sup_{b \in \mathcal{B}_{\mathcal{X}}} \sum_{x} p_x \sum_{y} p_{y|x} U\left(\frac{b_x}{q_x}\right) \tag{7.63}$$

$$= D_U(p_x\|q_x) \tag{7.64}$$

and

$$D_U(p_{x,y}\|q_{x,y}) = \sup_{b \in \mathcal{B}_{\mathcal{X} \times \mathcal{Y}}} \sum_{x,y} p_{x,y} U\left(\frac{b_{x,y}}{q_{x,y}}\right) \tag{7.65}$$

$$= \sup_{b \in \mathcal{B}_{\mathcal{X} \times \mathcal{Y}}} \sum_{x} p_x \sum_{y} p_{y|x} U\left(\frac{b_{x,y}}{q_{x,y}}\right) \tag{7.66}$$

$$\geq \sum_{x} p_x \sup_{b_{y|x} \in \mathcal{B}_{\mathcal{Y}}} \sum_{y} p_{y|x} U\left(\frac{b_{y|x} q_x}{q_{y|x} q_x}\right) \tag{7.67}$$

$$(\text{since } \tilde{b}_{x,y} = b_{y|x} q_x \in \mathcal{B}_{\mathcal{X} \times \mathcal{Y}}) \tag{7.68}$$

$$= D_U(p_{y|x}\|q_{y|x}). \quad \square \tag{7.69}$$

We have already discussed, in Corollary 7.5, Theorem 7.6, Corollary 7.6, and Corollary 7.7, various interpretations of the second law of thermodynamics, as our horizon recedes into the future. We now discuss an increase of a different type of uncertainty — we now show that more refined horse races have higher relative U-entropies.

Theorem 7.8 (*More refined horse races have higher relative U-entropies*) *Let $\tau : \mathcal{X} \to \mathcal{Y}$ be any onto function and p_y^τ, q_y^τ be the induced probabilities on $\mathcal{Y}$, i.e.,*

$$p_y^\tau = \sum_{\{x:\tau_x=y\}} p_x, \tag{7.70}$$

and

$$q_y^\tau = \sum_{\{x:\tau_x=y\}} q_x. \tag{7.71}$$

Then

$$D_U(p_y^\tau \| q_y^\tau) \le D_U(p_x \| q_x). \tag{7.72}$$

Proof:

$$D_U(p_y^\tau \| q_y^\tau) = \sup_{b \in \mathcal{B}_y} \sum_y p_y^\tau U\left(\frac{b_y}{q_y^\tau}\right) \tag{7.73}$$

$$= \sup_{b \in \mathcal{B}_y} \sum_y \left(\sum_{\{x:\tau_x=y\}} p_x\right) U\left(\frac{b_y}{\sum_{\{x:\tau_x=y\}} q_x}\right). \tag{7.74}$$

Define $\tilde{b}_x = \frac{q_x b_y}{\sum_{\{x':\tau(x')=y\}} q(x')}$. Then $\tilde{b}_x \in \mathcal{B}_{\mathcal{X}}$ and

$$\sum_y \left(\sum_{\{x:\tau_x=y\}} p_x\right) U\left(\frac{b_y}{\sum_{\{x:\tau_x=y\}} q_x}\right) = \sum_x p_x U\left(\frac{\tilde{b}_x}{q_x}\right) \tag{7.75}$$

$$\le D_U(p_x \| q_x). \tag{7.76}$$

□

Theorem 7.8 can also be regarded as a special case of Theorem 7.7, part (*i*).

We shall end this section with a statement of a generalized version of another of the pillars of information theory: the data processing inequality, which is expressed in terms of the mutual information of Definition 7.10. The data processing inequality tells us that for the Markov chain $X \to Y \to Z$, the mutual information between Y and X exceeds the mutual information of the "downstream" Z and X. A corollary tells us that no transformation of the variable Y will increase the information that we can derive about X. Before stating the data processing inequality, we state and prove the following Theorem.

Theorem 7.9 *The mutual information defined in Definition 7.10 has the following properties:*

(i) $I_U(X;Y) = I_U(Y;X)$, *and*

(ii) $I_U(X;Y) \leq I_U(X;Y,Z)$.

Proof: (i) is obvious. To prove (ii), note that by definition

$$I_U(X;Y) = \sup_{b \in \mathcal{B}_{X \times Y}} \sum_{x,y} p_{x,y} U\left(\frac{b_{x,y}}{p_x p_y}\right) \tag{7.77}$$

$$= \sup_{b \in \mathcal{B}_{X \times Y}} \sum_{x,y,z} p_{x,y,z} U\left(\frac{b_{x,y} p_{z|y}}{p_x p_{y,z}}\right) \tag{7.78}$$

$$\leq \sup_{b \in \mathcal{B}_{X \times Y \times Z}} \sum_{x,y,z} p_{x,y,z} U\left(\frac{b_{x,y,z}}{p_x p_{y,z}}\right) \tag{7.79}$$

$$= I_U(X;Y,Z). \tag{7.80}$$

□

We now state and prove a relative U-entropy generalized version of the data processing inequality.

Theorem 7.10 *(Data processing inequality): For the Markov chain* $X \rightarrow Y \rightarrow Z$, $I_U(X;Y) \geq I_U(X;Z)$.

Proof: We will first show that $I_U(X;Y) = I_U(X;Y,Z)$. From the previous result,

$$I_U(X;Y) \leq I_U(X;Y,Z) \tag{7.81}$$

On the other hand,

$$I_U(X;Y,Z) = \sup_{b \in \mathcal{B}_{\mathcal{X} \times \mathcal{Y} \times \mathcal{Z}}} \sum_{x,y,z} p_{x,y,z} U\left(\frac{b_{x,y,z}}{p_x p_{y,z}}\right) \tag{7.82}$$

$$= \sup_{b \in \mathcal{B}_{\mathcal{X} \times \mathcal{Y} \times \mathcal{Z}}} \sum_{x,y} p_{x,y} \sum_z p_{z|x,y} U\left(\frac{b_{x,y,z}}{p_x p_{y,z}}\right) \tag{7.83}$$

$$\leq \sup_{b \in \mathcal{B}_{\mathcal{X} \times \mathcal{Y} \times \mathcal{Z}}} \sum_{x,y} p_{x,y} U\left(\sum_z p_{z|x,y} \frac{b_{x,y,z}}{p_x p_{y,z}}\right) \tag{7.84}$$

(by Jensen's inequality) $\tag{7.85}$

$$= \sup_{b \in \mathcal{B}_{\mathcal{X} \times \mathcal{Y} \times \mathcal{Z}}} \sum_{x,y} p_{x,y} U\left(\frac{\sum_z b_{x,y,z}}{p_x p_y}\right) \tag{7.86}$$

(since $p_{z|x,y} = p_{z|y}$, by the Markov property) $\tag{7.87}$

$$\leq \sup_{b \in \mathcal{B}_{\mathcal{X} \times \mathcal{Y}}} \sum_{x,y} p_{x,y} U\left(\frac{b_{x,y}}{p_x p_y}\right) \tag{7.88}$$

$$= I_U(X;Y) \tag{7.89}$$

so $I_U(X;Y) = I_U(X;Y,Z) \geq I_U(X;Z)$. $\square$

Thus the information that Y encodes with respect to X is greater than the information that the "downstream" Z encodes with respect to X.

Corollary 7.8 *For the Markov chain $X \to Y \to Z$, if $Z = g(Y)$, we have $I_U(X;Y) \geq I_U(X;g(Y))$.*

Thus, no transformation of Y will provide more information about X than Y itself.

7.4.3   Power Utility

Consider the power utility, from Example 4.3 and Section 5.1.6, given by

$$U_\kappa(W) = \frac{W^{1-\kappa} - 1}{1 - \kappa}. \tag{7.90}$$

In this section, we compute the associated U-entropy and relative U-entropy, explore limiting behavior, and compare with Tsallis entropy. We recall from Section 5.1.6 that the power utility is commonly used, has desirable optimality properties, and the constant relative risk aversion property. We note that for this utility function, the (constant) relative risk aversion is given by

$$R(W) = -W\frac{U''(W)}{U'(W)} = \kappa. \tag{7.91}$$

From Section 5.1.6, with $\mathcal{O}_y = \frac{1}{q_y}$, we have

$$b_y^*(p) = q_y \left(\frac{p_y}{\lambda q_y} \right)^{\frac{1}{\kappa}} \tag{7.92}$$

where

$$\lambda = \left(\sum_y q_y \left(\frac{p_y}{q_y} \right)^{\frac{1}{\kappa}} \right)^{\kappa}. \tag{7.93}$$

After some algebra, we obtain

$$D_{U_\kappa}(p\|q) = \frac{\left(\sum_y p_y^{\frac{1}{\kappa}} q_y^{1-\frac{1}{\kappa}} \right)^{\kappa}}{1-\kappa} - \frac{1}{1-\kappa}. \tag{7.94}$$

7.4.3.1 U-Entropy for Large Relative Risk Aversion

Let us consider the limit of infinite risk aversion, which for the power utility corresponds to $\kappa \to \infty$, as can be seen from (7.91). It follows from (7.92) and (7.93) that a power utility investor with infinite risk aversion invests all his money in the bank account, i.e., allocates according to

$$b^*(p)_y = \frac{B}{\mathcal{O}_y} = \frac{1}{\mathcal{O}_y}, \tag{7.95}$$

no matter what his belief-measure, p, is, so that his after-bet wealth is $B = 1$ with certainty. Such an investor makes no use of the information provided by the model, so no model can outperform another and, therefore,

$$\lim_{\kappa \to \infty} D_{U_\kappa}(p\|q) = 0, \tag{7.96}$$

for any measures p, q and any utility function U.

What happens if the relative risk aversion of a power utility investor, i.e.,

κ, is large but finite? To answer this question, we expand (7.94) as follows.

$$D_{U_\kappa}(p\|q) = \frac{1}{1-\kappa}\left[\sum_y \left(\frac{p_y}{q_y}\right)^{\frac{1}{\kappa}} q_y\right]^\kappa - \frac{1}{1-\kappa}$$

$$= \frac{1}{1-\kappa}\left[\sum_y \left(1 + \frac{1}{\kappa}\log\frac{p_y}{q_y} + \frac{1}{2\kappa^2}\left(\log\frac{p_y}{q_y}\right)^2 + O(\kappa^{-3})\right)q_y\right]^\kappa$$

$$- \frac{1}{1-\kappa} \quad (\text{ since } z^{\frac{1}{\kappa}} = 1 + \frac{1}{\kappa}\log(z) + \frac{1}{2\kappa^2}(\log z)^2 + O(\kappa^{-3}) \text{ })$$

$$= \frac{1}{1-\kappa}\left[1 + \frac{1}{\kappa}\sum_y q_y\log\frac{p_y}{q_y} + \frac{1}{2\kappa^2}\sum_y q_y\left(\log\frac{p_y}{q_y}\right)^2 + O(\kappa^{-3})\right]^\kappa$$

$$- \frac{1}{1-\kappa}$$

$$= \frac{1 - e^{\sum_y q_y\log\frac{p_y}{q_y}}}{\kappa} + \frac{1 - e^{\sum_y q_y\log\frac{p_y}{q_y}}\left[1 + \frac{1}{2}var_q\left(\log\frac{p_y}{q_y}\right)\right]}{\kappa^2}$$

$$+ O\left(\kappa^{-3}\right),$$

where the last equality follows from the fact that

$$\frac{1}{1-\kappa}\left(1 + \frac{a}{\kappa} + \frac{b}{2\kappa^2} + \frac{c}{\kappa^3}\right)^\kappa - \frac{1}{1-\kappa} = \frac{1-e^a}{\kappa} + \frac{1 - e^a\left(1 + \frac{b-a^2}{2}\right)}{\kappa^2}$$

$$+ O\left(\kappa^{-3}\right).$$

So

$$D_{U_\kappa}(p\|q) = \frac{1 - e^{-D(q\|p)}}{\kappa} + \frac{1 - e^{-D(q\|p)}\left[1 + \frac{1}{2}var_q\left(\log\frac{p_y}{q_y}\right)\right]}{\kappa^2} + O\left(\kappa^{-3}\right).$$

$$(7.97)$$

Thus, we have related, for large κ, $D_{U_\kappa}(p\|q)$ to the Kullback-Leibler relative entropy.

7.4.3.2 Relation with Tsallis and Rényi Relative Entropies and the f-Divergence

In this section, we briefly describe other generalizations of relative entropy:

(*i*) Tsallis relative entropy,

(*ii*) Rényi relative entropy, and

(*iii*) the $f-$divergence,

and relate these quantities to the relative $U-$entropy in the case of a power utility function.

The Tsallis entropy (see Tsallis (1988)) provides a theoretical framework for a nonextensive thermodynamics, i.e., for a thermodynamics for which the entropy of a system made up of two independent subsystems is not simply the sum of the entropies of the subsystems. Examples for physical application of such a theory might be astronomical self-gravitating systems (see, for example, Plastino and Plastino (1999)).

The Tsallis entropy is given by

$$H_\alpha^T(X) = \frac{\sum_y p_x^\alpha - 1}{1 - \alpha} \qquad (7.98)$$

and can be generalized to the Tsallis relative entropy (see, for example, Tsallis and Brigatti (2004)),

$$D_\alpha^T(p\|q) = \frac{\sum_y p_y^\alpha q_y^{1-\alpha} - 1}{\alpha - 1}. \qquad (7.99)$$

It turns out that there is a simple monotonic relationship between the relative U-entropy and the relative Tsallis entropies:

$$D_{U_\kappa}(p\|q) = \frac{\left(D_{\frac{1}{\kappa}}^T(p\|q)\left(\frac{1}{\kappa} - 1\right) + 1 \right)^\kappa - 1}{1 - \kappa}. \qquad (7.100)$$

It is easy to see that the Rényi generalized divergence of order α (see Rényi (1961))

$$\frac{1}{\alpha - 1} log \left(\sum_y \frac{p_y^\alpha}{q_y^{\alpha-1}} \right) \qquad (7.101)$$

and the Tsallis relative entropy are continuous and monotone increasing functions of each other.

Csiszár (1972) defines another generalization of relative entropy, the f-divergence

$$\sum_y q_y f \left(\frac{p_y}{q_y} \right), \qquad (7.102)$$

where $f : R^+ \to R$ is a strictly convex function. We note that in the case that $f(u) = u^\alpha$, it is easy to see that the f-divergence is a monotone increasing function of the Tsallis relative entropy.

Thus, we see that the power utility, the relative U-entropy, the Tsallis and Rényi relative entropies, as well as the f-divergence, where f is a power function, are related via monotone transformations.

7.5 Exercises

1. Express $D_U(p\|q)$, $H_{U,\mathcal{O}}(p)$, and $H_U(p)$ in terms of $G_{U,\mathcal{O}}(p^2, p^1; p)$.

2. Show that a horse race investor operating in a horse race governed by the probability measure, p, but allocating to maximize his expected utility under the (misspecified) measure q, attains the expected utility $E_p[U(b^*(q), O)]$, where

$$E_p[U(b^*(q), O)] = E_p[U(O)] - H_{U,\mathcal{O}}(p) - D_{U,\mathcal{O}}(p\|q). \qquad (7.103)$$

3. Interpret (7.103) for a Kelly investor.

4. Prove Theorem 7.3.

5. State and prove an analog of Exercise (2) for the case of discrete conditional distributions.

6. Show that for the Markov chain $X \to Y \to Z$, we must have

$$p(x, y, z) = p(x)p(y|x)p(z|y). \qquad (7.104)$$

7. Show that for the Markov chain $X \to Y \to Z$, we must have

$$p(x, z|y) = p(x|y)p(z|y). \qquad (7.105)$$

8. Show that for the discrete time approximation of Example 7.1,

$$S_{n+1} = S_n(1 + \mu\Delta T + \sigma\sqrt{\Delta T}\epsilon_n), \qquad (7.106)$$

the stochastic process S_n is not stationary, but the stochastic process $\frac{S_{n+1}}{S_n}$ is stationary.

9. Prove Corollary 7.7: The conditional entropy $H_U(X_n|X_1)$ increases with n for a stationary Markov process.

Chapter 8

Utility-Based Model Performance Measurement

In Chapters 3 and 5, we introduced the horse race investor as an idealization of a decision maker in an uncertain environment. We have analyzed the usefulness of a probabilistic model to this investor, assuming that the investor aims at maximizing his expected wealth growth rate or expected utility. In this chapter we reconsider such an investor who is trying to measure the performance of a probabilistic model. This will lead us to utility-based performance measures for probabilistic models. The main ideas underlying these performance measures can be summarized in the following principle, which is depicted in Figure 8.1.

Principle 8.1 *(Model Performance Measurement) Given*

(i) an investor with a utility function, and

(ii) a market setting (in this chapter, a horse race or a conditional horse race) in which the investor can allocate,

the investor will allocate according to the model (so as to maximize his expected utility under the model).

We will then measure the performance of the candidate model for this investor via the average utility attained by the investor on an out-of-sample test dataset.

This performance measure that results from this principle can be used to measure the performance of any, parametric or nonparametric, probabilistic model. It does not rely on any specific functional form of the model, any assumptions about model parameters, or any real-world interpretation of probabilities (see, for example, Jaynes (2003), de Finetti (1974), Savage (1954), Bernardo and Smith (2000), or Grünwald (2002) for various interpretations of probabilities); it directly estimates the usefulness of the model to an investor in a horse race. In this sense, the above model performance measure is more closely related to predictive accuracy measures used in the machine learning community than the traditional statistical tests (for a discussion of these different approaches, see, for example, Breiman (2001) and the comments by Cox (2001), Efron (2001), Hoadley (2001), and Parzen (2001)).

The above performance measure should provide a useful tool for practical model building purposes; however, it might have to be balanced with other

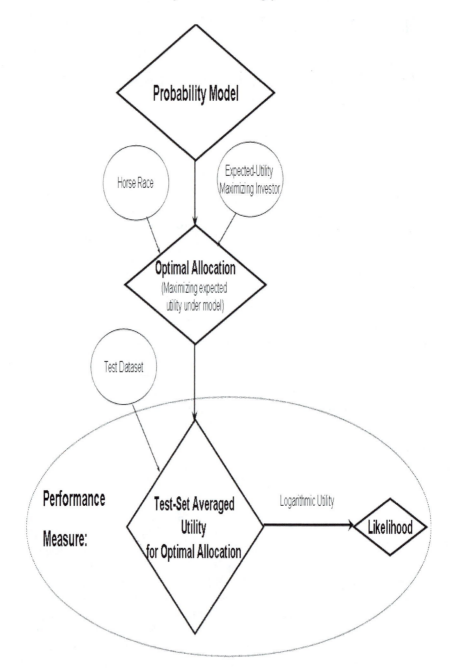

FIGURE 8.1: Model performance measurement principle.

objectives, such as model interpretability. It might also be useful to consider multiple test datasets instead of a single test dataset and take into account the variation of the performance measure over these test datasets, when building a model. We will briefly touch upon such considerations in Chapters 10 and 12, but they lie outside the scope of this book.

We shall see that our performance measure takes a particularly simple form for an investor who has a generalized logarithmic utility function. For such an investor, and only for such an investor, the relative performance of two models can essentially be measured by the likelihood ratio of the two models, and is therefore independent of the odds ratios in the horse race. Moreover, any investor who ranks models for variables with more than two states according to their likelihood ratio must have a generalized logarithmic utility function, and therefore uses the likelihood ratio not only to rank but also to measure model performance.

In Sections 8.1 and 8.2, we explicitly discuss the above model performance measures for discrete probability models, which is the simplest setting of practical interest and clearly illustrates the approach. In the subsequent sections, we generalize the performance measures to conditional probability models (see Section 8.3) and to probability density models (see Sections 8.4 and 8.5).

Based on the idea of the certainty equivalent, one can formulate model performance measures that are similar to the utility-based ones, but can be expressed as monetary values. We will introduce such performance measures in Section 8.6 and discuss its properties. In particular, we shall see that these monetary performance measures are monotone functions of the corresponding utility-based performance measures, and therefore lead to the same model rankings.

8.1 Utility-Based Performance Measures for Discrete Probability Models

In this section, we define decision-theoretic performance measures for discrete probability models, and discuss some of their properties.

We consider a random variable Y with the finite state space $\mathcal{Y}$. We denote model probability measures for Y by $q, q^{(1)}, q^{(2)}$ etc. and the empirical measure on a test dataset by $\tilde{p}$. In order to measure the performance of probability measures for Y, we adapt the viewpoint of an investor, as defined in Definition 3.4, in a horse race, as defined in Definition 3.1, and assume that the investor has a utility function that is compatible with the horse race, as discussed in Chapter 5. We recall from Definitions 3.1 and 3.4 that, for each $y \in \mathcal{Y}$, our investor bets the amount b_y on the event $Y = y$ such that $\sum_{y \in \mathcal{Y}} = 1$, and that each bet pays $\mathcal{O}_y$ if horse y wins and nothing otherwise. We have denoted the investor's allocation by $b = \{b_y,\ y \in \mathcal{Y}\}$.

In order to measure the performance of the probability measure q, we assume, as in Chapter 5, that our investor believes the model q and therefore allocates his \$1 so as to maximize his expected utility under q. Recall that our investor would then allocate according to

$$b^*(q) = \arg \max_{\{b:\sum_{y \in \mathcal{Y}} b_y = 1\}} \sum_{y \in \mathcal{Y}} q_y U(b_y \mathcal{O}_y), \tag{8.1}$$

which implies that

$$b_y^*(q) = \frac{1}{\mathcal{O}_y} (U')^{-1} \left(\frac{\lambda}{q_y \mathcal{O}_y} \right), \tag{8.2}$$

where λ is the solution of

$$\sum_{y \in \mathcal{Y}} \frac{1}{\mathcal{O}_y} (U')^{-1} \left(\frac{\lambda}{q_y \mathcal{O}_y} \right) = 1. \tag{8.3}$$

It is natural to measure model performance by the average utility over a test set under the expected utility-optimal allocation. Making the further assumption that such a test (out-of-sample) dataset is available, we define the following model performance measure.

Definition 8.1 *(Performance measure) For an investor with utility function U, the expected utility model-performance measure for the model q is*

$$E_{\tilde{p}}[U(b^*(q), \mathcal{O})] = \sum_{y \in \mathcal{Y}} \tilde{p}_y U(b_y^*(q) \mathcal{O}_y) \tag{8.4}$$

where $\tilde{p}$ is the empirical probability measure of the test set.

This performance measure belongs to the larger class of decision-theoretic performance measures that can be constructed within the so-called $\mathcal{M}$-completed framework (see Bernardo and Smith (2000), Section 6.1.3, for more details on the $\mathcal{M}$-completed framework). If we view the above performance measure from the $\mathcal{M}$-completed perspective, the empirical measure of the test set plays the role of the "actual belief model," as it does in the cross-validation method discussed in Bernardo and Smith (2000), Section 6.1.6. The performance measure from Definition 8.1 is specific to the horse race setting, which describes a fairly, but not completely, general situation. The horse race specification allows us to explicitly express the performance measure in terms of the payoffs associated with the states.

The following lemma follows directly from the definition of b^* (see 8.1) and Definition 8.1.

Lemma 8.1 *For any probability measure q,*

$$E_{\tilde{p}}[U(b^*(q), \mathcal{O})] \leq E_{\tilde{p}}[U(b^*(\tilde{p}), \mathcal{O})]\ ,\ \text{with equality if } q = \tilde{p}. \tag{8.5}$$

This lemma states that the best performance is achieved by a model that accurately predicts the frequency distribution of the test set. In the language of Bernardo and Smith (2000), this means that the above approach leads to a proper score function for probability measures. The property (8.5) holds for an investor with an arbitrary utility function; all investors agree on what is the perfect probability measure; however they may disagree (if they have different utility functions) on the ranking of imperfect probability measures.

In light of Definition 8.1, it is natural to compare two models by means of the following relative performance measures.

Definition 8.2 *(Relative performance measure) For an investor with utility function U, the expected utility relative model-performance measure for the models $q^{(1)}$ and $q^{(2)}$ is*

$$\Delta_U\left(q^{(1)},q^{(2)},\mathcal{O}\right) = E_{\tilde{p}}\left[U\left(b^*\left(q^{(2)}\right),\mathcal{O}\right)\right] - E_{\tilde{p}}\left[U\left(b^*\left(q^{(1)}\right),\mathcal{O}\right)\right].$$

Although the performance measure $\Delta_U\left(q^{(1)},q^{(2)},\mathcal{O}\right)$ depends on the empirical measure, $\tilde{p}$, for the sake of brevity, we didn't indicate this in the notation; this shouldn't cause any confusion.

8.1.1 The Power Utility

As an example, we explicitly derive $\Delta_U\left(q^{(1)},q^{(2)},\mathcal{O}\right)$ for the power utility function. Recall, from Chapter 5, that an investor with a power utility function given by

$$U_\kappa(W) = \frac{W^{1-\kappa}-1}{1-\kappa} \to log(W) \text{ as } \kappa \to 1, \tag{8.6}$$

where $\kappa \geq 0$, allocates according to

$$b_y^*(q) = \frac{\frac{1}{\mathcal{O}_y}(q_y\mathcal{O}_y)^{\frac{1}{\kappa}}}{\sum_{y'}\frac{1}{\mathcal{O}_{y'}}(q_{y'}\mathcal{O}_{y'})^{\frac{1}{\kappa}}}. \tag{8.7}$$

It follows that

$$U_\kappa(b_y^*(q)\mathcal{O}_y) = \frac{1}{1-\kappa}\left[\left(\frac{(q_y\mathcal{O}_y)^{\frac{1}{\kappa}}}{\sum_{y'}\frac{1}{\mathcal{O}_{y'}}(q_{y'}\mathcal{O}_{y'})^{\frac{1}{\kappa}}}\right)^{1-\kappa} - 1\right] \equiv \Phi_y^\kappa(q), \tag{8.8}$$

and

$$\Delta_{U_\kappa}\left(q^{(1)},q^{(2)},\mathcal{O}\right) = \sum_y \tilde{p}_y\left[\Phi_y^\kappa(q^{(2)}) - \Phi_y^\kappa(q^{(1)})\right]. \tag{8.9}$$

We shall consider a specific numerical example in Section 8.1.7.

8.1.2 The Kelly Investor

In Sections 3.3 and 5.1.4 we have discussed the Kelly investor who allocates his assets so as to maximize his expected wealth growth rate (see Definition 3.6). Next, we compute the relative performance measure from Definition 8.2 for such an investor.

According to Definition 3.5, the expected wealth growth rate corresponding to a probability measure q and a betting strategy b is given by

$$W(b, q) = E_q \left[\log \left(b, \mathcal{O} \right) \right] = \sum_{y \in \mathcal{Y}} q_y \log(b_y \mathcal{O}_y) . \qquad (8.10)$$

Comparing this to the expected utility, $E_q \left[U \left(b, \mathcal{O} \right) \right]$, we see that the Kelly investor is an expected utility-maximizing investor with the utility function

$$U(W) = \log W . \qquad (8.11)$$

Recall from Section 5.1.4 that in this setting,

$$b_y^*(q) = q_y , \qquad (8.12)$$

which is consistent with Theorem 3.1. It follows then from Definition 8.2 that our relative performance measure for the Kelly investor is the log-likelihood ratio (see Definition 6.2), i.e., that the following theorem holds.

Theorem 8.1 (Δ *for Kelly investor and log-likelihood ratio*) *The Kelly investor has the relative model performance measure*

$$\Delta_{log} \left(q^{(1)}, q^{(2)} \right) = l \left(q^{(1)}, q^{(2)} \right) . \qquad (8.13)$$

Recall that $l \left(q^{(1)}, q^{(2)} \right)$ denotes the log-likelihood ratio (see Definition 6.1).

It also follows that, for an expected utility maximizing investor with the logarithmic utility function, our relative performance measure, $\Delta_{log} \left(q^{(1)}, q^{(2)} \right)$, can be interpreted as the estimated gain in expected wealth growth rate for the investor who uses the model $q^{(2)}$, rather than model $q^{(1)}$, under the empirical probability measure $\tilde{p}$.

In Theorem 8.2 in Section 8.1.4 below, we shall generalize Theorem 8.1 and recover the log-likelihood ratio as a relative performance measure for a whole family of utility functions.

8.1.3 Horse Races with Homogeneous Returns

As another example, we compute the performance measure of the homogeneous expected return measure from Definition 3.3, which is given by

$$p^{(h)} = \left\{ p_y^{(h)} = \frac{B}{\mathcal{O}_y}, \ y \in \mathcal{Y} \right\} , \qquad (8.14)$$

where B is the bank account payoff. Recall that, under this measure, the expected payoff for a bet placed on a single horse, y, is always B, independent of y, i.e., that

$$p_y^{(h)} \mathcal{O}_y + \left(1 - p_y^{(h)}\right) 0 = B, \ \forall y \in \mathcal{Y} . \tag{8.15}$$

In order to compute the optimal allocation for our investor, we substitute $p_y^{(h)} \mathcal{O}_y = B$ into (8.2), and obtain

$$b_y^* \left(p^{(h)}\right) = \frac{1}{\mathcal{O}_y} (U')^{-1} \left(\frac{\lambda}{B}\right) = \frac{p_y^{(h)}}{B} (U')^{-1} \left(\frac{\lambda}{B}\right) .$$

Summing and using (8.3), we see that

$$1 = \frac{1}{B} (U')^{-1} \left(\frac{\lambda}{B}\right), \tag{8.16}$$

so

$$\lambda = BU'(B), \tag{8.17}$$

$$b_y^* \left(p^{(h)}\right) = p_y^{(h)} \text{ (as for the Kelly investor!), and} \tag{8.18}$$

$$b_y^* \left(p^{(h)}\right) \mathcal{O}_y = B . \tag{8.19}$$

It follows from (8.19) that the model performance measure from Definition 8.1 is

$$E_{\tilde{p}} \left[U \left(b^* \left(p^{(h)}\right), \mathcal{O}\right)\right] = B . \tag{8.20}$$

That is, under the homogeneous expected return measure, $p^{(h)}$, any investor, regardless of his utility function, will allocate according to $b^* = p^{(h)}$, which leads to the payoff, B, in any state. Thus, the expected utility performance measure, for any investor, regardless of utility function, leads to the same performance, B, under any empirical probability measure, $\hat{p}$.

8.1.4 Generalized Logarithmic Utility Function and the Log-Likelihood Ratio

In this section, we discuss model performance measurement for an investor with a generalized logarithmic utility function and relate it to the likelihood ratio. Recall that the generalized logarithmic utility function is given by

$$U(W) = \alpha \log(W - \gamma B) + \beta , \ \alpha > 0 . \tag{8.21}$$

This utility function was suggested by Bernoulli (1738) in his solution to the St. Petersburg paradox. It is often used in financial modeling (see, for example, Rubinstein (1976)), and seems to well describe the behavior of some fund managers (see Ziemba and MacLean (2005)).[1]

The utility function (5.63) has some very interesting properties. In particular, as the following theorem shows, it leads to the likelihood ratio as a model performance measure.

Theorem 8.2 *For a utility function of the form*

$$U(W) = \alpha \log(W - \gamma B) + \beta \tag{8.22}$$

with $\alpha > 0$ and $\gamma < 1$, the relative performance measure, $\Delta_U\left(q^{(1)}, q^{(2)}, \mathcal{O}\right)$, is given by

$$\Delta_U\left(q^{(1)}, q^{(2)}, \mathcal{O}\right) = \alpha \Delta_{\log}\left(q^{(1)}, q^{(2)}\right) \tag{8.23}$$

$$= \alpha\, l\left(q^{(1)}, q^{(2)}\right) , \tag{8.24}$$

i.e., is proportional to the log-likelihood ratio. Moreover, the optimal betting weights are given by

$$b_y^*(q) = q_y\left[1 - \gamma\right] + \frac{\gamma B}{\mathcal{O}_y} . \tag{8.25}$$

We note that the above relative performance measure

(*i*) is independent of the odds ratios, $\mathcal{O}$,

(*ii*) does not depend on the constants β and γ, and

(*iii*) depends on the constant α only in a trivial way.

Proof: From Section 5.1.5, we have (8.25). From (8.25), we have

$$U(b_y^*(q)\mathcal{O}_y) = \alpha \log q_y + \alpha \log\left(\mathcal{O}_y\left[1 - \gamma\right]\right) + \beta . \tag{8.26}$$

By inserting this expression into Definition 8.2 one can see that our relative performance measure is

$$\Delta_U\left(q^{(1)}, q^{(2)}, \mathcal{O}\right) = \alpha \sum_y \tilde{p}_y \log\left(\frac{q_y^{(2)}}{q_y^{(1)}}\right) . \tag{8.27}$$

Equation (8.24) follows from Definition 6.2 and (8.23) then follows from Theorem 8.1. $\square$

The condition $\gamma < 1$ in Theorem 8.2 insures compatibility between the utility function and the horse race, as discussed in Section 5.1.5.

We shall consider a specific example in Section 8.1.7.

---

[1]For a criticism of the logarithmic utility function, see Samuelson (1971) and Samuelson (1979); for a justification of nearly, but not quite, logarithmic utility functions, see Luenberger (1998) pp. 426-427. Janeček (2002) discusses drawdown consequences of the logarithmic and power utilities; he shows that the logarithmic utility function can lead to investment strategies that are quite aggressive. Thorp (1997) discusses his long term investment experience using the Kelly criterion, i.e., using a special case of the utility function (5.63).

8.1.5 Approximating the Relative Model Performance Measure with the Log-Likelihood Ratio

We have seen in Section 5.1.5 that for an investor with a generalized logarithmic utility function, the relative model performance measure is essentially the likelihood ratio. Obviously, this is not the case for an arbitrary utility function. However, since the generalized logarithmic utility function has three parameters, it can locally approximate any other utility function up to second order. This is illustrated in Figure 8.2. So we might expect that, if expected returns are sufficiently homogeneous, the relative performance measure for an investor with an arbitrary utility function can be approximated by the likelihood ratio. This is indeed the case, as the following theorem states.

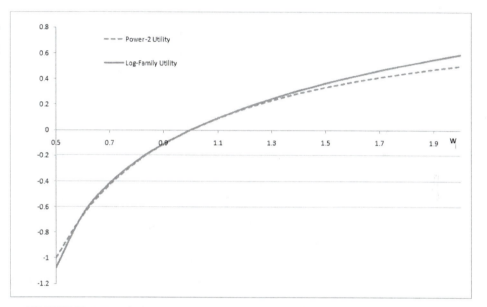

FIGURE 8.2: Approximation of the power utility with power 2.

Theorem 8.3 *Let us assume that the two models $q^{(1)}$ and $q^{(2)}$ define nearly homogeneous expected returns, i.e., that*

$$q_y^{(i)} \mathcal{O}_y = B \left(1 + \epsilon_y^{(q^{(i)})} \right) \ , \ i = 1, 2 \ . \tag{8.28}$$

Then, in the limit $\max_{y,i} |\epsilon_y^{(q^{(i)})}| \to 0$,

$$\Delta_U \left(q^{(1)}, q^{(2)}, \mathcal{O} \right) = \frac{BU'(B)}{R(B)} \Delta_{\log} \left(q^{(1)}, q^{(2)} \right) + o\left(\epsilon'\right) \qquad (8.29)$$

$$= \frac{BU'(B)}{R(B)} l \left(q^{(1)}, q^{(2)} \right) + o\left(\epsilon'\right) , \qquad (8.30)$$

where

$$\epsilon' = \max_{y \in \mathcal{Y}, i=1,2} |\epsilon_y^{(q^{(i)})}| , \qquad (8.31)$$

and

$$R(B) = -\frac{BU''(B)}{U'(B)} \qquad (8.32)$$

is the investor's relative risk aversion at the wealth level B.

Proof: See Section 8.7.1.

We emphasize that, to leading order in ϵ', these performance measures do not depend on the odds ratios.

8.1.6   Odds Ratio Independent Relative Performance Measure

In general, our relative performance measure Δ depends on the odds ratios. This can be an encumbrance in practical application; the odds ratios are often unknown or they can depend on time. For this reason, it may be desirable to have a relative performance measure that doesn't depend on the odds ratios. We ask the following question: Are there any utility functions for which the relative performance measure is independent of the odds ratios? As we have seen, the answer to this questions is "Yes." In the following theorem, we state the general form that the utility function must have in order to lead to a relative performance measure that is independent of the odds ratios:

Theorem 8.4 *If, for any empirical measure, $\tilde{p}$, and candidate model measures, $q^{(1)}$, $q^{(2)}$, the relative model performance measure, $\Delta_U \left(q^{(1)}, q^{(2)}, \mathcal{O} \right)$, is independent of the odds ratios, $\mathcal{O}$, then the utility function, U, has to have the form*

$$U(W) = \alpha \log(W - \gamma B) + \beta , \qquad (8.33)$$

where α, β, and γ are constants.

Proof: See Section 8.7.2.

We note that we have encountered again the generalized logarithmic utility function. We have seen in Theorem 8.2 that, for this utility function, the relative performance measure, Δ, is indeed independent of the odds ratios.

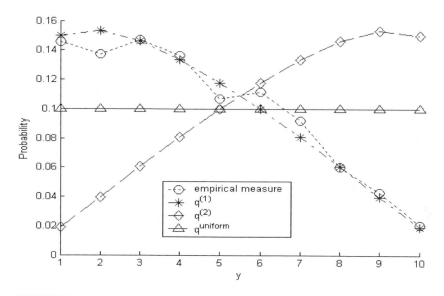

FIGURE 8.3: Example: Three probability measures.

8.1.7 A Numerical Example

As an example, let us consider a random variable Y with the state space $\mathcal{Y} = \{1, 2, ..., 10\}$ and the probability measures $q^{(1)}$, $q^{(2)}$, and $q^{uniform}$, which are shown in Figure 8.3. We assume that we have observed data with the empirical measure $\tilde{p}$, which is also shown in Figure 8.3. Let us further assume that the odds ratios are given by $\mathcal{O}_y = \frac{B}{\tilde{p}_y}$ with $B = 1.2$ (these odds ratios might have been set by a bookie who believes the empirical measure, $\tilde{p}$, and expects the same return for each horse, no matter what the associated risk is).

We compute relative performance measures between these probability measures

(*i*) for an investor with the power utility function (5.69) with $\kappa = 2$, and

(*ii*) for an investor with the generalized logarithmic utility function, with $\alpha = \frac{1}{B\kappa}$ chosen[2] such that (8.30) gives the appropriate approximation for the power utility investor in the nearly homogeneous return case (the choice of β and γ does not affect our relative performance measures).

The results are shown in Table 8.1. We first note that the performance difference between $q^{(1)}$ and $\tilde{p}$ is small (compared to the other values in the table) and, within the displayed precision, the same for both investors. The reason

---

[2]To see that this choice is appropriate, see Exercise 5.

for the smallness is that the two probability measures are fairly similar; the reason why both investors have basically the same Δ is that the returns are homogeneous under $\tilde{p}$ and nearly homogeneous under $q^{(1)}$, so that we can apply the nearly homogeneous returns approximation from Theorem 8.3.

The other relative performance measures in Table 8.1 are all taken with respect to the uniform measure $q^{uniform}$, which might be chosen by an unbiased investor before he sees the data. For both investors, the largest among these relative performance measures is $\Delta_U\left(q^{uniform}, \tilde{p}, \mathcal{O}\right)$. This is so because $\tilde{p}$ is the best performing probability measure (see Lemma 8.1). The second best outperforming probability measure is $q^{(1)}$ for both investors. In fact $q^{(1)}$ outperforms $q^{uniform}$ almost as well as $\tilde{p}$, which, of course, is no surprise since $\Delta_U\left(q^{(1)}, \tilde{p}\right)$ is small and Δ_U is transitive. The probability measure $q^{(2)}$ considerably underperforms the uniform measure, which, again, is no surprise considering Figure 8.3.

We note that, because $q^{uniform}$ doesn't correspond to nearly homogeneous returns, the approximation from Theorem 8.3 doesn't work too well here, i.e., the two investors have different relative performance measures between $q^{uniform}$ and any of the other three probability measures.

TABLE 8.1:   Relative performance measure Δ_U for the specific models from Figure 8.3. In this example, $\mathcal{O}_y = \frac{B}{\tilde{p}_y}$, $B = 1.2$, and $\kappa = 2$. The choices for β and γ do not affect the performance measures.

	$U(W) = \frac{W^{1-\kappa}-1}{1-\kappa}$	$U(W) = \frac{1}{B\kappa}\log(W+\gamma)+\beta$
$\Delta_U\left(q^{(1)}, \tilde{p}, \mathcal{O}\right):$	0.0013	0.0013
$\Delta_U\left(q^{uniform}, \tilde{p}, \mathcal{O}\right):$	0.0342	0.0456
$\Delta_U\left(q^{uniform}, q^{(1)}, \mathcal{O}\right):$	0.0329	0.0443
$\Delta_U\left(q^{uniform}, q^{(2)}, \mathcal{O}\right):$	-0.1868	-0.1581

8.2   Revisiting the Likelihood Ratio

We have seen in Theorem 8.2 that, for the generalized logarithmic utility, our relative model performance measure, Δ, reduces, up to a positive multiplicative constant, to the log-likelihood ratio. The following question arises: are there any other utility functions for which our relative performance measure, Δ, essentially reduces to the log-likelihood ratio? In other words, what kind of utility function can an investor have if he evaluates relative model performance via the log-likelihood ratio? It turns out that we are able to answer this question by answering the following more general question: what kind of utility function can an investor have if he ranks (but not necessarily evaluates) relative model performance via the log-likelihood ratio? To elaborate the difference between ranking and evaluating models, let us consider an investor who has three models, $q^{(1)}, q^{(2)}$, and $q^{(3)}$, and aims at ordering his model preferences, i.e., would like to make a statement of the type "$q^{(2)}$ is better than $q^{(1)}$, which is better than $q^{(3)}$," but does not necessarily make statements of the type "the performance difference between the models $q^{(1)}$ and $q^{(2)}$ is bigger than the one between $q^{(2)}$ and $q^{(3)}$." Such an investor ranks models only, but doesn't completely evaluate their performance. If he uses the log-likelihood ratio for ranking models, then his model performance measure can be any monotone transformation of the log-likelihood ratio. Obviously, if he uses the log-likelihood ratio for measuring model performance, i.e., for evaluating models, he also ranks models according to the log-likelihood ratio.

In order to answer our question, we first state and prove the following theorem.

Theorem 8.5 *Let us assume that*

(i) the utility function U is defined for all arguments $W \geq W_0$, where W_0 is some positive number,

(ii) for any number of states, $|\mathcal{Y}| \geq 3$, odds ratios, $\mathcal{O}$, and candidate model measures, q, there exists a λ that solves (5.18), and

(iii) for any empirical measure, $\tilde{p}$, number of states, $|\mathcal{Y}| \geq 3$, odds ratios, $\mathcal{O}$, and candidate model measures, $q^{(1)}$ and $q^{(2)}$, the relative model performance measure can be expressed as

$$\Delta_U\left(q^{(1)}, q^{(2)}, \mathcal{O}\right) = h_{\mathcal{O},m,\tilde{p}}\left(\sum_{y \in \mathcal{Y}} \tilde{p}_y g_{\mathcal{O},m,y}\left(q_y^{(1)}, q_y^{(2)}\right)\right) , \quad (8.34)$$

where $h_{\mathcal{O},m,\tilde{p}}$ and $g_{\mathcal{O},m,y}$ are real functions that are parameterized as the notation indicates.

Then the utility function, U, must have the form

$$U(W) = \alpha \log(W - \gamma B) + \beta , \quad (8.35)$$

with some constants α, β, and γ, and the relative performance measure is

$$\Delta_U\left(q^{(1)}, q^{(2)}\right) = \alpha l\left(q^{(1)}, q^{(2)}\right) , \quad (8.36)$$

where l is the log-likelihood ratio as defined in Definition 6.2.

Proof: see Section 8.7.3

The following corollary is a straightforward consequence of the above theorem.

Corollary 8.1 *If Assumptions (i) and (ii) of Theorem 8.5 hold and for any empirical measure, $\tilde{p}$, number of states, $|\mathcal{Y}| \geq 3$, odds ratios, $\mathcal{O}$, and candidate model measures, $q^{(1)}$, $q^{(2)}$, the relative model performance measure can be expressed as*

$$\Delta_U \left(q^{(1)}, q^{(2)}, \mathcal{O} \right) = h_{\mathcal{O},m,\tilde{p}} \left(l(q^{(1)}, q^{(2)}) \right) , \tag{8.37}$$

where l is the observed log-likelihood ratio from Definition 6.2 and $h_{\mathcal{O},m,\tilde{p}}$ is a real function that is parameterized as the notation indicates, then the utility function, U, must have the form

$$U(W) = \alpha \log(W - \gamma B) + \beta , \tag{8.38}$$

with some constants α, β, and γ, and the relative performance measure is

$$\Delta_U(q^{(1)}, q^{(2)}, \mathcal{O}) = \alpha \, l \left(q^{(1)}, q^{(2)} \right) . \tag{8.39}$$

This corollary answers our question: In an uncertain environment with more than two states, under some mild technical assumptions, even an investor who uses the log-likelihood ratio only for ranking models must have a generalized logarithmic utility function for his ranking criterion to be consistent with our decision-theoretic framework.

8.3 Utility-Based Performance Measures for Discrete Conditional Probability Models

In this section, we will generalize the model performance measures from Section 8.1 to conditional probability models. We shall see that most of the definitions and results from Section 8.1 can be easily generalized.

We consider probabilities of a random variable Y with state space $\mathcal{Y}$ given values of the random variable X with state space $\mathcal{X}$, and denote probability measures simply by q (there is no need for a more explicit notation such as $q^{Y|X}$ in this section, since we don't simultaneously consider conditional and unconditional measures here). We take the point of view of a conditional investor, as defined in Definition 3.11, in a conditional horse race, as defined in Definition 3.7, and assume that the investor has a utility function and the horse race is compatible, as discussed in Chapter 5.

We recall from Chapter 5 that the optimal allocation for a conditional investor who believes the (conditional probability) model q is

$$b^*(q) = \arg \max_{\{b: \sum_{y \in \mathcal{Y}} b_{y|x}=1 \,,\, \forall x \in \mathcal{X}\}} \sum_{y \in \mathcal{Y}} q_{y|x} U(b_{y|x} \mathcal{O}_{y|x}) . \tag{8.40}$$

Other definitions are possible, but Definition 5.6 seems to be most useful when we want to evaluate (or construct) conditional probabilities $q_{y|x}$, as opposed to joint probabilities $q_{x,y}$.

We recall from Chapter 5 that when the conditional discrete horse race and utility are compatible,

$$\sum_{y \in \mathcal{Y}} \frac{1}{\mathcal{O}_{y|x}} (U')^{-1} \left(\frac{\lambda_x}{q_{y|x} \mathcal{O}_y} \right) \sum_{y \in \mathcal{Y}} \frac{1}{\mathcal{O}_{y|x}} = 1 \qquad (8.41)$$

has a solution for λ_x for each $x \in \mathcal{X}$, and

$$b^*_{y|x}(q) = \frac{1}{\mathcal{O}_{y|x}} (U')^{-1} \left(\frac{\lambda_x}{q_{y|x} \mathcal{O}_{y|x}} \right) , \qquad (8.42)$$

where the λ_x is the solutions of (8.41).

Next, we generalize the model performance measure from Definition 8.1.

Definition 8.3 *For a conditional investor with utility function U, the expected utility model-performance measure for the (conditional probability) model q is*

$$E_{\tilde{p}} [U(b^*(q), \mathcal{O})] = \sum_{x \in \mathcal{X}} \tilde{p}_x \sum_{y \in \mathcal{Y}} \tilde{p}_{y|x} U \left(b^*_{y|x}(q) \mathcal{O}_{y|x} \right) \qquad (8.43)$$

where $\tilde{p}$ is the empirical probability measure of the test set.

For the sake or brevity, we have used the same notation as in Definition 8.1 here; it will always be clear from the context whether we refer to conditional or unconditional measures.

As in the unconditional case, Lemma 8.1 holds, i.e., the best performance is achieved by a model that accurately predicts the frequency distribution of the test set. This is the case for an investor with an arbitrary utility function; all investors agree on what is the perfect probability measure; however they may disagree (if they have different utility functions) on the ranking of imperfect probability measures.

We generalize the relative model performance measure from Definition 8.2 as follows.

Definition 8.4 *For a conditional investor with utility function U, the expected utility relative model-performance measure for the models $q^{(1)}$ and $q^{(2)}$ is*

$$\Delta_U \left(q^{(1)}, q^{(2)}, \mathcal{O} \right) = E_{\tilde{p}} \left[U \left(b^* \left(q^{(2)} \right), \mathcal{O} \right) \right] - E_{\tilde{p}} \left[U \left(b^* \left(q^{(1)} \right), \mathcal{O} \right) \right].$$

This definition might appear the same as the definition in Definition 8.2, but the interpretation is different, since $q^{(1)}$ and $q^{(2)}$ are conditional probability models in this section.

8.3.1 The Conditional Kelly Investor

In Section 5.1.4, Theorem 8.1, we have seen that for a Kelly investor the relative performance measure for unconditional probability models is the likelihood ratio. This can be generalized to the conditional Kelly investor from Definition 3.13.

Theorem 8.6 *(Δ for conditional Kelly investor and log-likelihood ratio) The conditional Kelly investor has the relative model performance measure*

$$\Delta_{log}\left(q^{(1)}, q^{(2)}\right) = l\left(q^{(1)}, q^{(2)}\right) , \qquad (8.44)$$

where $q^{(1)}$ and $q^{(2)}$ denote conditional probability measures, Δ_{log} is the relative model performance measure from Definition 8.4 for the case $U(W) = \log W$, and l denotes the conditional log-likelihood ratio from Definition 6.4.

Proof: The proof is essentially the same as the one for Theorem 8.1.

As in the unconditional case, for an expected utility maximizing investor with the logarithmic utility function, our relative performance measure, $\Delta_{log}\left(q^{(1)}, q^{(2)}\right)$, can be interpreted as the estimated gain in expected wealth growth rate for the investor who uses the model $q^{(2)}$, rather than model $q^{(1)}$, under the empirical probability measure $\tilde{p}$.

8.3.2 Generalized Logarithmic Utility Function, Likelihood Ratio, and Odds Ratio Independent Relative Performance Measure

We generalize Theorem 8.2 as follows.

Theorem 8.7 *For a utility function of the form*

$$U(W) = \alpha \log(W - \gamma B) + \beta \qquad (8.45)$$

with $\alpha > 0$ and $\gamma < 1$, where B denotes the worst conditional bank account defined in Definition 3.9, the relative performance measure, $\Delta_U\left(q^{(1)}, q^{(2)}, \mathcal{O}\right)$, between the two conditional probability measures $q^{(1)}$ and $q^{(2)}$ is given by

$$\Delta_U\left(q^{(1)}, q^{(2)}, \mathcal{O}\right) = \alpha \Delta_{log}\left(q^{(1)}, q^{(2)}\right) \qquad (8.46)$$

$$= \alpha l\left(q^{(1)}, q^{(2)}\right) , \qquad (8.47)$$

where Δ and l are defined by Definitions 8.4 and 6.4, respectively, i.e., is proportional to the (conditional) log-likelihood ratio. Moreover, the optimal betting weights are given by

$$b^*_{y|x}(q) = q_{y|x}\left[1 - \frac{\gamma B}{B_x}\right] - \frac{\gamma B}{\mathcal{O}_{y|x}} . \qquad (8.48)$$

Proof: The proof is essentially the same as the one for Theorem 8.2.

The above relative performance measure is independent of the odds ratios, $\mathcal{O}$, does not depend on the constants β and γ, and depends on the constant α only in a trivial way.

Theorem 8.3 can be generalized as follows.

Theorem 8.8 *Let us assume that* $B_x = \dfrac{1}{\sum_{y \in \mathcal{Y}} \frac{1}{\mathcal{O}_{y|x}}} = B$, $\forall x \in \mathcal{X}$, *for some*
B, and that the two models $q^{(1)}$ *and* $q^{(2)}$ *define nearly homogeneous returns,*
i.e., that

$$p_x q^{(i)}_{y|x} \mathcal{O}_{y|x} = B \left(1 + \epsilon^{(q^{(i)})}_{y|x} \right) \ , \ i = 1, 2 \ . \tag{8.49}$$

Then, in the limit $\max_{y,i} |\epsilon^{(q^{(i)})}_{y}| \to 0$,

$$\Delta_U \left(q^{(1)}, q^{(2)}, \mathcal{O} \right) = \frac{BU'(B)}{R(B)} \Delta_{\log} \left(q^{(1)}, q^{(2)} \right) + o\left(\epsilon' \right) \tag{8.50}$$

$$= \frac{BU'(B)}{R(B)} l \left(q^{(1)}, q^{(2)} \right) + o\left(\epsilon' \right) \ , \tag{8.51}$$

where

$$\epsilon' = \max_{x \in \mathcal{X}, y \in \mathcal{Y}, i=1,2} |\epsilon^{(q^{(i)})}_{y|x}| \ , \tag{8.52}$$

$$R(B) = -\frac{BU''(B)}{U'(B)} \tag{8.53}$$

is the investor's relative risk aversion at the wealth level B, and Δ *and the*
log-likelihood ratio, l, are defined by Definitions 8.4 and 6.4, respectively.

Proof: The proof is a straightforward generalization of the proof of Theorem 8.3.

Similar to Theorem 8.4, we have the following theorem.

Theorem 8.9 *If, for any empirical measure, $\tilde{p}$, and candidate measures, $q^{(1)}$,*
$q^{(2)}$, *the relative model performance measure, $\Delta_U \left(q^{(1)}, q^{(2)}, \mathcal{O} \right)$, is indepen-*
dent of the odds ratios, $\mathcal{O}$, then the utility function, U, has to have the form

$$U(W) = \alpha \log(W - \gamma B) + \beta \ , \tag{8.54}$$

where α, β, and γ are constants.

Proof: The proof is essentially the same as the one for Theorem 8.4.

8.4 Utility-Based Performance Measures for Probability Density Models

In this chapter, until now, we have discussed models for a random variable Y with a discrete (finite) state space. In many practical applications, however, we are interested in random variables that can take any value in a volume element. In this section, we will measure the performance of models for such random variables, i.e., for probability density models. We do so by taking the small-size limit of the discrete-Y model performance measure of a series of discretization of the support of Y as discussed in Chapter 5.

8.4.1 Performance Measures and Properties

Similar to Definition 8.1 for discrete probabilities, we define the performance measures for probability densities as follows.

$$E_{\tilde{p}}[U(b^*[q], \mathcal{O})] = \sum_i \tilde{p}_{y_i} U(b^*[q](y_i)\mathcal{O}(y_i)) , \qquad (8.55)$$

where $\tilde{p}$ is the empirical probability measure and the y_i are the observed Y-values on the test set. We note that $\tilde{p}$ is always a discrete measure, and not a density, since any test set is finite. The argument $b^*[q](y_i)\mathcal{O}(y_i)$ of U in (8.55) is consistent with (5.97) and (5.99), i.e., it is obtained in the limit $n \to \infty$ of the after-bet wealth when the investor bets on the discretized horse race described above.

It follows from Lemma 5.3, that, at least under the assumptions of the lemma, the performance measure (8.55) has the same form as the one from Definition 8.1 in Section 8.1. Thus, we are free to apply all results from Section 8.1 in the context of probability density model performance measures. In particular, Definition 8.2 and Theorems 8.4 and 8.2 apply in this context.

8.5 Utility-Based Performance Measures for Conditional Probability Density Models

In this section, we construct a performance measure for conditional probability density models of the form $p(y|x)$, where $y \in \mathcal{Y} \subset \mathbf{R}^{d_y}$ and $x \in \mathcal{X} \subset \mathbf{R}^{d_x}$. Since our performance measures will be evaluated on a real dataset, there will be at most a finite number of discrete points in $\mathcal{X}$.

We proceed by taking ideas from Sections 5.3 and 8.3. We discretize by forming "chunks" in the cross product space $X \times Y$. For the allocation $b(y|x)$, with

$$\int_{\mathcal{Y}} b(y|x)dy = 1 \text{ for all } x, \tag{8.56}$$

we find that, under a conditional model q and the empirical X-probabilities, the expected utility of these discretization schemes converges to

$$E[U] = \sum_{x_i} \tilde{p}_{x_i} \int_{\mathcal{Y}} q(y|x)U(b(y|x)\mathcal{O}_{y|x})dy . \tag{8.57}$$

So we define the model performance measure

$$E_{\tilde{p}}[U(b^*[q], \mathcal{O})] = \sum_i \tilde{p}_{x_i, y_i} U(b^*[q](y_i|x_i)\mathcal{O}(y_i|x_i)) , \tag{8.58}$$

which has the same form as the one from Definition 8.3 in Section 8.3.

8.6 Monetary Value of a Model Upgrade

The expected utility performance measure from the previous sections can be interpreted in a financial context; however, it is not a monetary value, but rather an expected utility. In practice, on the other hand, probabilistic models are often built or purchased at a cost — measured in dollars. Fortunately, expected utilities can be related to monetary values based on the idea of a certainty equivalent (see, for example, Luenberger (1998)). In this section, which presents the material from Sandow et al. (2007), we use this notion to construct model performance measures that are, in fact, monetary values. In particular, we shall define the monetary value, V, of an upgrade from the model $q^{(1)}$ to the model $q^{(2)}$ and show that this monetary value is a monotone increasing function of the expected utility gain, Δ, corresponding to the same model upgrade, i.e., that both of the performance measures, V and Δ, rank models the same way.

Although the performance measures V and Δ rank models the same way, the numerical performance differences they assign to models are not the same. However, we shall show in this section that, in the case of nearly homogeneous returns, these numerical differences are approximately proportional to each other, with a model independent proportionality factor; in effect, this proportionality factor serves as an exchange rate between utils and dollars.

We shall also derive explicit expressions for the monetary value, V, of a model upgrade for investors with generalized logarithmic and power utility functions. For the generalized logarithmic utility investor, V can be expressed in terms of the likelihood ratio. This fact can be used to find a particularly tractable approximation for V for an investor with an arbitrary utility function. For the power utility investor, we shall derive a general expression for

V and a simple approximation that holds in the limit of large risk aversion. We shall discuss the latter limit also for a general utility function and demonstrate that the value of a model upgrade has an upper limit proportional to the coefficient of (relative or absolute) risk aversion.

For ease of exposition, we shall focus our attention on the simplest possible setting — unconditional probabilities. However, we shall also briefly discuss the monetary value of a model upgrade for conditional probabilities.

8.6.1 General Idea and Definition of Model Value

Let us consider an investor who owns an additional (beyond his \$1) amount V that he doesn't invest. The optimal allocation for this investor is

Definition 8.5 *(Optimal allocation withholding V)*

$$b^*(q, V) = \arg \max_{\{b:\sum_{y\in\mathcal{Y}} b_y=1\}} \sum_{y\in\mathcal{Y}} q_y U(b_y \mathcal{O}_y + V) . \tag{8.59}$$

We note that we have used the same notation, b, we have used previously; this should not lead to any confusion, since the number of arguments and the context should always make clear what we are referring to.

The following lemma gives an explicit expression for the above optimal allocation.

Lemma 8.2 *(Optimal allocation withholding V)*

(i) If

$$\sum_{y\in\mathcal{Y}} \frac{1}{\mathcal{O}_y}(U')^{-1}\left(\frac{\lambda}{q_y\mathcal{O}_y}\right) - \frac{V}{B} = 1 \tag{8.60}$$

has a solution for λ, then

$$b_y^*(q, V) = \frac{1}{\mathcal{O}_y}(U')^{-1}\left(\frac{\lambda}{q_y\mathcal{O}_y}\right) - \frac{V}{\mathcal{O}_y} , \tag{8.61}$$

where λ is the solution of (8.60).

(ii) If

$$\frac{1}{(U')^{-1}(0) - V} < B^{-1} < \frac{1}{max\{0, (U')^{-1}(\infty) - V\}} , \tag{8.62}$$

then (8.60) has a solution for λ.

Proof: The proof is a straightforward generalization of the proof of Lemma 5.1. It can be found in Sandow et al. (2007).

One can easily show that (8.62) holds for the logarithmic, and power utility function, for any odds ratios; it holds for the quadratic utility function if

$B^{-1} > \frac{1}{(U')^{-1}(0)-V}$. (8.62) holds for the utility function $U(W) = \alpha \log(W - \gamma B) + \beta, \alpha > 0$, for all odds ratios, if $\gamma + V \geq 0$, and, if $\gamma + V < 0$, for odds ratios that satisfy the constraint $B + \gamma + V > 0$. $\square$

Borrowing and modifying the idea of the certainty equivalent, we next define the (relative) monetary value of a model.

Definition 8.6 *(Monetary value of a model upgrade)* *The monetary value,* $V_U\left(q^{(1)}, q^{(2)}, \mathcal{O}\right)$*, of upgrading from model* $q^{(1)}$ *to model* $q^{(2)}$ *is the solution for V of the following equation:*

$$\sum_{y\in\mathcal{Y}} \tilde{p}_y U\left(b_y^*(q^{(2)}, 0)\mathcal{O}_y\right) = \sum_{y\in\mathcal{Y}} \tilde{p}_y U\left(b_y^*(q^{(1)}, V)\mathcal{O}_y + V\right), \qquad (8.63)$$

where $\tilde{p}$ is the empirical probability measure of the test set.

$V_U\left(q^{(1)}, q^{(2)}, \mathcal{O}\right)$ is the amount of money the investor who owns model $q^{(1)}$ would have to be compensated with so that he is indifferent to

(*i*) the compensation, and

(*ii*) the model upgrade to model $q^{(2)}$.

8.6.2 Relationship between V and Δ

Definitions 8.2 and 8.6 define two relative model performance measures. This raises the question which of those measures is more useful. The measure Δ expresses the benefit of model 2 over model 1 in terms of an expected utility. For an investor with logarithmic utility, Δ is essentially equal to a wealth growth rate and it is also essentially equal to the logarithm of the likelihood ratio (see, for example, Cover and Thomas (1991)). However, utils are not, in general, the most natural currency for those facing model purchase decisions. Model vendors sell models for dollars; model buyers need to assess model value in the same units to make informed purchase decisions. The model performance measure V, on the other hand, is a monetary value. So both measures, Δ and V, seem to be useful. It would be desirable, if both measures gave the same answer to model selection problems. The two performance measures are generally not identical; however, both performance measures rank candidate models the same way, which follows from the theorem below.

Theorem 8.10 *For fixed $\tilde{p}, q^{(0)}, U$, and $\mathcal{O}$, $\Delta_U(q^{(0)}, q, \mathcal{O})$ is a strictly monotone increasing function of $V_U(q^{(0)}, q, \mathcal{O})$.*

Proof: See Section 8.7.4.

8.6.3 Best Upgrade Value

The following lemma follows directly from Theorem 8.10, Lemma 8.1, and Definition 8.2.

Lemma 8.3 *For any probability measures $q^{(0)}$ and q,*

$$V_U(q^{(0)}, q, \mathcal{O}) \leq V_U(q^{(0)}, \tilde{p}, \mathcal{O}) , \quad \text{with equality if } q = \tilde{p} . \tag{8.64}$$

That is, the highest monetary gain is achieved by upgrading to a model that accurately predicts the frequency distribution of the test set.

8.6.4 Investors with Power Utility Functions

As an example, we consider an investor with a power utility function. For such an investor, we have the following corollary.

Corollary 8.2 *If our investor has the power utility function*

$$U_\kappa(w) = \frac{w^{1-\kappa} - 1}{1 - \kappa} , \quad \kappa > 0 , \tag{8.65}$$

the monetary value of an upgrade from model $q^{(1)}$ to model $q^{(2)}$ is

$$V_{U_\kappa}(q^{(1)}, q^{(2)}, \mathcal{O}) = B \left[\left(\frac{A_\kappa(q^{(2)}, \mathcal{O})}{A_\kappa(q^{(1)}, \mathcal{O})} \right)^{\frac{1}{1-\kappa}} \frac{S_\kappa(q^{(1)}, \mathcal{O})}{S_\kappa(q^{(2)}, \mathcal{O})} - 1 \right] , \tag{8.66}$$

where $S_\kappa(q, \mathcal{O}) = \sum_y \frac{1}{\mathcal{O}_y} (q_y \mathcal{O}_y)^{\frac{1}{\kappa}}$ \tag{8.67}

and $A_\kappa(q, \mathcal{O}) = \sum_y \tilde{p}_y (q_y \mathcal{O}_y)^{\frac{1-\kappa}{\kappa}} .$ \tag{8.68}

Proof: See Section 8.7.5.

From Corollary 8.2, we can derive the following approximation for an investor with large risk aversion, i.e., with a large value for κ.

Corollary 8.3 *If our investor has the power utility function*

$$U_\kappa(w) = \frac{w^{1-\kappa} - 1}{1 - \kappa} , \quad \kappa > 0 , \tag{8.69}$$

then, in the limit $\kappa \to \infty$, the value of an upgrade from model $q^{(1)}$ to model $q^{(2)}$ is

$$V_{U_\kappa, 0}(q^{(1)}, q^{(2)}) = \frac{B}{\kappa} \left\{ E_{p^{(h)}} \left[\log \left(\frac{q^{(1)}}{q^{(2)}} \right) \right] + \log \left(\frac{E_{\tilde{p}} [(q^{(1)} \mathcal{O})^{-1}]}{E_{\tilde{p}} [(q^{(2)} \mathcal{O})^{-1}]} \right) \right\}$$
$$+ o \left(\frac{1}{\kappa} \right) ,$$

where $p^{(h)}$ is the homogeneous return measure from Definition 3.3.

Proof: See Section 8.7.5.

The following corollary follows immediately from Corollary 8.3 and Definition 3.3.

Corollary 8.4 *If our investor has the power utility function*

$$U(w) = \frac{w^{1-\kappa} - 1}{1 - \kappa} , \tag{8.70}$$

then, in the limit $\kappa \to \infty$, the value of an upgrade from the empirical measure, $\tilde{p}$, to the homogeneous return measure $p^{(h)}$ is

$$V_{U_\kappa, \mathcal{O}} \left(p^{(h)}, \tilde{p} \right) = \frac{B}{\kappa} D \left(p^{(h)} \middle\| \tilde{p} \right) + o \left(\frac{1}{\kappa} \right) , \tag{8.71}$$

where $D(q\|p)$ is the Kullback-Leibler relative entropy between q and p.

8.6.5 Approximating V for Nearly Homogeneous Expected Returns

In general, the two performance measures, Δ and V, are different. Although they rank models the same way, the numerical performance differences they assign to models are not the same. However, it turns out that, in the case of approximately homogeneous expected returns, these numerical differences are approximately proportional to each other, with a model independent proportionality factor; effectively this proportionality factor serves as an exchange rate between utils and dollars. The following theorem makes this statement precise.

Theorem 8.11 *Let us assume that the two models $q^{(1)}$ and $q^{(2)}$ define nearly homogeneous expected returns, i.e., that*

$$q_y^{(i)} \mathcal{O}_y = B \left(1 + \epsilon_y^{(q^{(i)})} \right) , \quad i = 1, 2 . \tag{8.72}$$

Then, in the limit $\max_{y \in \mathcal{Y}, i} \left| \epsilon_y^{(q^{(i)})} \right| \to 0$, the following holds:

(i) The monetary value of an upgrade from model $q^{(1)}$ to $q^{(2)}$ is given by

$$V_U \left(q^{(1)}, q^{(2)}, \mathcal{O} \right) = \frac{B}{R(B)} \Delta_{\log} \left(q^{(1)}, q^{(2)} \right) + o(\epsilon') \tag{8.73}$$

$$= \frac{B}{R(B)} l \left(q^{(1)}, q^{(2)} \right) + o(\epsilon') , \tag{8.74}$$

where

$$\epsilon' = \max_{y \in \mathcal{Y}, i=1,2} \left| \epsilon_y^{(q^{(i)})} \right| , \tag{8.75}$$

$$R(B) = -\frac{B U''(B)}{U'(B)} \tag{8.76}$$

is the investor's relative risk aversion at the wealth level B, and the log-likelihood ratio, l, is defined by Definition 6.4.

(ii)

$$\Delta_U\left(q^{(1)}, q^{(2)}, \mathcal{O}\right) = U'(B)V_U\left(q^{(1)}, q^{(2)}, \mathcal{O}\right) + o\left(\epsilon'\right) . \tag{8.77}$$

Proof: See Section 8.7.6.

8.6.6   Investors with Generalized Logarithmic Utility Functions

As stated in the following theorem, for an investor with a generalized logarithmic utility function, the value of a model upgrade is odds-ratio-independent and is related to the likelihood ratio.

Theorem 8.12 *If our investor has the following utility function*

$$U(W) = \alpha \log(W - \gamma B) + \beta , \tag{8.78}$$

the value of a model upgrade is

$$V_U\left(q^{(1)}, q^{(2)}, \mathcal{O}\right) = B(1-\gamma)\left(e^{\Delta_{\log}\left(q^{(1)}, q^{(2)}\right)} - 1\right) \tag{8.79}$$

$$= B(1-\gamma)\left(e^{l\left(q^{(1)}, q^{(2)}\right)} - 1\right) . \tag{8.80}$$

Proof: Let $\gamma' = -\gamma B$. We use Lemma 8.2 to derive the following explicit expression for $b^*(q)$.

$$b_y^*(q, V) = q_y\left[1 + \frac{\gamma' + V}{B}\right] - \frac{\gamma' + V}{\mathcal{O}_y} . \tag{8.81}$$

Hence

$$U[b_y^*(q, V)\mathcal{O}_y + V] = \alpha \log\left(q_y \mathcal{O}_y\left[1 + \frac{\gamma' + V}{B}\right]\right) + \beta, \tag{8.82}$$

where we have assumed, for the moment, that $B + \gamma' + V > 0$. So (8.63) in Definition 8.6 becomes

$$0 = \sum_{y \in \mathcal{Y}} \tilde{p}_y\left(\log \frac{q_y^{(2)}}{q_y^{(1)}} - \log \frac{B + \gamma' + V}{B + \gamma'}\right) . \tag{8.83}$$

Solving this equation, we obtain

$$V_U\left(q^{(1)}, q^{(2)}, \mathcal{O}\right) = (B + \gamma')\left(e^{\sum_{y \in \mathcal{Y}} \tilde{p}_y \log \frac{q_y^{(2)}}{q_y^{(1)}}} - 1\right) , \tag{8.84}$$

which, in conjunction with Definition 6.2 and Theorem 8.1, results in (8.79) and (8.80). We note that (8.84) implies that indeed $B + \gamma' + V > 0$. This concludes the proof of the theorem. □

8.6.7 The Example from Section 8.1.7

In this section, we revisit the example from Section 8.1.7, i.e., we consider a random variable Y with the state space $\mathcal{Y} = \{1, 2, ..., 10\}$ and the probability measures $q^{(1)}$, $q^{(2)}$, and $q^{uniform}$, which are shown in Figure 8.3. We assume that we have observed data with the empirical measure $\tilde{p}$, which is also shown in Figure 8.3. We assume further that the odds ratios are given by $\mathcal{O}_y = \frac{B}{\tilde{p}_y}$ with $B = 1.2$.

We compute relative performance measures between these probability measures

(*i*) for an investor with the power utility function (5.69) with $\kappa = 2$, and

(*ii*) for an investor with a generalized logarithmic utility function, with $\gamma = \frac{\kappa-1}{\kappa}$ (γ was chosen such that the generalized logarithmic utility function locally, around B, approximates the power utility function; the choice of α and β does not affect our relative performance measures, which follows from Theorem 8.12).

The results are shown in Table 8.2. We can make the same observations as we have made in Section 8.1.7 for the expected utility performance measures. The upgrade value from $q^{(1)}$ to $\tilde{p}$ is small (compared to the other values in the table) and, within the displayed precision, the same for both investors. The reason for the smallness is that the two probability measures are fairly similar; the reason why both investors have basically the same V is that the returns are homogeneous under $\tilde{p}$ and nearly homogeneous under $q^{(1)}$, so that we can apply the nearly homogeneous returns approximation from Theorem 8.3.

The other upgrade values in Table 8.2 are all taken with respect to the uniform measure $q^{uniform}$. For both investors, the largest among these relative performance measures is $V_{U,\mathcal{O}}\left(q^{uniform}, \tilde{p}\right)$, which is a consequence of Lemma 8.3. The second best outperforming probability measure is $q^{(1)}$ for both investors. In fact $q^{(1)}$ outperforms $q^{uniform}$ almost as well as $\tilde{p}$. The probability measure $q^{(2)}$ considerably underperforms the uniform measure, in line with Figure 8.3 and Table 8.1.

8.6.8 Extension to Conditional Probabilities

So far we have computed monetary values of model upgrades for unconditional models for a single variable Y. Practitioners, on the other hand, are often interested in evaluating conditional probability measures for a variable Y given X; so we would like to generalize the ideas from the previous sections. In Section 8.3, we have introduced the expected utility performance measure for conditional probability measures, using the notion of the conditional horse

TABLE 8.2: Value of a model upgrade $V_{U,\mathcal{O}}$ for the specific models from Figure 8.3. In this example, $\mathcal{O}_y = \frac{B}{\tilde{p}_y}$, $B = 1.2$, $\kappa = 2$, and $\gamma = B\frac{1-\kappa}{\kappa}$. The choices for α and β do not affect the upgrade values.

	$U(W) = \frac{W^{1-\kappa}-1}{1-\kappa}$	$U(W) = \alpha \log(W+\gamma) + \beta$
$V_U\left(q^{(1)}, \tilde{p}, \mathcal{O}\right):$	0.0019	0.0019
$V_U\left(q^{uniform}, \tilde{p}, \mathcal{O}\right):$	0.0492	0.0693
$V_U\left(q^{uniform}, q^{(1)}, \mathcal{O}\right):$	0.0473	0.0672
$V_U\left(q^{uniform}, q^{(2)}, \mathcal{O}\right):$	-0.2126	-0.1894

race. This performance measure is given by

$$\Delta_U\left(q^{(1)}, q^{(2)}, \mathcal{O}\right) = \sum_x \tilde{p}_x \sum_y \tilde{p}_{y|x} \left[U\left(b^*_{y|x}\left(q^{(2)}, 0\right)\mathcal{O}_{y|x}\right)\right.$$
$$\left. - U\left(b^*_{y|x}\left(q^{(1)}, 0\right)\mathcal{O}_{y|x}\right)\right],$$

where $\mathcal{O}_{y|x}$ is the odds ratio for a bet on horse y when x is known, and the $b^*_{y|x}(q, 0)$ are the optimal betting weights in the conditional horse race for an investor who believes the conditional probability measure q and has the withholding amount $V = 0$.

We can define, in a manner similar to that in Definition 8.6, the value, $V_U\left(q^{(1)}, q^{(2)}, \mathcal{O}\right)$, of a model upgrade as the solution for V of the following equation.

$$\sum_x \tilde{p}_x \sum_y \tilde{p}_{y|x} \ U\left(b^*_{y|x}(q^{(2)}, 0)\mathcal{O}_{y|x}\right)$$
$$= \sum_x \tilde{p}_x \sum_y \tilde{p}_{y|x} U\left(b^*_{y|x}(q^{(1)}, V)\mathcal{O}_{y|x} + V\right).$$

One can derive an explicit expression for the value of a model upgrade in the case of a generalized logarithmic utility, i.e., for $U(W) = \alpha \log(W - \gamma B) + \beta$, under the additional assumption that

$$\frac{1}{\sum_y \frac{1}{\mathcal{O}_{y|x}}} = B, \ \forall x. \tag{8.85}$$

In this case we obtain

$$V_U\left(q^{(1)}, q^{(2)}, \mathcal{O}\right) = B(1-\gamma)\left(e^{\Delta_{\log}\left(q^{(1)}, q^{(2)}\right)} - 1\right), \tag{8.86}$$

where

$$\Delta_{\log}\left(q^{(1)}, q^{(2)}\right) = \sum_{x,y} \tilde{p}_{x,y} \log \frac{q^{(2)}_{y|x}}{q^{(1)}_{y|x}} \tag{8.87}$$

is the conditional log-likelihood ratio.

8.7   Some Proofs

8.7.1   Proof of Theorem 8.3

We follow the lines of the proof of Theorem 2 in Sandow et al. (2007). First we state and prove the following lemma.

Lemma 8.4 *Let us assume that our investor believes a model, q, with nearly homogeneous returns, i.e., with*

$$q_y \mathcal{O}_y = B \left(1 + \epsilon_y^{(q)} \right) . \tag{8.88}$$

Then, in the limit $\max_{y \in \mathcal{Y}} \left| \epsilon_y^{(q)} \right| \to 0$, the investor allocates according to

$$b_y^*(q) = \frac{1}{\mathcal{O}_y} \left(B + \frac{B}{R(B)} \epsilon_y^{(q)} \right) + o \left(\bar{\epsilon}^{(q)} \right) , \tag{8.89}$$

where

$$\bar{\epsilon}^{(q)} = \max_{y \in \mathcal{Y}} \left| \epsilon_y^{(q)} \right| , \tag{8.90}$$

and

$$R(B) = -\frac{B U''(B)}{U'(B)} \tag{8.91}$$

is the investor's relative risk aversion at the wealth level B.

Proof: First we note that it follows from $\sum_{y \in \mathcal{Y}} q_y = 1$, (8.88), and the definition of the bank account, Definition 3.2, that

$$\sum_{y \in \mathcal{Y}} \frac{\epsilon_y^{(q)}}{\mathcal{O}_y} = 0 . \tag{8.92}$$

Next, we use Lemma 5.1. We first solve (5.18), which after inserting (8.88) reads

$$
\begin{aligned}
1 &= \sum_{y \in \mathcal{Y}} \frac{1}{\mathcal{O}_y} (U')^{-1} \left(\frac{\lambda}{B \left(1 + \epsilon_y^{(q)} \right)} \right) \\
&= \sum_{y \in \mathcal{Y}} \frac{1}{\mathcal{O}_y} \left[(U')^{-1} \left(\frac{\lambda}{B} \right) - \epsilon_y^{(q)} \frac{\lambda}{B} \left((U')^{-1} \right)' \left(\frac{\lambda}{B} \right) + o(\bar{\epsilon}^{(q)}) \right] \\
&= \frac{1}{B} (U')^{-1} \left(\frac{\lambda}{B} \right) + o(\bar{\epsilon}^{(q)}) \text{ (by (8.92) and Definition 3.2)} , \quad (8.93)
\end{aligned}
$$

for λ. The solution is

$$\lambda = B U'(B) + o(\bar{\epsilon}^{(q)}) . \tag{8.94}$$

Inserting (8.94) and (8.88) into (5.19), i.e., into

$$b_y^*(q) = \frac{1}{\mathcal{O}_y}(U')^{-1}\left(\frac{\lambda}{q_y \mathcal{O}_y}\right),$$

we obtain

$$
\begin{aligned}
b_y^*(q)\mathcal{O}_y &= (U')^{-1}\left(\frac{U'(B)}{1 + \epsilon_y^{(q)}} + o(\bar{\epsilon}^{(q)})\right) \\
&= (U')^{-1}\left(U'(B) - \epsilon_y^{(q)}U'(B) + o(\bar{\epsilon}^{(q)})\right) \\
&= B - \frac{U'(B)}{U''(B)}\epsilon_y^{(q)} + o(\bar{\epsilon}^{(q)}),
\end{aligned}
\tag{8.95}
$$

which, in conjunction with (8.91), results in (8.89). This concludes the proof of Lemma 8.4. □

Next, we prove Theorem 8.3. To this end, we expand, using Lemma 8.4,

$$U\left(b_y^*(q^{(i)})\mathcal{O}_y\right) = U(B) + \frac{B}{R(B)}\epsilon_y^{(q^{(i)})}U'(B) + o(\epsilon'), \tag{8.96}$$

where

$$\epsilon' = \max_{y \in \mathcal{Y}, i=1,2}|\epsilon_y^{(q^{(i)})}|. \tag{8.97}$$

We insert (8.96) into the definition of Δ, Definition 8.2, and obtain

$$\Delta_U\left(q^{(1)}, q^{(2)}, \mathcal{O}\right) = \frac{BU'(B)}{R(B)}\sum_{y \in \mathcal{Y}}\tilde{p}_y\left[\epsilon_y^{(q^{(2)})} - \epsilon_y^{(q^{(1)})}\right] + o(\epsilon'). \tag{8.98}$$

Next, we use

$$
\begin{aligned}
\log\frac{q_y^{(2)}}{q_y^{(1)}} &= \log\frac{q_y^{(2)}\mathcal{O}_y}{q_y^{(1)}\mathcal{O}_y} \\
&= \log\frac{B\left(1 + \epsilon_y^{(q^{(2)})}\right)}{B\left(1 + \epsilon_y^{(q^{(1)})}\right)} + o(\epsilon') \text{ by (8.28))} \\
&= \epsilon_y^{(q^{(2)})} - \epsilon_y^{(q^{(1)})} + o(\epsilon')
\end{aligned}
\tag{8.99}
$$

to write (8.98) as

$$\Delta_U\left(q^{(1)}, q^{(2)}, \mathcal{O}\right) = \frac{BU'(B)}{R(B)}\sum_{y \in \mathcal{Y}}\tilde{p}_y\log\frac{q_y^{(2)}}{q_y^{(1)}} + o(\epsilon'), \tag{8.100}$$

which, in conjunction with Definition 6.2 and Theorem 8.1, proves (8.29) and (8.30). This completes the proof of Theorem 8.3. □

8.7.2 Proof of Theorem 8.4

In this section, following the lines of Friedman and Sandow (2003b), we prove Theorem 8.4, which states that if, for any empirical measure, $\tilde{p}$, $\Delta_U\left(q^{(1)}, q^{(2)}, \mathcal{O}\right)$ is independent of the odds ratios, $\mathcal{O}$, then the utility function, U, has to have the form

$$U(W) = \alpha \log(W + \gamma') + \beta , \tag{8.101}$$

where α, β, and γ are constants, $\gamma' = -\gamma B$, and $\gamma < 1$.

We prove the theorem by considering the markets of the form

$$\mathcal{O}_y = \overline{\mathcal{O}} , \ \forall y \in \mathcal{Y} , \tag{8.102}$$

and by considering the two models

$$q_y^{(1)} = \frac{1}{m} , \ \forall y \in \mathcal{Y} , \tag{8.103}$$

$$\text{where } m = |\mathcal{Y}| , \tag{8.104}$$

$$q_y^{(2)} = \frac{1}{m}, y \notin \{y_1, y_2\} , \tag{8.105}$$

$$q_{y_1}^{(2)} = \frac{1}{m} + \epsilon , \tag{8.106}$$

$$\text{and } q_{y_2}^{(2)} = \frac{1}{m} - \epsilon . \tag{8.107}$$

Furthermore, we choose $\overline{\mathcal{O}}$ such that

$$\exists \delta \text{ such that } \mathcal{N}_\delta\left(\frac{\overline{\mathcal{O}}}{m}\right) \subset (0, \infty) \cap dom(U) , \tag{8.108}$$

where $\mathcal{N}_\delta(x)$ denotes a δ-neighborhood of the point x. As we shall see below, this condition ensures that, for both of the triples $(U, \mathcal{O}, q^{(1)})$ and $(U, \mathcal{O}, q^{(2)})$, we can find the optimal betting weights according to Lemma 5.1.

Before we prove Theorem 8.4, we state and prove two lemmas.

Lemma 8.5 *Under the market assumption (8.102), the model assumptions (8.103)-(8.107), and the condition (8.108), the optimal betting weights for $y = y_1$ for the two models are*

$$b_{y_1}^*(q^{(1)}) = \frac{1}{\overline{\mathcal{O}}}(U')^{-1}\left(\frac{\lambda^{(1)}}{\frac{1}{m}\overline{\mathcal{O}}}\right) , \tag{8.109}$$

$$b_{y_1}^*(q^{(2)}) = \frac{1}{\overline{\mathcal{O}}}(U')^{-1}\left(\frac{\lambda^{(2)}}{\left(\frac{1}{m} + \epsilon\right)\overline{\mathcal{O}}}\right) , \tag{8.110}$$

and the Lagrange multipliers (defined in equations (5.19) and (5.18)) are related as

$$\frac{\lambda^{(2)}}{\left(\frac{1}{m} + \epsilon\right)\overline{\mathcal{O}}} = \frac{m\lambda^{(1)}}{\overline{\mathcal{O}}} - \epsilon \frac{m^2\lambda^{(1)}}{\overline{\mathcal{O}}} + o(\epsilon) . \tag{8.111}$$

Proof: First, we consider $\lambda^{(1)}$. Substituting (8.102) and (8.103) into Lemma 5.1, (5.18), we obtain

$$1 = \sum_{y \in \mathcal{Y}} \frac{1}{\overline{\mathcal{O}}} (U')^{-1} \left(\frac{\lambda^{(1)}}{\frac{1}{m}\overline{\mathcal{O}}} \right) . \tag{8.112}$$

Simplifying, we have

$$\frac{m\lambda^{(1)}}{\overline{\mathcal{O}}} = U' \left(\frac{\overline{\mathcal{O}}}{m} \right). \tag{8.113}$$

Note that, by (8.108), this equation has a solution for $\lambda^{(1)}$. Therefore (5.18) has a solution, and, by Lemma 5.1, the optimal betting weights are given by (5.19), which, by (8.103), leads to (8.109).

Next, we consider $\lambda^{(2)}$. Substituting (8.102) and (8.105)-(8.107) into Lemma 5.1, (5.18), we have

$$\overline{\mathcal{O}} = \sum_{y \notin \{y_1, y_2\}} (U')^{-1} \left(\frac{\lambda^{(2)}}{\frac{1}{m}\overline{\mathcal{O}}} \right) + (U')^{-1} \left(\frac{\lambda^{(2)}}{\left(\frac{1}{m}+\epsilon\right)\overline{\mathcal{O}}} \right) + (U')^{-1} \left(\frac{\lambda^{(2)}}{\left(\frac{1}{m}-\epsilon\right)\overline{\mathcal{O}}} \right) . \tag{8.114}$$

Note that

$$\frac{\lambda^{(2)}}{\left(\frac{1}{m} \pm \epsilon\right)\overline{\mathcal{O}}} = \frac{m\lambda^{(2)}}{\overline{\mathcal{O}}} \mp \epsilon \frac{m^2\lambda^{(2)}}{\overline{\mathcal{O}}} + o(\epsilon). \tag{8.115}$$

Substituting this equation into (8.114), we obtain

$$\overline{\mathcal{O}} = (m-2)(U')^{-1} \left(\frac{m\lambda^{(2)}}{\overline{\mathcal{O}}} \right)$$
$$+ (U')^{-1} \left(\frac{m\lambda^{(2)}}{\overline{\mathcal{O}}} - \epsilon \frac{m^2\lambda^{(2)}}{\overline{\mathcal{O}}} + o(\epsilon) \right)$$
$$+ (U')^{-1} \left(\frac{m\lambda^{(2)}}{\overline{\mathcal{O}}} + \epsilon \frac{m^2\lambda^{(2)}}{\overline{\mathcal{O}}} + o(\epsilon) \right)$$
$$= (m-2)(U')^{-1} \left(\frac{m\lambda^{(2)}}{\overline{\mathcal{O}}} \right)$$
$$+ (U')^{-1} \left(\frac{m\lambda^{(2)}}{\overline{\mathcal{O}}} \right) - \epsilon \frac{m^2\lambda^{(2)}}{\overline{\mathcal{O}}} \left((U')^{-1} \right)' \left(\frac{m\lambda^{(2)}}{\overline{\mathcal{O}}} \right) + o(\epsilon)$$
$$+ (U')^{-1} \left(\frac{m\lambda^{(2)}}{\overline{\mathcal{O}}} \right) + \epsilon \frac{m^2\lambda^{(2)}}{\overline{\mathcal{O}}} \left((U')^{-1} \right)' \left(\frac{m\lambda^{(2)}}{\overline{\mathcal{O}}} \right) + o(\epsilon)$$
$$= m(U')^{-1} \left(\frac{m\lambda^{(2)}}{\overline{\mathcal{O}}} \right) + o(\epsilon). \tag{8.116}$$

By (8.108), this equation has a solution for $\lambda^{(2)}$ if ϵ is small enough. Therefore (5.18) has a solution, and, by Lemma 5.1, the optimal betting weights are given by (5.19), which, by (8.106), leads to (8.110).

We note that, in (8.116), the terms of first order in ϵ have canceled each other. Rearranging (8.116) and using (8.113), we obtain

$$\frac{m\lambda^{(2)}}{\overline{\mathcal{O}}} = U'\left(\frac{\overline{\mathcal{O}}}{m}\right) + o(\epsilon) = \frac{m\lambda^{(1)}}{\overline{\mathcal{O}}} + o(\epsilon). \tag{8.117}$$

Combining this equation with (8.115), we obtain

$$\frac{\lambda^{(2)}}{\left(\frac{1}{m}+\epsilon\right)\overline{\mathcal{O}}} = \frac{m\lambda^{(1)}}{\overline{\mathcal{O}}} - \epsilon\frac{m^2\lambda^{(1)}}{\overline{\mathcal{O}}} + o(\epsilon) , \tag{8.118}$$

i.e., (8.111). This completes the proof of the lemma. $\square$

Lemma 8.6 *Let U be a strictly concave, strictly increasing function. Let*

$$f(x) = U\left((U')^{-1}(x)\right) \tag{8.119}$$

and suppose that

$$\frac{d}{dx}\left(xf'(x)\right) = 0 \ , \ \forall\, x \in (U'(\overline{W}), U'(\underline{W})) \ , \tag{8.120}$$

for some constants $\underline{W} \le \overline{W}$. Then we must have

$$U(W) = \alpha \log(W + \gamma') + \beta \ , \forall\, W \in (\underline{W}, \overline{W}) \ , \tag{8.121}$$

where α, β, and γ' are constants.

Proof: It follows from (8.120) that

$$0 = f'(x) + xf''(x) \ , \ \forall\, x \in (U'(\overline{W}), U'(\underline{W})) \ . \tag{8.122}$$

All solutions of this differential equation have the form

$$f(x) = \alpha' \log x + \beta' \ , \ \forall\, x \in (U'(\overline{W}), U'(\underline{W})) \ , \tag{8.123}$$

with constants α' and β'.

What is the form of the function U then? U is related to f by means of (8.119), so we have

$$U\left((U')^{-1}(x)\right) = \alpha' \log x + \beta' \ , \ \forall\, x \in (U'(\overline{W}), U'(\underline{W})) \ . \tag{8.124}$$

Let

$$W = (U')^{-1}(x) \ . \tag{8.125}$$

We notice that, since x can take all values in $(U'(\overline{W}), U'(\underline{W}))$, W can take all values in $(\underline{W}, \overline{W})$. Therefore (8.124) reads

$$U(W) = \alpha' \log x + \beta', \tag{8.126}$$

which we exponentiate to obtain

$$e^{\frac{(U(W)-\beta')}{\alpha'}} = x. \tag{8.127}$$

Comparing (8.125) and (8.127), we see that

$$U'(W) = e^{\frac{(U(W)-\beta')}{\alpha'}} \, , \;\; \forall \, W \in (\underline{W}, \overline{W}) \, . \tag{8.128}$$

The solution to this nonlinear first order equation in W is known to exist and be unique, given initial conditions (see Braun (1975), Theorem 2', p. 106). It is given by

$$U(W) = -\alpha' \log \left(\frac{W + \gamma'}{\alpha'} \right) + \beta' \, , \;\; \forall \, W \in (\underline{W}, \overline{W}) \, , \tag{8.129}$$

i.e., it has the form

$$U(W) = \alpha \log(W + \gamma') + \beta \, , \;\; \forall \, W \in (\underline{W}, \overline{W}) \tag{8.130}$$

where α, β, and γ' are constants. $\square$

We now prove Theorem 8.4. Our performance measure is

$$\Delta_U \left(q^{(1)}, q^{(2)}, \mathcal{O} \right) = \sum_{y \in \mathcal{Y}} \tilde{p}_y \left(U \left(b_y^* \left(q^{(2)} \right) \mathcal{O}_y \right) - U \left(b_y^* \left(q^{(1)} \right) \mathcal{O}_y \right) \right) \, . \tag{8.131}$$

We note that, in particular, by the assumptions of the theorem, Δ has to be independent of the payoffs for a set of empirical measures that assign probability one to a particular state and zero to all others. This leads to the requirement that

$$U \left(b_y^* \left(q^{(2)} \right) \mathcal{O}_y \right) - U \left(b_y^* \left(q^{(1)} \right) \mathcal{O}_y \right) \tag{8.132}$$

does not depend on $\mathcal{O}$, for any $y \in \mathcal{Y}$. Using our market assumption, (8.102), for $y = y_1$, we see that

$$U \left(b_{y_1}^* \left(q^{(2)} \right) \overline{\mathcal{O}} \right) - U \left(b_{y_1}^* \left(q^{(1)} \right) \overline{\mathcal{O}} \right) \tag{8.133}$$

does not depend on $\overline{\mathcal{O}}$. It follows from Lemma 8.5 (see (8.110)) that

$$b_{y_1}^* (q^{(2)}) \overline{\mathcal{O}} = (U')^{-1} \left(\frac{\lambda^{(2)}}{\overline{\mathcal{O}} \left(\frac{1}{m} + \epsilon \right)} \right) . \tag{8.134}$$

Also from Lemma 8.5 (see (8.109)), it follows that

$$b_{y_1}^* (q^{(1)}) \overline{\mathcal{O}} = (U')^{-1} \left(\frac{\lambda^{(1)}}{\overline{\mathcal{O}} \frac{1}{m}} \right) . \tag{8.135}$$

Substituting these expression into (8.133), we see that

$$U\left((U')^{-1}\left(\frac{\lambda^{(2)}}{\overline{\mathcal{O}}\left(\frac{1}{m}+\epsilon\right)}\right)\right) - U\left((U')^{-1}\left(\frac{\lambda^{(1)}}{\overline{\mathcal{O}}\frac{1}{m}}\right)\right) \qquad (8.136)$$

does not depend on $\overline{\mathcal{O}}$. Hence

$$f\left(\frac{\lambda^{(2)}}{\overline{\mathcal{O}}\left(\frac{1}{m}+\epsilon\right)}\right) - f\left(\frac{\lambda^{(1)}}{\overline{\mathcal{O}}\frac{1}{m}}\right), \qquad (8.137)$$

where

$$f(x) = U\left((U')^{-1}(x)\right) \qquad (8.138)$$

does not depend on $\overline{\mathcal{O}}$.

Applying (8.111) from Lemma 8.5, we see that

$$f\left(\frac{m\lambda^{(1)}}{\overline{\mathcal{O}}} - \epsilon\frac{m^2\lambda^{(1)}}{\overline{\mathcal{O}}} + o(\epsilon)\right) - f\left(\frac{m\lambda^{(1)}}{\overline{\mathcal{O}}}\right) = -\epsilon\frac{m^2\lambda^{(1)}}{\overline{\mathcal{O}}}f'\left(\frac{m\lambda^{(1)}}{\overline{\mathcal{O}}}\right) + o(\epsilon)$$
$$(8.139)$$

does not depend on $\overline{\mathcal{O}}$. We have (after dividing by m and ϵ), to leading order,

$$\frac{m\lambda^{(1)}}{\overline{\mathcal{O}}}f'\left(\frac{m\lambda^{(1)}}{\overline{\mathcal{O}}}\right) \qquad (8.140)$$

does not depend on $\overline{\mathcal{O}}$. Recall that

$$\frac{m\lambda^{(1)}}{\overline{\mathcal{O}}} = U'\left(\frac{\overline{\mathcal{O}}}{m}\right) \text{ (by (8.113))}; \qquad (8.141)$$

so, substituting into (8.140), we see that

$$U'\left(\frac{\overline{\mathcal{O}}}{m}\right)f'\left(U'\left(\frac{\overline{\mathcal{O}}}{m}\right)\right) \qquad (8.142)$$

does not depend on $\overline{\mathcal{O}}$. Letting

$$x = U'\left(\frac{\overline{\mathcal{O}}}{m}\right), \qquad (8.143)$$

it follows that, since $\frac{\overline{\mathcal{O}}}{m}$ can take all values in the interior of $(0, \infty) \cap dom(U)$ (see (8.108)), x takes all values in the interior of $(U'(\infty), U'(0)) \cap range(U')$. It follows that our first order condition (8.142) not depending on $\overline{\mathcal{O}}$ is equivalent to

$$x f'(x) \qquad (8.144)$$

not depending on x, i.e., our first order condition (8.142) is equivalent to

$$\frac{d}{dx}\left(x f'(x)\right) = 0 \ , \ \forall x \in (U'(\infty), U'(0)) \cap range(U') \ , \qquad (8.145)$$

where we have used the fact that, by (8.138) and Assumption 4.1, f' is differentiable. By Lemma 8.6, we must have

$$U(W) = \alpha \log(W + \gamma') + \beta \ , \ \forall \, W > \max\{0, -\gamma'\} \qquad (8.146)$$

where α, β, and γ' are constants. $\Box$.

8.7.3 Proof of Theorem 8.5

We follow the logic from Friedman and Sandow (2006b). First we state and prove (with one exception, for which we give a reference) the following lemmas.

Lemma 8.7 *Under assumptions (ii) and (iii) of Theorem 8.5, the relative model performance measure,* $\Delta_U \left(q^{(1)}, q^{(2)}, \mathcal{O} \right)$, *must have the form*

$$\Delta_U \left(q^{(1)}, q^{(2)}, \mathcal{O} \right) = \sum_{y \in \mathcal{Y}} \tilde{p}_y G_{\mathcal{O},m,y}(q_y^{(1)}, q_y^{(2)}) \ , \qquad (8.147)$$

where the function $G_{\mathcal{O},m,y}$ *is parameterized as indicated by the notation.*

Proof: By Definition 8.2, the relative performance measure is defined as

$$\Delta_U \left(q^{(1)}, q^{(2)}, \mathcal{O} \right) = \sum_{y \in \mathcal{Y}} \tilde{p}_y \left[U \left(b_y^* \left(q^{(2)} \right) \mathcal{O}_y \right) - U \left(b_y^* \left(q^{(1)} \right) \mathcal{O}_y \right) \right] \ , \qquad (8.148)$$

where b^* is given by Definition 5.3, so that assumption (iii) of Theorem 8.5 reads

$$\sum_{y \in \mathcal{Y}} \tilde{p}_y \left[U \left(b_y^* \left(q^{(2)} \right) \mathcal{O}_y \right) - U \left(b_y^* \left(q^{(1)} \right) \mathcal{O}_y \right) \right]$$

$$= h_{\mathcal{O},m,\tilde{p}} \left(\sum_{y \in \mathcal{Y}} \tilde{p}_y g_{\mathcal{O},m,y}(q_y^{(1)}, q_y^{(2)}) \right) \ ,$$

for all $\tilde{p}$. By choosing, for each y in turn, $\tilde{p} = \tilde{p}^y = \{\tilde{p}_{y'} = \delta_{y',y}, y' \in \mathcal{Y}\}$, we find

$$\left[U \left(b_y^* \left(q^{(2)} \right) \mathcal{O}_y \right) - U \left(b_y^* \left(q^{(1)} \right) \mathcal{O}_y \right) \right] = h_{\mathcal{O},m,\tilde{p}^y} \left(g_{\mathcal{O},m,y}(q_y^{(1)}, q_y^{(2)}) \right) \ , \qquad (8.149)$$

for all $y \in \mathcal{Y}$. Defining the function $G_{\mathcal{O},m,y}$ by means of

$$G_{\mathcal{O},m,y}(x_1, x_2) = h_{\mathcal{O},m,\tilde{p}^y}(g_{\mathcal{O},m,y}(x_x, x_2)), \qquad (8.150)$$

we obtain

$$\left[U \left(b_y^* \left(q^{(2)} \right) \mathcal{O}_y \right) - U \left(b_y^* \left(q^{(1)} \right) \mathcal{O}_y \right) \right] = G_{\mathcal{O},m,y} \left(q_y^{(1)}, q_y^{(2)} \right) \ . \qquad (8.151)$$

Inserting (8.151) into (8.148) results in (8.147), which completes the proof of the lemma.$\Box$

Lemma 8.8 *(Bernardo (1979)) Let a score function be a mapping that assigns a real number to each pair (q, y), where $q = \{q_y, y \in \mathcal{Y}\}$ is a probability measure for Y. If a score function u is*

(i) smooth, i.e., continuously differentiable as a function of the q_y,

(ii) proper, i.e., has the property

$$\sup_q \sum_{y \in \mathcal{Y}} p_y u(q, y) = \sum_{y \in \mathcal{Y}} p_y u(p, y) , \qquad (8.152)$$

and

(iii) local, in the sense that $u(q, y)$ is independent of $q_{y' \neq y}$,

and $m = |\mathcal{Y}| > 2$, then the score function must be of the form

$$u(q, y) = a \log q_y + c_y , \qquad (8.153)$$

where $a > 0$ and the c_y are arbitrary constants.

Proof: See Bernardo (1979), or Proposition 2.29 in Bernardo and Smith (2000).

Lemma 8.9 *Under the assumptions (i)-(iii) of Theorem 8.5, the score function*

$$u(q, y) = U(b_y^*(q)\mathcal{O}_y) \qquad (8.154)$$

is local in the sense that it is independent of $q_{y' \neq y}$, and the relative model performance measure, $\Delta_U\left(q^{(1)}, q^{(2)}, \mathcal{O}\right)$, must have the form

$$\Delta_U\left(q^{(1)}, q^{(2)}, \mathcal{O}\right) = A_m(\mathcal{O})\, l\left(q^{(1)}, q^{(2)}\right) , \qquad (8.155)$$

where l is the log-likelihood ratio from Definition 6.2, and A_m is a positive-valued function, possibly parameterized by m, of the odds ratios, $\mathcal{O}$.

Proof: We will use Lemma 8.8 to prove Lemma 8.9. To this end, we first prove that the assumptions of Lemma 8.8 hold.

The score function, u, as defined by (8.154), is smooth, which follows from our assumptions about U, Assumption *(ii)* of Theorem 8.5, and (5.19).

Furthermore, the score function, u, is proper, which follows from the definition of the score function, (8.154), and the definition of b^*, Definition 5.3.

Next, we show that the score function defined by (8.154) is local. We write the relative performance measure in terms of the score function as

$$\Delta_U\left(q^{(1)}, q^{(2)}, \mathcal{O}\right) = \sum_{y \in \mathcal{Y}} \tilde{p}_y \left\{ u(q^{(2)}, y) - u(q^{(1)}, y) \right\} . \qquad (8.156)$$

In order to see that the score function is local, we use Assumption (iii) of Lemma 8.9 (Theorem 8.5) and Lemma 8.7 to write

$$\Delta_U\left(q^{(1)},q^{(2)},\mathcal{O}\right)=\sum_{y\in\mathcal{Y}}\tilde{p}_y G_{\mathcal{O},m,y}(q_y^{(1)},q_y^{(2)})\ ,\ \forall\tilde{p}\ , \tag{8.157}$$

which, combined with (8.156), results in

$$\sum_{y\in\mathcal{Y}}\tilde{p}_y[u(q^{(2)},y)-u(q^{(1)},y)]=\sum_{y\in\mathcal{Y}}\tilde{p}_y G_{\mathcal{O},m,y}(q_y^{(1)},q_y^{(2)}). \tag{8.158}$$

By choosing, for each y in turn, $\tilde{p}=\tilde{p}^y=\{\tilde{p}_{y'}=\delta_{y',y},y'\in\mathcal{Y}\}$, we find

$$u(q^{(2)},y)-u(q^{(1)},y)=G_{\mathcal{O},m,y}(q_y^{(1)},q_y^{(2)})\ , \tag{8.159}$$

which proves that $u(q,y)$ is independent of the $q_{y'\neq y}$, i.e., that u is local.

Having shown that the score function is smooth, proper, and local, it follows from Lemma 8.8 that the score function can be written as

$$u(q,y)=A_m(\mathcal{O})\log(q_y)+B_{m,y}(\mathcal{O})\ ,\ A_m(\mathcal{O})>0\ . \tag{8.160}$$

(At this point, we have used the assumption that the number of states is greater than two, since Lemma 8.8 holds only under this condition.) Lemma 8.9 follows then from (8.156), (8.160), and Definition 6.2.□

Lemma 8.10 *If a real continuous function F has the property*

$$F\left(\frac{y}{m}\right)=a\left(y\right)\ ,\ \forall y\geq mW_0,m=M,M+1,\dots, \tag{8.161}$$

for some real function a, some positive integer M, and some real positive number W_0, then $F(y)$ and $a(y)$ are independent of y for all $y\geq W_0$.

Proof: It follows from (8.161) that

$$F\left(\frac{y}{m}\right)=F\left(\frac{y}{M}\right)\ ,\ \forall y\geq mW_0,m=M,M+1,\dots\ . \tag{8.162}$$

We now choose two integers m and n with

$$n\geq m\geq M\ . \tag{8.163}$$

Choosing $y=nW_0$ implies $y\geq mW_0$ and, by means of (8.162),

$$F\left(\frac{n}{m}W_0\right)=F\left(W_0\frac{n}{M}\right)\ , \tag{8.164}$$

and choosing $m=n$ in (8.164) implies

$$F\left(W_0\right)=F\left(W_0\frac{n}{M}\right)\ . \tag{8.165}$$

Combining (8.164) with (8.165), we have

$$F\left(\frac{n}{m}W_0\right) = F(W_0) \ , \ \forall m = M, M+1, \dots, \ n = M, M+1, \dots \text{ with } n \geq m \ . \tag{8.166}$$

The independence of $F(y)$ of y then follows from the continuity of F and the fact that any real number larger than one can be obtained as the limit of a ratio $\frac{n}{m}$ with $m \in \{M, M+1, \dots\}$, $n \in \{M, M+1, \dots\}$ and $n \geq m$. It follows from (8.161) that $a(y)$ is independent of y too, which completes the proof of the lemma.$\square$

Next, we prove Theorem 8.5. Under the assumptions of the theorem (which are identical with those of Lemma 8.9), it follows from Lemma 8.9 that

$$\Delta_U\left(q^{(1)}, q^{(2)}, \mathcal{O}\right) = A_m(\mathcal{O})\, l\left(q^{(1)}, q^{(2)}\right) \ , \text{ i.e.,} \tag{8.167}$$

$$\sum_{y \in \mathcal{Y}} \tilde{p}_y \left(U\left(b_y^*\left(q^{(2)}\right)\mathcal{O}_y\right) - U\left(b_y^*\left(q^{(1)}\right)\mathcal{O}_y\right) \right) = A_m(\mathcal{O}) \sum_{y \in \mathcal{Y}} \tilde{p}_y \log\left(\frac{q_y^{(2)}}{q_y^{(1)}}\right) . \tag{8.168}$$

The preceding equation must hold for any choice of the odds ratios; in particular it must hold if

$$\mathcal{O}_y = \overline{\mathcal{O}} > mW_0 \ , \ \forall y \ . \tag{8.169}$$

(The constraint on the admissible values for $\overline{\mathcal{O}}$ ensures that the performance measures exist, at least for certain models, as we shall see below.) In this case, (8.168) reads

$$\sum_{y=1}^{m} \tilde{p}_y \left(U(b_y^*(q^{(2)})\overline{\mathcal{O}}) - U(b_y^*(q^{(1)})\overline{\mathcal{O}}) \right) = a_m(\overline{\mathcal{O}}) \sum_{y=1}^{m} \tilde{p}_y \log\left(\frac{q_y^{(2)}}{q_y^{(1)}}\right) , \tag{8.170}$$

$$\text{where } a_m(\overline{\mathcal{O}}) = A(\mathcal{O}_1, \dots, \mathcal{O}_m)|_{\mathcal{O}_1 = \overline{\mathcal{O}}, \dots, \mathcal{O}_m = \overline{\mathcal{O}}} \ .$$

According to the above equation, the function $a_m(\overline{\mathcal{O}})$ is indexed by the number, m, of states. However, it turns out that it is independent of m. In order to see this, we add an additional state, $y = m+1$, to the existing states, which we denote by $y = 1, \dots, m$, with the property $q_{m+1}^{(1)} = q_{m+1}^{(2)}$. For each of the original states we multiply $\tilde{p}_y$ by the factor $1 - \kappa$, $0 < \kappa < 1$; the state $y = m+1$ gets $\tilde{p}_{m+1} = \kappa$. Under this setup, (8.170) reads

$$(1-\kappa) \sum_{y=1}^{m} \tilde{p}_y \left(U(b_y^*(q^{(2)})\overline{\mathcal{O}}) - U(b_y^*(q^{(1)})\overline{\mathcal{O}}) \right)$$

$$+ \kappa \left(U(b_{m+1}^*(q^{(2)})\overline{\mathcal{O}}) - U(b_{m+1}^*(q^{(1)})\overline{\mathcal{O}}) \right)$$

$$= a_{m+1}(\overline{\mathcal{O}})(1-\kappa) \sum_{y=1}^{m} \tilde{p}_y \log\left(\frac{q_y^{(2)}}{q_y^{(1)}}\right) + \kappa \log\left(\frac{q_{m+1}^{(2)}}{q_{m+1}^{(1)}}\right) . \tag{8.171}$$

On both sides of this equation, the terms corresponding to $y = m+1$ are zero because $q_{m+1}^{(1)} = q_{m+1}^{(2)}$ and the score function $U(b_y^*(q)\overline{\mathcal{O}})$ is local (see Lemma 8.9). Hence, we obtain

$$\sum_{y=1}^{m} \tilde{p}_y \left(U(b_y^*(q^{(2)})\overline{\mathcal{O}}) - U(b_y^*(q^{(1)})\overline{\mathcal{O}}) \right) = a_{m+1}(\overline{\mathcal{O}}) \sum_{y=1}^{m} \tilde{p}_y \log \left(\frac{q_y^{(2)}}{q_y^{(1)}} \right).$$

(8.172)

This equation holds for any choice of $q_{m+1}^{(1)} = q_{m+1}^{(2)}$; in particular it holds for $q_{m+1}^{(1)} = q_{m+1}^{(2)} = 0$, in which case we recover the original m-state setup. In the latter case, (8.170) holds as well. Comparing (8.172) with (8.170) results in $a_{m+1}(\overline{\mathcal{O}}) = a_m(\overline{\mathcal{O}})$. Hence, the function a_m does not depend on m; for the sake of convenience we will drop the index m.

(8.170) must hold for any empirical measure, $\tilde{p}$, in particular for a measure that is concentrated at a particular value y. Therefore, we must have

$$U\left(b_y^*\left(q^{(2)}\right)\overline{\mathcal{O}}\right) - U\left(b_y^*\left(q^{(1)}\right)\overline{\mathcal{O}}\right) = a(\overline{\mathcal{O}})\log\left(\frac{q_y^{(2)}}{q_y^{(1)}}\right).$$

(8.173)

The above equation must hold for any choice of models; in particular it must hold for

$$q_y^{(1)} = \frac{1}{m}, \ \forall y,$$

(8.174)

$$\text{where } m = |\mathcal{Y}|,$$

(8.175)

$$q_y^{(2)} = \frac{1}{m}, \ y \notin \{y_1, y_2\},$$

(8.176)

$$q_{y_1}^{(2)} = \frac{1}{m} + \epsilon,$$

(8.177)

$$\text{and } q_{y_2}^{(2)} = \frac{1}{m} - \epsilon.$$

(8.178)

For the above models, (8.173) reads for $y = y_1$:

$$U\left(b_{y_1}^*\left(q^{(2)}\right)\overline{\mathcal{O}}\right) - U\left(b_{y_1}^*\left(q^{(1)}\right)\overline{\mathcal{O}}\right) = a(\overline{\mathcal{O}})\log\left(1 + m\epsilon\right).$$

(8.179)

It follows from Lemma 8.5 that

$$b_{y_1}^*(q^{(2)})\overline{\mathcal{O}} = (U')^{-1}\left(\frac{\lambda^{(2)}}{\overline{\mathcal{O}}\left(\frac{1}{m}+\epsilon\right)}\right)$$

(8.180)

$$\text{and } b_{y_1}^*(q^{(1)})\overline{\mathcal{O}} = (U')^{-1}\left(\frac{\lambda^{(1)}}{\overline{\mathcal{O}}\frac{1}{m}}\right).$$

(8.181)

Substituting these expressions into (8.179), we must have

$$f\left(\frac{\lambda^{(2)}}{\overline{\mathcal{O}}\left(\frac{1}{m}+\epsilon\right)}\right) - f\left(\frac{\lambda^{(1)}}{\overline{\mathcal{O}}\frac{1}{m}}\right) = a(\overline{\mathcal{O}})\log\left(1 + m\epsilon\right),$$

(8.182)

where

$$f(x) = U\left((U')^{-1}(x)\right). \tag{8.183}$$

Applying again Lemma 8.5, we see that

$$f\left(\frac{\lambda^{(2)}}{\overline{\mathcal{O}}\left(\frac{1}{m}+\epsilon\right)}\right) - f\left(\frac{\lambda^{(1)}}{\overline{\mathcal{O}}\frac{1}{m}}\right) = f\left(\frac{m\lambda^{(1)}}{\overline{\mathcal{O}}} - \epsilon\frac{m^2\lambda^{(1)}}{\overline{\mathcal{O}}} + o(\epsilon)\right) - f\left(\frac{m\lambda^{(1)}}{\overline{\mathcal{O}}}\right)$$

$$= -\epsilon\frac{m^2\lambda^{(1)}}{\overline{\mathcal{O}}}f'\left(\frac{m\lambda^{(1)}}{\overline{\mathcal{O}}}\right) + o(\epsilon). \tag{8.184}$$

Therefore, (8.182) can be rewritten as

$$m\epsilon\frac{m\lambda^{(1)}}{\overline{\mathcal{O}}}f'\left(\frac{m\lambda^{(1)}}{\overline{\mathcal{O}}}\right) + o(\epsilon) = -a(\overline{\mathcal{O}})m\epsilon. \tag{8.185}$$

It follows from (5.18) and (8.174) that

$$\frac{m\lambda^{(1)}}{\overline{\mathcal{O}}} = U'\left(\frac{\overline{\mathcal{O}}}{m}\right). \tag{8.186}$$

Substituting the above equation into (8.185), we see that

$$m\epsilon\, U'\left(\frac{\overline{\mathcal{O}}}{m}\right)f'\left(U'\left(\frac{\overline{\mathcal{O}}}{m}\right)\right) + o(\epsilon) = -a(\overline{\mathcal{O}})m\epsilon\,, \ \forall \epsilon > 0\,, \ m = 3,4,\dots$$

$$\text{and } \overline{\mathcal{O}} \geq mW_0\,, \tag{8.187}$$

i.e., that

$$U'\left(\frac{\overline{\mathcal{O}}}{m}\right)f'\left(U'\left(\frac{\overline{\mathcal{O}}}{m}\right)\right) + \frac{o(\epsilon)}{m\epsilon} = -a(\overline{\mathcal{O}})\,, \ \forall \epsilon > 0\,, \ m = 3,4,\dots$$

$$\text{and } \overline{\mathcal{O}} \geq mW_0\,. \tag{8.188}$$

Taking the limit of the left hand side as $\epsilon \to 0+$, we obtain

$$U'\left(\frac{\overline{\mathcal{O}}}{m}\right)f'\left(U'\left(\frac{\overline{\mathcal{O}}}{m}\right)\right) = -a(\overline{\mathcal{O}})\,, \ \forall m = 3,4,\dots \text{ and } \overline{\mathcal{O}} \geq mW_0\,. \tag{8.189}$$

It follows then from Lemma 8.10 that the function a must be constant. Next, we define

$$x = U'\left(\frac{\overline{\mathcal{O}}}{m}\right); \tag{8.190}$$

we see that, since $\frac{\overline{\mathcal{O}}}{m}$ can take all values $\geq W_0$, x takes all values in the interior of $(U'(\infty), U'(W_0)]$. It follows from (8.189) and the fact that a is a constant function that

$$\frac{d}{dx}(xf'(x)) = 0\,, \ \forall x \in (U'(\infty), U'(W_0)]\,, \tag{8.191}$$

where we have used the fact that f' is differentiable. By Lemma 8.6 and (8.183), we must have

$$U(W) = \alpha \log(W + \gamma) + \beta \ , \ \forall \, W > max\{W_0, -\gamma\} \tag{8.192}$$

where α, β, and γ are constants. This is the same as (8.35).

Finally, (8.36) follows from Theorem 8.2. $\square$

8.7.4 Proof of Theorem 8.10

Following the logic from Sandow et al. (2007), we first state and prove the following lemma.

Lemma 8.11 *The expression*

$$U \left[b_y^* \left(q^{(0)}, V \right) \mathcal{O}_y + V \right] \tag{8.193}$$

is strictly monotone increasing in V.

Proof: Since U is strictly concave, U' and therefore $(U')^{-1}$ are strictly monotone decreasing. It follows from (8.60) that λ decreases when V increases.

Next, we use (8.61) to write

$$U \left[b_y^* \left(q^{(0)}, V \right) \mathcal{O}_y + V \right] = U \left((U')^{-1} \left(\frac{\lambda}{q_y^{(0)} \mathcal{O}_y} \right) \right). \tag{8.194}$$

The function

$$f(x) = (U')^{-1}(x) \tag{8.195}$$

is strictly monotone decreasing, since its derivative

$$f'(x) = \frac{1}{U''\left((U')^{-1}(x)\right)} < 0 \text{ (because } U \text{ is strictly concave)} . \tag{8.196}$$

Since furthermore U is strictly monotone increasing, the function

$$U \left((U')^{-1}(x) \right) \tag{8.197}$$

is strictly monotone decreasing too. It follows then from (8.194) and the fact that λ decreases when V increase that $U \left[b_y^* \left(q^{(0)}, V \right) \mathcal{O}_y + V \right]$ is a strictly monotone increasing function of V. $\square$

Next, we prove Theorem 8.10. To this end, we have to show that, for fixed $\tilde{p}, q^{(0)}$ and $\mathcal{O}$, $\Delta_U \left(q^{(0)}, q, \mathcal{O} \right)$ is a strictly monotone increasing function of $V_U \left(q^{(0)}, q, \mathcal{O} \right)$. For fixed $\tilde{p}, q^{(0)}$ and $\mathcal{O}$, we can view Δ and V as functions of

$$\overline{U}(q) = \sum_y \tilde{p}_y U \left[b_y^* \left(q, 0 \right) \mathcal{O}_y \right] . \tag{8.198}$$

We have

$$\Delta_U \left(q^{(0)}, q, \mathcal{O} \right) = \overline{U}(q) - c \,, \tag{8.199}$$

where

$$c = \sum_y \tilde{p}_y U \left[b_y^* \left(q^{(0)}, 0 \right) \mathcal{O}_y \right] \tag{8.200}$$

does not depend on q, and

$$V_U \left(q^{(0)}, q, \mathcal{O} \right) = F^{-1}(\overline{U}(q)) \,, \tag{8.201}$$

where the function F is defined as

$$F(V) = \sum_y \tilde{p}_y U \left[b_y^* \left(q^{(0)}, V \right) \mathcal{O}_y + V \right] \,. \tag{8.202}$$

Combining (8.199) and (8.201), we obtain

$$V_U \left(q^{(0)}, q, \mathcal{O} \right) = F^{-1} \left(\Delta_U \left(q^{(0)}, q, \mathcal{O} \right) + c \right) \,. \tag{8.203}$$

It follows from Lemma 8.11 and (8.202) that F, and consequently F^{-1}, are strictly monotone increasing in V. Theorem 8.10 follows then from (8.203). $\square$

8.7.5 Proof of Corollary 8.2 and Corollary 8.3

The proofs that follow are the ones from Sandow et al. (2007).

Proof of Corollary 8.2: In order to prove (8.66), we first use Lemma 5.1. We see from (8.60) that

$$\lambda = \left(\sum_{y'} \frac{1}{\mathcal{O}_{y'}} (q_{y'} \mathcal{O}_{y'})^{\frac{1}{\kappa}} \right)^{\kappa} \left(1 + \frac{V}{B} \right)^{-\kappa} \,. \tag{8.204}$$

We see, from (8.61), that

$$b_y^*(q, V) \mathcal{O}_y = \left(1 + \frac{V}{B} \right) \frac{(q_y \mathcal{O}_y)^{\frac{1}{\kappa}}}{S_\kappa(q, \mathcal{O})} - V \,, \tag{8.205}$$

where $S_\kappa(q, \mathcal{O}) = \sum_y \frac{1}{\mathcal{O}_y} (q_y \mathcal{O}_y)^{\frac{1}{\kappa}}$ (see (8.67)). It follows that

$$U(b_y^*(q, V) \mathcal{O}_y + V) = \frac{1}{1 - \kappa} \left[\left(1 + \frac{V}{B} \right)^{1-\kappa} \left(\frac{(q_y \mathcal{O}_y)^{\frac{1}{\kappa}}}{S_\kappa(q, \mathcal{O})} \right)^{1-\kappa} - 1 \right] \,. \tag{8.206}$$

Inserting this equation into (8.63) from Definition 8.6, we obtain

$$0 = \frac{A_\kappa(q^{(2)}, \mathcal{O})}{\left(S_\kappa(q^{(2)}, \mathcal{O}) \right)^{1-\kappa}} - \frac{A_\kappa(q^{(1)}, \mathcal{O})}{\left(S_\kappa(q^{(1)}, \mathcal{O}) \right)^{1-\kappa}} \left(1 + \frac{V}{B} \right)^{1-\kappa} \,, \tag{8.207}$$

where $A_\kappa(q, \mathcal{O}) = \sum_y \tilde{p}_y \, (q_y \mathcal{O}_y)^{\frac{1-\kappa}{\kappa}}$ (see (8.68)). Solving for V results in (8.66). □

Proof of Corollary 8.3: Using (8.67), the expansion

$$(q_y \mathcal{O}_y)^{\frac{1}{\kappa}} = 1 + \frac{1}{\kappa} \log(q_y \mathcal{O}_y) + o\left(\frac{1}{\kappa}\right) ,$$

and Definition 3.3, we obtain

$$S_\kappa(q, \mathcal{O}) = \frac{1}{B} + \frac{1}{B\kappa} E_{p^{(h)}} \left[\log(q\mathcal{O})\right] + o\left(\frac{1}{\kappa}\right) . \qquad (8.208)$$

This leads to

$$\frac{S_\kappa(q^{(1)}, \mathcal{O})}{S_\kappa(q^{(2)}, \mathcal{O})} = 1 - \frac{1}{\kappa} E_{p^{(h)}} \left[\log\left(\frac{q^{(2)}}{q^{(1)}}\right)\right] + o\left(\frac{1}{\kappa}\right) . \qquad (8.209)$$

Next, we expand the term $\left(\frac{A_\kappa(q^{(2)}, \mathcal{O})}{A_\kappa(q^{(1)}, \mathcal{O})}\right)^{\frac{1}{1-\kappa}}$. We substitute κ by $\frac{1}{\alpha}$ into this term, and define it as a function of α:

$$f(\alpha) = \left(\frac{A_{\frac{1}{\alpha}}(q^{(2)}, \mathcal{O})}{A_{\frac{1}{\alpha}}(q^{(1)}, \mathcal{O})}\right)^{\frac{\alpha}{\alpha-1}} . \qquad (8.210)$$

Then

$$\log f(\alpha) = \frac{\alpha}{\alpha-1} \left[\log(A_{\frac{1}{\alpha}}(q^{(2)}, \mathcal{O})) - \log(A_{\frac{1}{\alpha}}(q^{(1)}, \mathcal{O}))\right] , \qquad (8.211)$$

and we obtain

$$\begin{aligned}
\frac{f'(\alpha)}{f(\alpha)} &= (\log(f(\alpha)))' \\
&= -\frac{1}{(\alpha-1)^2} \left[\log(A_{\frac{1}{\alpha}}(q^{(2)}, \mathcal{O})) - \log(A_{\frac{1}{\alpha}}(q^{(1)}, \mathcal{O}))\right] \\
&\quad + \frac{\alpha}{\alpha-1} \left[\log(A_{\frac{1}{\alpha}}(q^{(2)}, \mathcal{O})) - \log(A_{\frac{1}{\alpha}}(q^{(1)}, \mathcal{O}))\right]' . \qquad (8.212)
\end{aligned}$$

Based on

$$\lim_{\alpha \to 0} f(\alpha) = 1 , \qquad (8.213)$$

$$\lim_{\alpha \to 0} A_{\frac{1}{\alpha}}(q^{(1)}, \mathcal{O}) = \sum_y \tilde{p}_y (q_y^{(1)} \mathcal{O}_y)^{-1} , \qquad (8.214)$$

$$\lim_{\alpha \to 0} A_{\frac{1}{\alpha}}(q^{(2)}, \mathcal{O}) = \sum_y \tilde{p}_y (q_y^{(2)} \mathcal{O}_y)^{-1} , \qquad (8.215)$$

and

$$\lim_{\alpha \to 0} \frac{\alpha}{\alpha - 1} [\log(A_{\frac{1}{\alpha}}(q^{(2)}, \mathcal{O})) - \log(A_{\frac{1}{\alpha}}(q^{(1)}, \mathcal{O}))]' = 0 , \tag{8.216}$$

(8.212) leads to

$$f'(0) = \log \left(\sum_y \tilde{p}_y (q_y^{(1)} \mathcal{O}_y)^{-1} \right) - \log \left(\sum_y \tilde{p}_y (q_y^{(2)} \mathcal{O}_y)^{-1} \right) . \tag{8.217}$$

So we get the expansion

$$\left(\frac{A_\kappa(q^{(2)}, \mathcal{O})}{A_\kappa(q^{(1)}, \mathcal{O})} \right)^{\frac{1}{1-\kappa}} = 1 + \frac{1}{\kappa} \left\{ \log \left(\sum_y \tilde{p}_y (q_y^{(1)} \mathcal{O}_y)^{-1} \right) - \log \left(\sum_y \tilde{p}_y (q_y^{(2)} \mathcal{O}_y)^{-1} \right) \right\}$$
$$+ o \left(\frac{1}{\kappa} \right) .$$

Combining this equation with (8.66) and (8.209), we get

$$V_{U_\kappa} \left(q^{(1)}, q^{(2)}, \mathcal{O} \right) = \frac{B}{\kappa} \left\{ E_{p^{(h)}} \left[\log \left(\frac{q^{(1)}}{q^{(2)}} \right) \right] + \log \left(\frac{E_{\tilde{p}} \left[(q^{(1)} \mathcal{O})^{-1} \right]}{E_{\tilde{p}} \left[(q^{(2)} \mathcal{O})^{-1} \right]} \right) \right\} + o \left(\frac{1}{\kappa} \right) . \tag{8.218}$$

This completes the proof of the corollary . □

8.7.6   Proof of Theorem 8.11

Following the logic from Sandow et al. (2007), we first state and prove the following lemma.

Lemma 8.12 *Let us assume that our investor*

(i) has initial capital of $\$(1 + V)$ *and allocates* $\$1$ *to the horse race, and*

(ii) believes a model, q, with nearly homogeneous returns, i.e., with

$$q_y \mathcal{O}_y = B(1 + \epsilon_y^{(q)}) . \tag{8.219}$$

Then, in the limit $\max_{y \in \mathcal{Y}} \left| \epsilon_y^{(q)} \right| \to 0$, *the investor allocates according to*

$$b_y^*(q, V) = \frac{1}{\mathcal{O}_y} \left(B + \frac{B + V}{R(B + V)} \epsilon_y^{(q)} \right) + o \left(\bar{\epsilon}^{(q)} \right) , \tag{8.220}$$

where

$$\bar{\epsilon}^{(q)} = \max_{y \in \mathcal{Y}} \left| \epsilon_y^{(q)} \right| , \tag{8.221}$$

and

$$R(W) = -\frac{W U''(W)}{U'(W)} \tag{8.222}$$

is the investor's relative risk aversion at the wealth level W.

Proof: First we note that it follows from $\sum_{y \in \mathcal{Y}} q_y = 1$, (8.219), and Definition 3.2 that

$$\sum_{y \in \mathcal{Y}} \frac{\epsilon_y^{(q)}}{\mathcal{O}_y} = 0 .$$

(8.223)

Next, we solve (8.60), which after inserting (8.219) reads

$$
\begin{aligned}
1 + \frac{V}{B} &= \sum_{y \in \mathcal{Y}} \frac{1}{\mathcal{O}_y} (U')^{-1} \left(\frac{\lambda}{B(1 + \epsilon_y^{(q)})} \right) \\
&= \sum_{y \in \mathcal{Y}} \frac{1}{\mathcal{O}_y} \left[(U')^{-1} \left(\frac{\lambda}{B} \right) - \epsilon_y^{(q)} \frac{\lambda}{B} \left((U')^{-1} \right)' \left(\frac{\lambda}{B} \right) + o(\epsilon^{(q)}) \right] \\
&= \frac{1}{B} (U')^{-1} \left(\frac{\lambda}{B} \right) + o(\epsilon^{(q)})
\end{aligned}
$$

(8.224)

(by (8.223) and Definition 3.2) ,

for λ. The solution is

$$\lambda = B U'(B + V) + o(\epsilon^{(q)}) .$$

(8.225)

Inserting (8.225) and (8.219) into (8.61), we have

$$
\begin{aligned}
b_y^* \mathcal{O}_y &= (U')^{-1} \left(\frac{U'(B + V)}{1 + \epsilon_y^{(q)}} + o(\epsilon^{(q)}) \right) - V \\
&= B - \frac{U'(B + V)}{U''(B + V)} \epsilon_y^{(q)} + o(\epsilon^{(q)}) ,
\end{aligned}
$$

(8.226)

which, in conjunction with (8.222), results in (8.89). This concludes the proof of Lemma 8.12. $\square$

Next, we prove Theorem 8.11. To this end, we expand $V_U \left(q^{(1)}, q^{(2)}, \mathcal{O} \right)$ with respect to the $\epsilon_y^{(q^{(i)})}$. Let V_0 and V_1 denote the first two terms in this expansion; we have

$$V_U \left(q^{(1)}, q^{(2)}, \mathcal{O} \right) = V_0 + V_1 + o(\epsilon') ,$$

(8.227)

$$\text{where } \epsilon' = \max_{y \in \mathcal{Y}, i = 1, 2} \left| \epsilon_y^{(q^{(i)})} \right| \text{ (see (8.75)) },$$

(8.228)

$$V_0 = O(1) , \text{ and}$$

(8.229)

$$V_1 = O(\epsilon') .$$

(8.230)

Since $V_0 = \lim_{\epsilon' \to 0} V_U \left(q^{(1)}, q^{(2)}, \mathcal{O} \right)$ and the limit $\epsilon' \to 0$ corresponds to the case of two (identical) homogeneous-return models, we must have $V_0 = 0$, i.e., we obtain

$$V_U \left(q^{(1)}, q^{(2)}, \mathcal{O} \right) = V_1 + O(\epsilon') .$$

(8.231)

Next, we prove statement (i) of Theorem 8.11. It follows from Lemma 8.12 and $V = O(\epsilon')$ that

$$b_y^*(q^{(2)}, 0) = \frac{B}{\mathcal{O}_y}\left(1 + \frac{\epsilon_y^{(q^{(2)})}}{R(B)}\right) + o(\epsilon') \text{ , and}$$

$$b_y^*(q^{(1)}, V) = \frac{B}{\mathcal{O}_y}\left(1 + \frac{\epsilon_y^{(q^{(1)})}}{R(B)}\right) + o(\epsilon') \text{ .} \tag{8.232}$$

Now we expand

$$U\left(b_y^*(q^{(2)}, 0)\mathcal{O}_y\right) = U(B) + \frac{B}{R(B)}\epsilon_y^{(q^{(2)})}U'(B) + o(\epsilon') \text{ , and}$$

$$U\left(b_y^*(q^{(1)}, V)\mathcal{O}_y + V\right) = U(B) + \left[\frac{B}{R(B)}\epsilon_y^{(q^{(1)})} + V_1\right]U'(B) + o(\epsilon')$$

$$\text{(since } V_1 = O(\epsilon')\text{).}$$

We insert this into the definition of V, which is

$$\sum_{y\in\mathcal{Y}}\tilde{p}_y U\left(b_y^*(q^{(2)}, 0)\mathcal{O}_y\right) - \sum_{y\in\mathcal{Y}}\tilde{p}_y U\left(b_y^*(q^{(1)}, V)\mathcal{O}_y + V_1\right) = o(\epsilon') \text{ ,} \tag{8.233}$$

(see Definition 8.6), and obtain

$$\frac{B}{R(B)}\sum_{y\in\mathcal{Y}}\tilde{p}_y\epsilon_y^{(q^{(2)})} - \frac{B}{R(B)}\sum_{y\in\mathcal{Y}}\tilde{p}_y\epsilon_y^{(q^{(1)})} - V_1 = 0 + o(\epsilon') \text{ ,} \tag{8.234}$$

i.e.,

$$V_U\left(q^{(1)}, q^{(2)}, \mathcal{O}\right) = V_1 + o(\epsilon')$$

$$= \frac{B}{R(B)}\sum_{y\in\mathcal{Y}}\tilde{p}_y\left[\epsilon_y^{(q^{(2)})} - \epsilon_y^{(q^{(1)})}\right] + o(\epsilon') \text{ .} \tag{8.235}$$

We see from this equation that indeed $V_1 = O(\epsilon')$, as we have assumed. Next, we use

$$\log\frac{q_y^{(2)}}{q_y^{(1)}} = \log\frac{q_y^{(2)}\mathcal{O}_y}{q_y^{(1)}\mathcal{O}_y}$$

$$= \log\frac{B\left(1 + \epsilon_y^{(q^{(2)})}\right)}{B\left(1 + \epsilon_y^{(q^{(1)})}\right)} + o(\epsilon') \text{ (from (8.72))}$$

$$= \epsilon_y^{(q^{(2)})} - \epsilon_y^{(q^{(1)})} + o(\epsilon') \tag{8.236}$$

to write (8.235) as

$$V_U\left(q^{(1)}, q^{(2)}, \mathcal{O}\right) = \frac{B}{R(B)} \sum_{y \in \mathcal{Y}} \tilde{p}_y \log \frac{q_y^{(2)}}{q_y^{(1)}} + o\left(\epsilon'\right) . \qquad (8.237)$$

The above equation, together with Definition 6.2 and Theorem 8.1, proves statement (i) of the theorem. Statement (ii) follows then from Theorem 8.3, which concludes the proof of Theorem 8.11. □

8.8   Exercises

1. Construct an example for which side information improves the expected utility of an investor who aims at maximizing the latter.

2. A coin has probability $\tilde{p}_1 = 0.55$ of turning up heads and probability $\tilde{p}_2 = 0.45$ of turning up tails. The odds ratios for a bet on a toss of this coin are $\mathcal{O}_1 = \mathcal{O}_2 = \$2.1$. Compute for an expected utility maximizing investor, the gain in expected utility with respect to an investor who believes the uniform measure. Assume that the investor has a generalized logarithmic utility function and that he believes the probability measure

   (a) $q_1 = 1 - q_2 = 0.55$, or
   (b) $q_1 = 1 - q_2 = 0.9$.

   Compute, for the same investor, the monetary value of upgrading from the uniform measure to the above model q.

3. Assume the same setting as in the previous problem, except that the investor has a power utility now. Plot the investor's gain in expected utility and the monetary value of the model upgrade as functions of the investor's risk aversion.

4. A hypothetical, expected wealth growth rate-maximizing trader places bets on whether companies default ($Y = 1$) or survive ($Y = 0$) over the course of a year, given side information $x \in R^d$ (the financials of the firm, economic conditions, etc.). He gets his probabilities from the conditional probability model $q(Y = y|x), y \in \{0, 1\}$.

   (a) What is expected wealth growth rate cost of making Kelly (i.e., expected growth rate optimal) bets according to the model $q(Y = y|x)$ rather than the "true" model $p(Y = y|x)$, measured on the equally likely points $\{x_1, \ldots, x_N\}$?

(b) If the model $q(Y = y|x)$ was produced by a maximum-likelihood logistic regression

$$q(Y = 1|x) = \frac{1}{1 + exp(-\beta_0 + \beta \cdot x)} \qquad (8.238)$$

trained on the set $\{(x_i, y_i), i = 1, \ldots, N\}$, prove that

$$\frac{1}{N} \sum_i q(Y = 1|x_i) = \text{percentage of defaulters in the training set.}$$

$$(8.239)$$

(Hint: Derive the dual for the MRE problem subject to the above equation as a constraint.)

5. Which value must be chosen for α, so that the generalized logarithmic utility function

$$U(W) = \alpha \log(W - \gamma B) + \beta \qquad (8.240)$$

approximates a power κ utility function locally near wealth level B?

6. Derive an explicit expression for the performance measure Δ in the case of a linear utility function.

Chapter 9

Select Methods for Estimating Probabilistic Models

In Chapters 6 and 8, we have measured the performance of probabilistic models from a statistical and from a decision-theoretic point of view. Obviously, before we can evaluate a model, we have to build it, preferably such that it performs well according to our chosen evaluation criterion. How one can do this is the topic of this and the following chapter. In this chapter we review some commonly used methods, and in the next chapter we will recast these methods in a decision-theoretic framework. We note that the review in this chapter is far from complete, but rather restricted to those methods that can be related to the decision-theoretic framework that is the main topic of this book.

In this chapter, we shall restrict ourselves to the simplest possible setting, which is the estimation of unconditional probabilities of a discrete random variable. We have chosen this setting for the sake of convenience; it allows us to clarify the main ideas. A generalization to conditional probabilities and continuous random variables is in most cases straightforward; we will address this issue in later chapters.

In Section 9.1, we shall review classical parametric inference, introducing the main ideas of parametric inference and discussing maximum likelihood inference in some detail. In maximum likelihood inference, the parameters of a model are chosen so as to maximize the likelihood. This approach is consistent with measuring model performance via the likelihood; we have discussed various motivations for doing so in Sections 6.2, 8.1, and 8.2. Some of these motivations are of a decision-theoretic nature.

In Section 9.2, we discuss the problem of overfitting and regularized maximum likelihood inference. Overfitting is the often-observed phenomenon of an estimated model matching the training data very well but out-of-sample data poorly. Regularization is a way to mitigate overfitting. In the regularized maximum-likelihood method, one doesn't maximize the log-likelihood itself but an objective function that is the sum of the likelihood and a term that penalizes nonsmooth distribution. The resulting distributions are smoother than the ones obtained from a maximum-likelihood estimation, and are therefore less likely to overfit.

Another powerful method for estimating probabilistic models is Bayesian inference. In Section 9.3, we review the main ideas of this approach, such as prior and posterior measures. We shall demonstrate that, for large sample sizes, it is related to maximum-likelihood inference. Regularization of the latter method can then be interpreted as the consequence of choosing a nonuniform prior.

An alternative approach to estimating probabilities is the minimum relative entropy (MRE) method. We review this method in Section 9.4 in this chapter, and, in the next chapter, we shall relate it to the decision-theoretic framework from Chapter 8. In Section 9.4, we discuss the MRE method with equality constraints, i.e., the standard MRE problem, and with relaxed inequality constraints, i.e., the relaxed MRE problem. We shall see that the dual of the standard MRE problem is a maximum-likelihood problem and the dual of the relaxed MRE problem is a regularized maximum-likelihood problem for a family of exponential distributions.

9.1 Classical Parametric Methods

In the classical parametric approach to probabilistic model building one assumes that the probability measure we are to estimate is a member of parametric family, and one infers the "true" parameter value from data. In this section, we will briefly discuss the main ideas behind this approach and the maximum-likelihood method as the most commonly used method for parameter inference. For a more detailed review of this topic we refer the reader to textbooks on classical statistics, such as the one by Davidson and MacKinnon (1993).

To be specific, we set out to estimate a probability measure for the random variable Y with the finite state space $\mathcal{Y}$. We have chosen this setting for the sake of convenience; it is straightforward to generalize to conditional probabilities or to consider infinite state spaces.

9.1.1 General Idea

In classical parametric statistics, we assume that the random variable Y with the state space $\mathcal{Y}$ has a "true" probability measure of the form

$$p^\theta = \{p_y^\theta , \; y \in \mathcal{Y}\} , \qquad (9.1)$$

where θ is a vector of parameters. However, we don't know the "true" value, θ_0, of the parameter vector, θ, and, therefore, have to estimate it from the data. An example for a parametric measure of the type (9.1) is the binomial distribution, which assigns the probabilities $p_y^\theta = \binom{n}{y}\theta^y (1 - \theta)^{n-y}$, with the parameter θ to the random variable Y with the state space $\mathcal{Y} = \{0, 1, ..., n\}$.

We shall not discuss the validity of the above assumption here; interesting discussions of this topic can be found, for example, in Jaynes (2003). We would like to point out though that the existence of a "true" underlying probability measure, or, equivalently, of a data generating process, is not necessary in the decision-theoretic approaches that we shall introduce in Chapter 10.

There are various methods for estimating the parameter vector θ; the most widely used one is arguably the maximum-likelihood method. Alternatives to the maximum-likelihood method are M-estimators and the method of moments; these methods are beyond the scope of this book. We shall describe the maximum-likelihood method below, in Section 9.1.3. However, before we do so, we shall discuss some general desirable properties of parameter estimators.

9.1.2 Properties of Parameter Estimators

Let us consider an estimator $\hat{\theta}_N$ for θ, that was inferred from a sample of size N. It is often useful to know if this estimator has the properties defined below.

Bias

The bias of an estimator $\hat{\theta}_N$ for θ is

$$\|E[\hat{\theta}_N] - \theta_0\| ,$$

where $\|...\|$ denotes the norm of a vector, and θ_0 is the "true" parameter vector. An estimator $\hat{\theta}_N$ is said to be unbiased if

$$E[\hat{\theta}_N] = \theta_0$$

and asymptotically unbiased if

$$\lim_{N\to\infty} E[\hat{\theta}_N] = \theta_0 .$$

We generally prefer estimators with a small bias, since, everything else being equal, they are more likely to be close to the "true" parameter value.

Below we list some examples for biased and unbiased estimators. In all of these examples we denote the observed values for Y by $y_1, ..., y_N$.

(*i*) The sample mean

$$\overline{y}_N = \frac{1}{N} \sum_{i=1}^{N} y_i \qquad (9.2)$$

is an unbiased estimator for $E_{p^{\theta_0}}[Y]$ (see Exercise 1).

(*ii*) The sample variance

$$\hat{\sigma}_N^2 = \frac{1}{N} \sum_{i=1}^{N} (y_i - \overline{y}_N)^2 \qquad (9.3)$$

is a biased estimator for the variance of Y under p^{θ_0} (see Exercise 2).

(*iii*) The following variance estimator is unbiased

$$\frac{1}{N-1}\sum_{i=1}^{N}(y_i - \bar{y}_N)^2$$

(see Exercise 3).

Variance and efficiency

We generally prefer estimators with a low variance, since they are less likely to be far from their mean. This follows from the Chebyshev inequality, which, in the case of a one-dimensional θ, implies

$$P\left((\hat{\theta} - E[\hat{\theta}])^2 > c^2\right) \le \frac{Var(\hat{\theta})}{c^2}.$$

The following definition is useful for expressing the above preference: One unbiased estimator is said to be more efficient than another if the difference between their covariance matrices is a nonnegative definite matrix.

We note that, although low bias and low variance are both desirable, the two objectives cannot always be achieved at the same time. Therefore, we often have to consider the tradeoff between bias and variance.

Consistency

An estimator $\hat{\theta}_N$ for θ is consistent if

$$plim_{N\to\infty}\hat{\theta}_N = \theta.$$ (9.4)

Recall that

$$plim_{N\to\infty}a_N = a$$

means that, for any ϵ and $\delta > 0$, there exists an N such that for all $n > N$

$$P(||a_n - a|| > \epsilon) < \delta.$$

So consistency means that, at a given confidence level, we can, at least theoretically, get $\hat{\theta}_N$ arbitrarily close to θ_0 by increasing the sample size.

We note that although consistency and unbiasedness are closely related, they are different and neither one implies the other (see, for example, the discussion in Davidson and MacKinnon (1993), Section 4.5).

Asymptotic normality

An estimator $\hat{\theta}_N$ for θ (on a sample of size N) is said to be asymptotically normal-distributed if its distribution converges to a normal distribution as $N \to \infty$.

An estimator $\hat{\theta}_N$ for θ is usually asymptotically normal-distributed if the central limit theorem applies (see Davidson and MacKinnon (1993), Section

4.6). For example, the variance estimator

$$\hat{\sigma}_N^2 = \frac{1}{N} \sum_{i=1}^{N} (y_i - \bar{y}_N)^2$$

(see (9.3)) is asymptotically normal. This follows from the central limit theorem, since the above expression can be written as the sum of $N-1$ independent random variables (assuming that the y_i are i.i.d.).

If an estimator is asymptotically normal-distributed, we know its approximate distribution (in case of a large sample) whenever we can estimate its mean and variance. This is useful for hypothesis testing.

Invariance under reparameterization

Let us transform the model parameters (i.e., reparameterize the model) as

$$\theta' = g(\theta) , \tag{9.5}$$

where g is some differentiable function. Then we call the estimator $\hat{\theta}$ for θ invariant under the reparameterization if

$$\hat{\theta}' = g(\hat{\theta}) . \tag{9.6}$$

This means that we have the same probabilities for Y after reparameterization.

Since the parameterization of a model is completely arbitrary, we generally prefer estimators that are invariant under arbitrary reparameterization. We note that, however, invariance under reparameterization of an estimator implies that, in general, the estimator cannot be unbiased. In order to see this, suppose an estimator is unbiased for a given parameterization, i.e.,

$$E[\hat{\theta}] = \theta_0 , \tag{9.7}$$

and that we reparameterize with a nonlinear function g as

$$\theta' = g(\theta) . \tag{9.8}$$

Assuming invariance, i.e.,

$$\hat{\theta}' = g(\hat{\theta}) ,$$

we have

$$E[\hat{\theta}'] = E[g(\hat{\theta})] . \tag{9.9}$$

On the other hand, the "true" value of θ' is

$$\theta_0' = g(\theta_0)$$
$$= g(E[\hat{\theta}]) \text{ (by (9.7))} . \tag{9.10}$$

Since, in general, $E[g(\hat{\theta})] \neq g(E[\hat{\theta}])$, (9.9) and (9.10) imply

$$\theta_0' \neq E[\hat{\theta}'] \ .$$

This means that the transformed parameter estimator is biased.

Computability

For practical ends, it is important that an estimator can be computed, at least numerically. Typically, one has to solve an optimization problem to do so. Convex optimization problems are generally preferable to nonconvex problems, since they avoid the issue of local minima. So there is some advantage to estimators that are the solution of a convex optimization problem.

9.1.3   Maximum-Likelihood Inference

The perhaps most widespread method of parametric inference is the maximum-likelihood method. When following this method, we estimate the value of a parameter-vector θ by maximizing the likelihood. The following definition makes this formal.

Definition 9.1 *The maximum-likelihood estimator for the parameter θ of the probability measure p^θ is given by*

$$\hat{\theta}^{(ML)} = \arg \max_{\theta} \log L_{\mathcal{D}}\left(p^\theta\right) \ , \tag{9.11}$$

where $\mathcal{D} = \{y_1, y_2, ..., y_N\}$ are the observed data and $L_{\mathcal{D}}\left(p^\theta\right) = \prod_{i=1}^{N} p_{y_i}^\theta$ denotes the likelihood from Definition 6.1.

We note that maximum likelihood inference works the same way for conditional probability measure if we use Definition 6.3 for the likelihood and for probability densities if p^θ is such a density.

Maximum-likelihood inference is the logical consequence of using the likelihood as a model performance measure. In Sections 6.2, 8.1, and 8.2, we have discussed in some detail why the likelihood is a reasonable model performance measure; the reasons are the following.

(i) The likelihood is, by definition, the probability of the data under the model measure (see Definition 6.1).

(ii) The likelihood principle is equivalent to the conjunction of the conditionality principle and the sufficiency principle (see Birnbaum (1962), or Section 6.2.2).

(iii) The likelihood ratio provides a decision criterion for model selection that is optimal in the sense of the Neyman-Pearson lemma (see Neyman and Pearson (1933), or Section 6.2.3).

(*iv*) The likelihood principle is consistent with Bayesian logic (see Jaynes (2003), or Section 6.2.2).

(*v*) An investor who bets in a horse race so as to optimize his wealth growth rate measures relative model performance by means of the likelihood ratio (see Cover and Thomas (1991), or Section 6.2.4).

(*vi*) An expected utility maximizing investor with a utility function of the form $U(W) = \alpha \log(W - \gamma B) + \beta$ who bets in a horse race measures relative model performance by means of the likelihood ratio (Theorem 8.2).

Maximum-likelihood estimators have the following properties.

(*i*) They are generally biased. An example for a biased estimator is the maximum-likelihood estimator for θ of the probability distribution

$$p_y^\theta = \theta e^{-\theta y} \, , y \geq 0 \, , \, \theta > 0 \, ,$$

which can easily be shown to have the expectation

$$E_{p^{\theta_0}} \left[\hat{\theta}^{(ML)} \right] \approx \theta_0 \left(1 + \frac{1}{N} \right) \, , \text{ for large } N \, .$$

(*ii*) Maximum-likelihood estimators are, under some technical conditions, consistent (Wald's consistency theorem, see, for example, Davidson and MacKinnon (1993), Theorem 8.1).

(*iii*) Maximum-likelihood estimators are, under some technical conditions, asymptotically normal (see, for example, Davidson and MacKinnon (1993), Theorem 8.3).

(*iv*) Maximum-likelihood estimators are invariant under reparameterization (see, for example, Davidson and MacKinnon (1993), Section 8.3).

(*v*) Maximum-likelihood estimation generally leads to a nonconvex optimization problem. However, in certain interesting special cases, some of which we will encounter later in this book, the optimization problem is convex.

Properties *(ii)-(iv)* are generally viewed as desirable. These properties, along with the reasons discussed above, make the maximum-likelihood method the arguably most popular method of parameter estimation in classical statistics.

9.2   Regularized Maximum-Likelihood Inference

Maximum-likelihood inference tends to work well for well-specified parametric models as long as the model isn't too complex and there are enough observations available. If, on the other hand, the model specification is too complex for the dataset we train it on, the maximum-likelihood method can lead to so-called overfitting. A model is said to overfit, if it fits the training data well, but generalizes poorly. Somebody who measures model performance by means of the likelihood would consider a model overfit if it has an out-of-sample likelihood that is much lower than its in-sample likelihood. This oft-observed phenomenon occurs in many practical situations (see, for example, the discussion in Hastie et al. (2009), for more details); we will encounter some examples later in this book.

A commonly used approach to mitigating overfitting is regularization, also often called penalization. The main idea of this approach is to choose the model parameters such that they maximize the sum of the log-likelihood and an additional regularization term that penalizes nonsmoothness. The resulting model has, by construction, a lower in-sample likelihood than the unregularized maximum-likelihood model; however, its out-of-sample likelihood can be higher than the one of the maximum-likelihood model.

In order to define the regularized maximum-likelihood estimator, let us assume that the family of measures p^θ is parameterized such that lower values of the $|\theta_j|$ correspond to smoother measures than higher values, where the θ_j are the elements of the parameter vector $\theta = (\theta_1, ..., \theta_J)^T$. An example for such a parameterization is the measure $p_y^\theta \propto e^{-\theta_1|y| - \theta_2 y^2}$. We have used the term smoothness in a loose sense here; it can be defined precisely, for example, in terms of the sum of the squared (discretize) derivatives of p_y^θ with respect to y. We make the following definition.

Definition 9.2 *The ϕ-regularized maximum-likelihood estimator for the parameter vector θ of the probability measure p^θ is given by*

$$\hat{\theta}^{(\phi)}(\alpha) = \arg\max_{\theta \in \mathbf{R}^J} \left\{ \frac{1}{N} \log L_{\mathcal{D}}\left(p^\theta\right) - \alpha\phi(\theta) \right\}, \qquad (9.12)$$

where $\mathcal{D} = \{y_1, y_2, ..., y_N\}$ are the observed data, $L_{\mathcal{D}}\left(p^\theta\right) = \prod_{i=1}^{N} p_{y_i}^\theta$ denotes the likelihood from Definition 6.1, and ϕ is a continuous function on $\mathbf{R}^J$ that is increasing in $|\theta_j|$, $\forall j = 1...J$. The parameter $\alpha \geq 0$ is called the regularization parameter.

The larger the regularization parameter is chosen, the smoother is the probability measure $p^{\hat{\theta}^{(\phi)}(\alpha)}$, which is a consequence of the following lemma.

Lemma 9.1 *If $\alpha_1 > \alpha_2$, then $\left|\hat{\theta}_j^{(\phi)}(\alpha_1)\right| \leq \left|\hat{\theta}_j^{(\phi)}(\alpha_2)\right|$, $\forall j = 1...J$.*

Proof: We prove the lemma by contradiction. Suppose that $\alpha_1 > \alpha_2$ and

$$\left| \hat{\theta}_j^{(\phi)}(\alpha_1) \right| > \left| \hat{\theta}_j^{(\phi)}(\alpha_2) \right| \text{ for some } j . \qquad (9.13)$$

It follows from (9.12) that

$$L_{\mathcal{D}}\left(p^{\hat{\theta}(\alpha_1)} \right) - \alpha_1 \phi(\hat{\theta}(\alpha_1)) \geq L_{\mathcal{D}}\left(p^{\hat{\theta}(\alpha_2)} \right) - \alpha_1 \phi(\hat{\theta}(\alpha_2)) ,$$

$$\text{and } L_{\mathcal{D}}\left(p^{\hat{\theta}(\alpha_2)} \right) - \alpha_2 \phi(\hat{\theta}(\alpha_2)) \geq L_{\mathcal{D}}\left(p^{\hat{\theta}(\alpha_1)} \right) - \alpha_2 \phi(\hat{\theta}(\alpha_1)) ,$$

which implies

$$\alpha_1 \left[\phi(\hat{\theta}(\alpha_1)) - \phi(\hat{\theta}(\alpha_2)) \right] \leq L_{\mathcal{D}}\left(p^{\hat{\theta}(\alpha_1)} \right) - L_{\mathcal{D}}\left(p^{\hat{\theta}(\alpha_2)} \right)$$

$$\leq \alpha_2 \left[\phi(\hat{\theta}(\alpha_1)) - \phi(\hat{\theta}(\alpha_2)) \right] ,$$

i.e.,

$$\alpha_1 \left[\phi(\hat{\theta}(\alpha_1)) - \phi(\hat{\theta}(\alpha_2)) \right] \leq \alpha_2 \left[\phi(\hat{\theta}(\alpha_1)) - \phi(\hat{\theta}(\alpha_2)) \right] .$$

It follows then from (9.13) and the fact ϕ is increasing in the $|\theta_j|$ that $\phi(\hat{\theta}(\alpha_1)) - \phi(\hat{\theta}(\alpha_2)) > 0$ and, consequently,

$$\alpha_1 \leq \alpha_2 , \qquad (9.14)$$

which contradicts our assumption. $\square$

We note that

$$\left| \theta^{(\phi)}(\alpha) \right| \leq \left| \theta^{(\phi)}(0) \right| = \left| \theta^{(ML)} \right| \qquad (9.15)$$

follows from Lemma 9.1.

The quality of the parameter estimator, $\theta^{(\phi)}(\alpha)$, from Definition 9.2 depends obviously on the choice of the regularization-parameter α. Many practitioners simply try a variety of values for α, test each model via cross validation, and pick the model with the largest out-of-sample likelihood.

The idea of regularization is closely related to the principle of structural risk minimization (see Vapnik (1998)). According to this principle, one minimizes the guaranteed risk for a given dataset; this leads to a complexity penalization, i.e., a regularization, that is a function of a risk error estimate. The resulting regularization terms can be distribution-free, such as the ones based on the VC dimension (see Vapnik and Chernovenkis (1968) and Vapnik and Chernovenkis (1971)), or data-dependent (see, for example, Lugosi and Nobel (1999), or Massart (2000)). An overview over some of these methods is provided by Bartlett et al. (2000).

Regularization is also related to the various information criteria that can be used for model selection. These criteria usually consist of the sum of likelihood of a model and a complexity penalty. The best known of these criteria are perhaps the Akaike information criterion (AIC) from Akaike (1973) and the Bayesian information criterion (BIC) from Schwarz (1978). We shall not discuss these criteria in detail, but rather refer the reader to textbook by Burnham and Anderson (2002).

9.2.1 Regularization and Feature Selection

Often one assumes that the probability measure p^θ contains a number of so-called features, each of which corresponds to a component of the parameter vector θ, where features denote functions of y. A typical form, which we will encounter later in this book, is

$$p_y^\theta \propto e^{\sum_{j=1}^J \theta_j f_j(y)} \ , \tag{9.16}$$

where the f_j are the features. When, in later sections, we consider conditional probabilities, each feature could correspond to a particular explanatory variable.

A model builder often faces the question of which features should be included in the model. A fair amount of research is focused on this very interesting and important question (see, for example, Blum and Langley (1997), or Guyon and Elisseeff (2003), for reviews). We will not discuss feature selection in detail here, but we shall explore its link to regularization. In particular, we shall discuss whether and when regularization can induce feature selection. To this end, we state the following lemma.

Lemma 9.2 *Let $\log L_{\mathcal{D}}\left(p^\theta\right)$ be concave and differentiable with respect to θ, and let ϕ be convex. If, for some j,*

$$\alpha \left.\frac{\partial^- \phi(\theta)}{\partial^- \theta_j}\right|_{\theta=(0,\dots,0)^T} \le \frac{1}{N}\log \left.\frac{\partial L_{\mathcal{D}}\left(p^\theta\right)}{\partial \theta_j}\right|_{\theta=(0,\dots,0)^T} \le \alpha \left.\frac{\partial^+ \phi(\theta)}{\partial^+ \theta_j}\right|_{\theta=(0,\dots,0)^T} \ ,$$

where ∂^+ and ∂^- denote right-hand derivative and the left-hand derivative, respectively, then

$$\hat{\theta}_j^{(\phi)}(\alpha) = 0 \ , \tag{9.17}$$

where the $\hat{\theta}_j^{(\phi)}(\alpha)$ are the components of the estimated parameter vector (9.12).

Proof: The lemma follows directly from (9.12) after expanding $\log L_{\mathcal{D}}\left(p^\theta\right)$ and $\phi(\theta)$ around $\theta = (0,\dots,0)$ for a small positive and for a small negative θ_j and using the fact that, since $\log L_{\mathcal{D}}\left(p^\theta\right) - \alpha\phi(\theta)$ is concave, each local maximum is also a global one (see Theorem 2.7). $\square$

Lemma 9.2 relates regularization to feature selection: since we have assumed that each component θ_j of the parameter vector corresponds to a feature, setting $\theta_j = 0$ is equivalent to excluding, i.e., "unselecting", the feature from the model. Therefore, the regularized estimator from Definition 9.2 corresponds to a model, $p^{\theta^{(\phi)}(\alpha)}$, that may contain only some but not all the features that other models in the family contain; we have selected features.

9.2.2 ℓ_κ-Regularization, the Ridge, and the Lasso

A popular regularization method is the ℓ_κ-regularization, where the regularization function ϕ in Definition 9.2 is the ℓ_κ-norm. This specification leads to the following definition.

Definition 9.3 *The ℓ_κ-regularized maximum-likelihood estimator for the parameter θ of the probability measure p^θ is given by*

$$\hat{\theta}^{(\ell_\kappa)}(\alpha) = \arg \max_{\frac{1}{N}\log \theta \in \mathbf{R}^J} \left\{ L_\mathcal{D}\left(p^\theta\right) - \alpha \ell_\kappa(\theta) \right\} , \qquad (9.18)$$

where $\kappa \geq 1$, $\alpha \geq 0$, $\mathcal{D} = \{y_1, y_2, ..., y_N\}$ are the observed data, $L_\mathcal{D}\left(p^\theta\right) = \prod_{i=1}^N p_{y_i}^\theta$ denotes the likelihood from Definition 6.1, and $\ell_\kappa = \frac{1}{\kappa}\sum_j |\theta_j|^\kappa$ denotes the ℓ_κ-norm.

Two particularly often used powers in the ℓ_κ-regularization are $\kappa = 1$ and $\kappa = 2$; we shall discuss them next.

ℓ_1-Regularization

This regularization is a well-known tool in the context of linear regression, where it was introduced by Tibshirani (1996) and is known as the Lasso. It has also been used in the maximum-likelihood context (see, for example, Perkins et al. (2003), Riezler and Vasserman (2004), or Ng (2004)). We shall discuss examples for this regularization later in this book.

According to Lemma 9.2, ℓ_1-regularization can lead to feature selection: since $\frac{\partial^- \ell_1(\theta)}{\partial^- \theta_j}\Big|_{\theta=(0,...,0)^T} = -1$ and $\frac{\partial^+ \ell_1(\theta)}{\partial^+ \theta_j}\Big|_{\theta=(0,...,0)^T} = 1$, all features with

$$\frac{1}{N} \left| \frac{\partial \log L_\mathcal{D}\left(p^\theta\right)}{\partial \theta_j} \right|_{\theta=(0,...,0)^T} \leq \alpha \qquad (9.19)$$

have $\hat{\theta}_j^{(\ell_1)}(\alpha) = 0$, i.e., are not selected. If α is large enough, there will be such unselected features. In fact, there exists some value for α for which no feature is selected at all and the resulting distribution is uniform.

ℓ_2-Regularization

ℓ_2-Regularization is the same regularization that is used in a ridge regression, which is the minimization of an ℓ_2-regularized sum of square errors. The latter is a standard regression method (see, for example, Hastie et al. (2009)). ℓ_2-regularization is also often used in the maximum-likelihood context, i.e., for estimators of the form (9.18); we shall discuss a number of examples below.

ℓ_2-regularization doesn't lead to feature selection beyond the, usually nonexistent, selection by the maximum-likelihood method. This is consistent with

Lemma 9.2, since $\left.\frac{\partial^- \ell_2(\theta)}{\partial^- \theta_j}\right|_{\theta=(0,\dots,0)^T} = \left.\frac{\partial^- \ell_2(\theta)}{\partial^- \theta_j}\right|_{\theta=(0,\dots,0)^T} = 0$, so that, accord-

ing to Lemma 9.2, the condition of feature exclusion is $\left.\frac{\partial L_\mathcal{D}(p^\theta)}{\partial \theta_j}\right|_{\theta=(0,\dots,0)^T} =$

0.

Comparison of ℓ_1-regularization with ℓ_2-regularization

Both of these regularizations have been shown to be useful for practical applications. Empirical evidence and theoretical arguments seem to suggest that
ℓ_1-regularization is the better method in the sparse scenario, i.e., when many
of the features we have included are actually irrelevant, while ℓ_2-regularization
is the better method in the dense scenario, i.e., when most of the features are
actually important (see, Ng (2004), and Friedman et al. (2004)). The theoretical result from Ng (2004), which was proved in the context of logistic
regression but might apply to more general situations, is particularly interesting. Ng (2004) has shown that, using ℓ_1-regularization, the number of training
observations required to learn "well" grows only logarithmically in the number of irrelevant features, while, using ℓ_2-regularization, (at least in the worst
case) it grows at least linearly in the number of irrelevant features.

9.3 Bayesian Inference

In this section, we briefly discuss the Bayesian method for estimating probabilistic models. We limit our discussion to those aspects of Bayesian inference that relate to the decision-theoretic methods that are the subject of this
book. For a more comprehensive review, we refer the reader to the textbooks
by Bernardo and Smith (2000), Jaynes (2003), and Robert (1994) for more
detailed analysis of the Bayesian view on probabilities and their estimation,
and to Gelman et al. (2000) for a practical guide to Bayesian inference.

In this section, we restrict ourselves to the simplest setting, i.e., we aim at
estimating the (unconditional) probabilities of the (discrete) random variable
Y with a finite state space $\mathcal{Y}$. The generalization to continuous random variables and to conditional probabilities is straightforward, but we won't discuss
it here.

9.3.1 Prior and Posterior Measures

As we did in Sections 6.2.2, 9.1, and 9.2, we assume that the probability
measure of Y is a member of a family of parametric measures, p^θ, that is parameterized by the vector θ. This assumption is crucial to standard Bayesian
inference. Another crucial ingredient to Bayesian analysis is the view that the

parameter θ is a random variable itself. It is the objective of Bayesian inference to infer the probability distribution of θ from a set $\mathcal{D} = \{y_1, ..., y_N\}$ of independent observations for Y. (We have made the assumption of independence for the sake of convenience here; it is not essential.) To this end, we assume that the model builder, before seeing the data $\mathcal{D}$, assigns the probability density

$$P(\theta)$$

to the parameter vector θ. The above probability measure is called the *prior* measure, since it reflects the model builder's knowledge prior to inferring from the data $\mathcal{D}$. We have assumed here that θ is a continuous parameter, as it is the case in the majority of practical applications; if θ has a discrete state space, we have to interpret $P(\theta)$ as probabilities and replace the integrals below by sums, in order to develop the Bayesian framework.

The observed data, $\mathcal{D}$, provide the model builder with information about the parameter vector θ. As a consequence, he will update the probabilities for θ to the so-called *posterior* measure,

$$P(\theta|\mathcal{D}) \ .$$

As the notation indicates, the posterior probabilities are the probabilities of θ given the observed data, $\mathcal{D}$. This conditioning on the observed data is perhaps the most fundamental difference between Bayesian and classical statistics.

Bayes' law implies that the posterior measure is related to the prior measure as follows:

$$P(\theta|\mathcal{D}) = \frac{P(\theta)P(\mathcal{D}|\theta)}{P(\mathcal{D})} \ , \tag{9.20}$$

Since we have assumed that the probability measure of Y is a member of a family of parametric measures, p^θ, the probability $P(\mathcal{D}|\theta)$ of observing the data $\mathcal{D}$ given the parameter value θ is equal to the probability of $\mathcal{D}$ under the measure p^θ. The latter one is, according to Definition 6.1, the likelihood ratio $L_\mathcal{D}\left(p^\theta\right)$. Moreover, it follows from the same assumption that

$$P(\mathcal{D}) = \int_\Theta L_\mathcal{D}\left(p^\theta\right) P(\theta) d\theta \ ,$$

where Θ denotes the state space of θ. So we have

$$P(\theta|\mathcal{D}) = \frac{P(\theta)L_\mathcal{D}\left(p^\theta\right)}{\int_\Theta L_\mathcal{D}\left(p^{\theta'}\right) P(\theta')d\theta'} \ . \tag{9.21}$$

The denominator in the right hand side of the above equation is independent of θ; so it can be viewed as a normalization constant. One often writes the above equation as

$$P(\theta|\mathcal{D}) \propto Q(\theta, \mathcal{D}) \ , \tag{9.22}$$

where

$$Q(\theta, \mathcal{D}) = L_{\mathcal{D}} \left(p^{\theta} \right) P(\theta) , \qquad (9.23)$$

and "$\propto$" means that equality holds up to a θ-independent proportionality factor. The expression $Q(\theta, \mathcal{D})$ is often referred to as the unnormalized posterior measure. It plays an important role in practical applications.

9.3.2   Prior and Posterior Predictive Measures

In most practical applications, the model builder is interested in the probabilities of the random variable Y or in some expectations with respect to these probabilities. Before the data are considered, the model builder believes that Y has a probability measure of the form p^{θ} and that θ has the prior probability measure. So he assigns the probabilities

$$\begin{aligned} P(y) &= \int_{\Theta} P(y, \theta) \\ &= \int_{\Theta} P(\theta) P(y|\theta) d\theta \\ &= \int_{\Theta} P(\theta) p_y^{\theta} d\theta \end{aligned} \qquad (9.24)$$

to the possible values y of Y. We refer to the above measure as the prior predictive measure.

After we have observed the data $\mathcal{D}$, we update the probabilities of θ from the prior to the posterior ones, and we update the prior predictive measure to the posterior predictive measure, which is given by the following expression.

$$\begin{aligned} P(y|\mathcal{D}) &= \int_{\Theta} P(y, \theta|\mathcal{D}) d\theta \\ &= \int_{\Theta} P(\theta|\mathcal{D}) P(y|\theta, \mathcal{D}) d\theta \\ &= \int_{\Theta} P(\theta|\mathcal{D}) p_y^{\theta} d\theta \end{aligned} \qquad (9.25)$$

(since we have assumed that $P(y|\theta, \mathcal{D}) = P(y|\theta) = p_y^{\theta}$) .

Inserting (9.21) in the above equation, we obtain

$$P(y|\mathcal{D}) = \frac{\int_{\Theta} p_y^{\theta} P(\theta) L_{\mathcal{D}} \left(p^{\theta} \right) d\theta}{\int_{\Theta} L_{\mathcal{D}} \left(p^{\theta} \right) P(\theta) d\theta} \qquad (9.26)$$

$$\propto \int_{\Theta} p_y^{\theta} P(\theta) L_{\mathcal{D}} \left(p^{\theta} \right) d\theta , \qquad (9.27)$$

where "$\propto$" means here that equality holds up to a y-independent proportionality factor. This means that a model builder who has the prior $P(\theta)$ infers the probability measure (9.26) for Y from the data $\mathcal{D}$. This measure depends on the data only through the likelihood function, i.e., as we have discussed in Section 6.2.2, the Bayesian model builder abides by the likelihood principle.

9.3.3 Asymptotic Analysis

Asymptotic analysis plays an important role in classical (frequentist) statistics. In this type of analysis, one studies the properties of an estimator for a parameter in the limit $N \to \infty$, where N is the number of observations we use for inferring the parameters. The importance of these asymptotic properties derives from the fact that, in classical statistics, one assumes that there is an underlying "true" model, which is identical with the empirical measure, i.e., with the relative frequencies, that an infinite-size sample would have. In Bayesian inference, on the other hand, we usually don't work with the notion of a "true" measure and we always condition on the sample we have actually observed. Therefore, asymptotic analysis plays a less fundamental role in Bayesian inference; in fact, some textbooks, such as Robert (1994), don't discuss it at all. Even so, asymptotic analysis can help us building some intuition about what to expect for large sample sizes. For this reason, we briefly discuss it here, in an informal way. For more details, we refer the reader to Gelman et al. (2000), Appendix B, or Bernardo and Smith (2000), Section 5.3.

Let us assume in this section that we have observed the data $\mathcal{D}_N = \{y_1, ..., y_N\}$ (we added an index N to the notation here to indicate the sample-size dependence), and that these data are sampled from independent and identically distributed random variables, each of which has the probability measure f. We note that this assumption essentially amounts to assuming the existence of a "true" measure, i.e., to an assumption that we usually don't make in Bayesian analysis or in the decision-theoretic framework this book focuses on. We make it only for the sake of the asymptotic analysis; it is not needed for the remainder of this book.

An important result from Bayesian asymptotic analysis is the following.

Convergence: Let us assume that the relative entropy $D(f\|p^\theta)$ has the unique minimum

$$\theta_0 = \arg \min_\theta D(f\|p^\theta) \tag{9.28}$$

with respect to θ. If θ is defined on a compact set and A is a neighborhood of θ_0 with nonzero probability, then $P(\theta \in A|\mathcal{D}_N) \to 1$ as $N \to \infty$.

This statement points to an interesting connection between relative entropy and the asymptotic Bayes estimator, θ_0; the latter one minimizes the relative entropy from the "true" measure, f.

Sketch of proof of convergence: We follow the logic from Gelman et al. (2000), Appendix B, here. Unlike in the rest of our section on Bayesian analysis, we assume here that θ has a finite state space. Proving convergence for this simple case illuminates the general case, which we shall not prove here. Let us consider the posterior odds

$$\log\left(\frac{P(\theta|\mathcal{D}_N)}{P(\theta_0|\mathcal{D}_N)}\right) = \log\left(\frac{P(\theta)}{P(\theta_0)}\right) + \sum_{i=1}^{N}\log\left(\frac{p_{y_i}^{\theta}}{p_{y_i}^{\theta_0}}\right) \quad (9.29)$$

(by (9.21) and Definition 6.1) .

The second term on the right hand side is the sum of N i.i.d. random variables, each of which has the mean

$$E_f\left[\log\left(\frac{p_{y_i}^{\theta}}{p_{y_i}^{\theta_0}}\right)\right] = D(f\|p^{\theta_0})) - D(f\|p^{\theta})$$

$$\begin{cases} = 0 \text{ if } \theta = \theta_0 \\ < 0 \text{ if } \theta \neq \theta_0 \end{cases} \quad (9.30)$$

(by the definition of θ_0, (9.28)).

It follows that, if $\theta \neq \theta_0$, the second term in the right hand side of (9.29) is the sum of N i.i.d. random variables with negative mean. By the law of large numbers, this sum approaches $-\infty$ as $N \to \infty$. Therefore,

$$\lim_{N\to\infty} \log\left(\frac{P(\theta|\mathcal{D}_N)}{P(\theta_0|\mathcal{D}_N)}\right) = -\infty \quad \text{for } \theta \neq \theta_0 . \quad (9.31)$$

It follows that

$$\lim_{N\to\infty} P(\theta|\mathcal{D}_N) = 0 \quad \text{for } \theta \neq \theta_0 . \quad (9.32)$$

Since all probabilities sum to one, we must have $P(\theta_0|\mathcal{D}_N) = 1$. This completes the proof of the above convergence statement for the special case we have considered here. $\square$

Another important result from Bayesian asymptotic analysis is the following.

Asymptotic normality: Under some regularity assumptions, as $N \to \infty$, the posterior measure of θ approaches normality with mean θ_0 and variance $(NJ(\theta_0))^{-1}$, where θ_0 is the value that minimizes the relative entropy between p^{θ} and f and

$$J(\theta) = -E_f\left[\frac{\partial^2 \log p^{\theta}}{\partial\theta^2}\right] \quad (9.33)$$

is the Fisher information.

Before we sketch a proof of this statement, we would like to emphasize that the variance of the posterior measure decreases towards zero with N, i.e., that the posterior measure becomes increasingly localized as N increases, and that in the limit $N \to \infty$ the posterior measure depends solely on the data and not on the prior.

Sketch of proof of asymptotic normality: We roughly follow the logic from Gelman et al. (2000), Appendix B, here. In order to get an idea of how a proof would work, we consider the following special case.

(*i*) The parameter vector θ is one-dimensional and continuous,

(*ii*) θ_0 from (9.28) is unique, and

(*iii*) $f(y) = p^{\hat{\theta}}$, for some $\hat{\theta}$, i.e., f is in our parametric family of measures.

It follows from assumption (*ii*) and (*iii*), and the fact that

$$\arg \min_{q \in Q} D(f \| q) = f \ , \tag{9.34}$$

if $f \in Q$ (see Lemma 2.10, (*iv*)), that

$$\begin{aligned}
\theta_0 &= \arg \min_{\theta} D(f \| p^\theta) \ \text{(by (9.28))} \\
&= \arg \min_{\theta} D(p^{\hat{\theta}} \| p^\theta) \\
&= \hat{\theta} \ . \tag{9.35}
\end{aligned}$$

Next, we assume that, in accordance with the above convergence result, for large N, a small neighborhood of $\hat{\theta} = \theta_0$ has all the probability mass, and we expand the posterior distribution around θ_0:

$$\log P(\theta | \mathcal{D}_N) \approx \log P(\theta_0 | \mathcal{D}_N) + \frac{1}{2}(\theta - \theta_0)^2 \left. \frac{d^2}{d\theta^2} \log P(\theta | \mathcal{D}_N) \right|_{\theta = \theta_0} . \tag{9.36}$$

Using (9.20), i.e.,

$$\log P(\theta | \mathcal{D}_N) = \log P(\theta) + \sum_{i=1}^{N} \log p_{y_i}^\theta - \log P(\mathcal{D}_N) \ , \tag{9.37}$$

we obtain

$$\left. \frac{d^2}{d\theta^2} \log P(\theta | \mathcal{D}_N) \right|_{\theta = \theta_0} \approx \left. \frac{d^2}{d\theta^2} \log P(\theta) \right|_{\theta = \theta_0} + \sum_{i=1}^{N} \left. \frac{d^2}{d\theta^2} \log p_{y_i}^\theta \right|_{\theta = \theta_0} . \tag{9.38}$$

The second term on the right hand side is the sum of N i.i.d. (according to f) random variables; so, by the law of large numbers, we can approximate:

$$\begin{aligned}
\sum_{i=1}^{N} \left. \frac{d^2}{d\theta^2} \log p_{y_i}^\theta \right|_{\theta = \theta_0} &\approx N E_f \left[\left. \frac{d^2}{d\theta^2} \log p_{y_i}^\theta \right|_{\theta = \theta_0} \right] \\
&\approx -N J(\theta_0) \ \text{(by (9.33))} \ . \tag{9.39}
\end{aligned}$$

Combining (9.38) with (9.39), we obtain

$$\frac{d^2}{d\theta^2} \log P(\theta|\mathcal{D}_N)|_{\theta=\theta_0} \approx \frac{d^2}{d\theta^2} \log P(\theta)\bigg|_{\theta=\theta_0} - NJ(\theta_0)$$

$$\approx -NJ(\theta_0) \, ,$$

which, after combination with (9.36), results in

$$\log P(\theta|\mathcal{D}_N) \approx \log P(\theta_0|\mathcal{D}_N) - \frac{1}{2}NJ(\theta_0)(\theta - \theta_0)^2 - \log P(\mathcal{D}_N) \, . \quad (9.40)$$

This completes the sketch of the proof of the above asymptotic normality statement for the special case considered here.□

9.3.4 Posterior Maximum and the Maximum-Likelihood Method

For a large number of observations, the posterior measure is, under some technical conditions, fairly localized around its maximum (see Section 9.3.3). This suggests that the following approximation holds reasonably well if N is large.

$$P(\theta|\mathcal{D}) \approx \delta(\theta - \hat{\theta}) \, , \quad (9.41)$$

where

$$\hat{\theta} = \arg\max_{\theta} P(\theta|\mathcal{D})$$

is the maximum of posterior measure, and δ is Dirac's delta. It follows from the definition of Dirac's delta that

$$\int P(\theta|\mathcal{D})d\theta = 1; \, , \quad (9.42)$$

so the measure in the r.h.s of (9.41) is properly normalized.

Since the normalized and the unnormalized posterior measure differ only by a θ-independent factor, it follows from (9.42) that

$$\hat{\theta} = \arg\max_{\theta} Q(\theta, \mathcal{D}) \, , \quad (9.43)$$

where Q is the unnormalized posterior measure from (9.23).

The approximation (9.41) leads to the following posterior predictive measure for Y.

$$P(y|\mathcal{D}) = \int_{\Theta} p_y^{\theta} P(\theta|\mathcal{D})d\theta \text{ (by (9.25))}$$

$$\approx \int_{\Theta} p_y^{\theta} \delta(\theta - \hat{\theta})d\theta \text{ (by (9.41))}$$

$$\approx P(y|\hat{\theta}) \, . \quad (9.44)$$

Therefore, in the approximation (9.41), the Bayesian model builder infers a single model from the p^θ-family, just as a classical (non-Bayesian) model builder would do. His parameter estimator (in the classical sense) is given by (9.43), i.e., by

$$\hat{\theta} = \arg\max_{\theta} \left[L_{\mathcal{D}}\left(p^\theta\right) P(\theta) \right] \quad \text{(by (9.23) and (9.43))}$$

$$= \arg\max_{\theta} \left[\log L_{\mathcal{D}}\left(p^\theta\right) + \log P(\theta) \right] \quad \text{(since the log is monotone). (9.45)}$$

Let us consider some specific priors.

(*i*) *Uniform prior:* In this case,

$$\hat{\theta} = \arg\max_{\theta} \log L_{\mathcal{D}}\left(p^\theta\right) , \tag{9.46}$$

which is the maximum-likelihood estimator from Definition 9.1. The above equation, in conjunction with (9.44), states that a Bayesian model builder with a uniform prior who uses the approximation (9.41) infers the probability measure for Y by means of the (unregularized) maximum-likelihood method.

(*ii*) *Gaussian prior:* If

$$P(\theta) \propto e^{-(\theta^T - \bar{\theta}^T)\Sigma^{-1}(\theta - \bar{\theta})} , \tag{9.47}$$

where $\bar{\theta}$ is some mean parameter vector and Σ is a covariance matrix, then

$$\hat{\theta} = \arg\max_{\theta} \left[\log L_{\mathcal{D}}\left(p^\theta\right) - (\theta^T - \bar{\theta}^T)\Sigma^{-1}(\theta - \bar{\theta}) \right] . \tag{9.48}$$

The above estimator has, after the appropriate parameter transformation, the same form as the ℓ_2-regularized maximum-likelihood estimator from Definition 9.3 with $\kappa = 2$. It follows then, in conjunction with (9.44), that a Bayesian model builder with a Gaussian prior who uses the approximation (9.41) infers the probability measure for Y by means of the ℓ_2-regularized maximum-likelihood method.

(*iii*) *Exponential prior:* If

$$P(\theta) \propto e^{-\sum_j \alpha_j |\theta_j|} , \tag{9.49}$$

then

$$\hat{\theta} = \arg\max_{\theta} \left[\log L_{\mathcal{D}}\left(p^\theta\right) - \sum_j \alpha_j |\theta_j| \right] . \tag{9.50}$$

After an appropriate parameter rescaling, the above estimator has the same form as the ℓ_1-regularized maximum-likelihood estimator from Definition 9.3 with $\kappa = 1$. It follows then, in conjunction with (9.44), that a Bayesian model builder with a Gaussian prior who uses the approximation (9.41) infers the probability measure for Y by means of the ℓ_1-regularized maximum-likelihood method.

9.4   Minimum Relative Entropy (MRE) Methods

In the minimum relative entropy (MRE) approach, one estimates a probability measure such that it minimizes, under certain constraints, the relative entropy with respect to a prior measure. One rationale for this method is that we would like our probability measure to contain as little information beyond the prior information as possible while being consistent with our constraints, which might be the result of reliable observations. We will discuss a decision-theoretic motivation later in this book, in Chapter 10.

Historically, the MRE approach was introduced by Jaynes (1957a) as the maximum entropy (ME) method; the latter is a special case of the former for a uniform prior measure. Jaynes (1957a) used the ME approach to derive the canonical and grand-canonical probability distribution from statistical physics, thereby giving an information-theoretic interpretation to these probability measures. More recently, Globerson and Tishby (2004) have formulated a minimum mutual information (MMI) principle, which, it has been argued, is particularly appropriate for conditional probability estimation problems.

The ME and MRE approaches have since been used to infer probabilities for a variety of problems in numerous fields of science and engineering. Physicists have used MRE methods to restore images from space telescope data (see, for example, Wu (1997)) and nuclear magnetic resonance data (see, for example, Hore (1991)). MRE has been used for natural language processing (see, for example, Berger et al. (1996), or Chen and Rosenfeld (1999)), sentiment identification (see, for example, Mehra et al. (2002)), named entity recognition (see, for example, Chieu and Ng (2002)), and QA systems (see, for example, Ravichandran et al. (2003)). Biomedical applications of ME and MRE methods include gene ontology (see Raychaudhuri et al. (2002)) and gene selection for cancer classification (see, for example, Liu et al. (2005)). The MMI principle has been used in neural code analysis (see Globerson et al. (2009)). Financial theorists and practitioners have used these MRE and ME methods to calibrate option pricing models to market data (see, for example, Gulko (2002), Avellaneda (1998), Frittelli (2000), and Cont and Tankov (1999)) and estimate conditional probabilities from real-world data (see, for example Golan et al. (1996)). More examples for applications can be found in Buck and Macaulay (1991), Wu (1997), and below in this book.

In this section, we briefly review MRE methods, from the traditional perspective, in the simplest setting, i.e., for discrete unconditional probabilities. Then, in Chapter 10, we shall reinterpret these methods from a decision-theoretic point of view, and generalize them to conditional probabilities and to probability densities. In Chapter 10, we shall also formulate a generalized MRE method, which is based on the relative $(U, \mathcal{O})$-entropy.

9.4.1 Standard MRE Problem

We aim at finding a probability measure for the random variable Y with the finite state space $\mathcal{Y}$. In the MRE approach, we do so by minimizing, under certain constraints, the relative entropy with respect to a prior measure, q^0. The constraints are usually defined in terms of certain functions f_j , $j = 1...J$, of Y, the so-called feature functions.

Problem 9.1 *(Standard MRE problem) Find the MRE measure, which is given by*

$$q^* = \arg \min_{q \in \mathbf{R}^{|\mathcal{Y}|}} D(q\|q^0) \tag{9.51}$$

$$subject\ to\ 1 = \sum_{y \in \mathcal{Y}} q_y , \tag{9.52}$$

$$q_y \geq 0 , \ \forall y \in \mathcal{Y} , \tag{9.53}$$

$$and\ E_q[f_j] = E_{\tilde{p}}[f_j] , \ \forall j = 1, ..., J . \tag{9.54}$$

Here, $D(q\|q^0)$ is the relative entropy from Definition 2.7, and $\tilde{p}$ is the empirical measure corresponding to the observed data $\mathcal{D} = \{y_1, y_2, ..., y_N\}$.

The above MRE problem is a convex problem, and, as we shall show in the next theorem, it has the following dual.

Problem 9.2 *(Dual of Standard MRE problem)*

$$Find\ \beta^* = \arg \max_{\beta \in \mathbf{R}^J} \left[\frac{1}{N} \log L_D \left(\hat{q}^{(\beta)} \right) \right] , \tag{9.55}$$

$$where\ \hat{q}_y^{(\beta)} = \frac{1}{Z(\beta)} q_y^0 e^{\beta^T f(y)} , \tag{9.56}$$

$$Z(\beta) = \sum_{y \in \mathcal{Y}} q_y^0 e^{\beta^T f(y)} , \tag{9.57}$$

$\mathcal{D} = \{y_1, y_2, ..., y_N\}$ *is the observed dataset, and $L_D \left(\hat{q}^{(\beta)} \right)$ is the likelihood from Definition 6.1.*

The following theorem relates Problems 9.1 and 9.2 to each other and makes a statement about the solution to Problem 9.1.

Theorem 9.1 *If a finite β^* as defined by (9.55) exists, then the solution to Problem 9.1 is unique, and it is given by*

$$q* = \hat{q}^{(\beta^*)} , \tag{9.58}$$

where β^* is the solution to Problem 9.2.

Proof: See Section 9.4.4.

That is, the MRE measure q^* is an exponential distribution, the parameters of which are estimated by means of the maximum-likelihood method. The exact functional form of the measure q^* is determined by the choice of feature functions.

Like the primal problem, Problem 9.1, the dual problem, Problem 9.2, is a convex optimization problem. This follows from the general fact that duals of convex problems are convex problems (see Section 2.2.5).

The practical importance of Theorem 9.1 lies in the fact that the dual problem, i.e., Problem 9.2, has the dimension J, while the primal problem, i.e., Problem 9.1, has the dimension $|\mathcal{Y}|$, and that in most practical applications $|\mathcal{Y}| > J$.

The following corollary, which follows directly from Theorems 9.1 and 2.11, gives a sensitivity interpretation to the parameter vector β^* of the exponential distribution $\hat{q}^{(\beta^*)}$.

Corollary 9.1

$$\beta_j^* = \frac{\partial D\left(\hat{q}^{(\beta^*)}|q^0\right)}{\partial E_{\tilde{p}}[f_j]} \ . \tag{9.59}$$

9.4.2   Relation of MRE to ME and MMI

As we have mentioned, the ME problem can be obtained as a special case of the MRE problem, when the prior measure is constant. We shall say a bit more about the relation between MRE and MMI in Chapter 10, after introducing conditional probability estimation problems.

9.4.3   Relaxed MRE

It is well known that the standard MRE problem, i.e., Problem 9.1, can lead to overfitting models when there are relatively few data in the training set and the number of features is relatively large. This is not surprising, since the dual of Problem 9.1 is a maximum-likelihood estimation, and, as we have discussed in Section 9.2, the latter type of estimation can produce models that overfit.

One can mitigate overfitting by relaxing the equality constraints in the MRE problem. An oft-used way to do this is the following.

Problem 9.3 *(Relaxed MRE Problem) Find the relaxed MRE measure, which* *is given by*

$$q^* = \arg\min_{q \in \mathbf{R}^{|\mathcal{Y}|}} D(q\|q^0) \tag{9.60}$$

$$\text{subject to } 1 = \sum_{y \in \mathcal{Y}} q_y \,, \tag{9.61}$$

$$q_y \geq 0 \,, \; \forall y \in \mathcal{Y} \,, \tag{9.62}$$

$$E_q[f_j] - E_{\tilde{p}}[f_j] = c_j \,, \; \forall j = 1, 2, ..., J \,, \tag{9.63}$$

$$\text{and } \frac{1}{\omega}\ell_\omega^\omega(c) \leq \alpha \,. \tag{9.64}$$

Here, $D(q\|q^0)$ *is the relative entropy from Definition 2.7,* $\ell_\omega(c) = \left(\sum_{j=1}^{J} |c_j|^\omega\right)^{\frac{1}{\omega}}$ *denotes the* ℓ_ω*-norm with* $\omega > 1$*, and* α *is a positive number.*

For the sake of convenience, we have used the same notation as for the standard MRE problem here. It shall always become clear from the context which setting we refer to.

The set of feasible measures of Problem 9.3, i.e., the set of measures for which the constraints of the problem hold, is a superset of the set of feasible measures of Problem 9.1. In particular, the former set can be nonempty, even if the latter one is empty. So we can view the relaxation of the equality constraints also as a way to transform certain problems that have no solutions, because their constraints are too restrictive, into problems that do have solutions.

Alternatively, one could replace the constraint (9.64) by an additive term in the objective function that penalizes large values of the $|c_j|$ (see, for example, Lebanon and Lafferty (2001), or Friedlander and Gupta (2003)). Under certain conditions, this formulation leads to the same result as Problem 9.3 (see Exercise 17 in Section 2.4).

As we shall show in the next theorem, the dual of Problem 9.3 is the following.

Problem 9.4 *(Dual of relaxed MRE problem)*

$$\text{Find } \beta^* = \arg\max_{\beta \in \mathbf{R}^J} \left\{ \frac{1}{N} \log L_{\mathcal{D}}\left(\hat{q}^{(\beta)}\right) - \alpha^{\frac{1}{\omega}} q^{\frac{1}{\omega}} (\kappa - 1)^{-\frac{1}{\omega}} \ell_\kappa(\nu) \right\}, \tag{9.65}$$

$$\text{where } \hat{q}_y^{(\beta)} = \frac{1}{Z(\beta)} q_y^0 e^{\beta^T f(y)} \,, \tag{9.66}$$

$$Z(\beta) = \sum_{y \in \mathcal{Y}} q_y^0 e^{\beta^T f(y)} \,, \tag{9.67}$$

$\mathcal{D} = \{y_1, y_2, ..., y_N\}$ *is the observed dataset, and* $L_{\mathcal{D}}\left(\hat{q}^{(\beta)}\right)$ *is the likelihood from Definition 6.1,* $\ell_\kappa(\beta) = \left(\sum_{j=1}^{J} |\beta_j|^\kappa\right)^{\frac{1}{\kappa}}$ *denotes the* ℓ_κ*-norm, and* $\kappa = \frac{\omega}{\omega - 1}$*.*

The following theorem relates Problems 9.3 and 9.4 to each other.

Theorem 9.2 *If a finite β^* as defined by (9.65) exists, then the solution to Problem 9.3 is unique, and it is given by*

$$q* = \hat{q}^{(\beta^*)} , \tag{9.68}$$

where β^* is the solution to Problem 9.4.

Proof: Equation (9.68), i.e., the equivalence of Problems 9.3 and 9.4, follows from Theorems 2.13 and 9.1. The uniqueness of the solution follows from the strict convexity of the relative entropy (see Lemma 2.10, *(iii)*) and Theorem 2.8. $\square$

By virtue of Corollary 2.1, Theorem 9.2 also holds for the important case $\omega = \infty$ and $\kappa = 1$.

We note that, by Corollary 2.2, one can replace the second term in the objective function of Problem 9.4 by $\alpha \ell_\kappa^\kappa(\nu)$ and obtain the same family of solutions.

Like the dual of the standard MRE problem, Problem 9.4 is a convex optimization problem.

As in the standard MRE case, the practical importance of Theorem 9.2 lies in the fact that the dual problem, i.e., Problem 9.4, has the dimension J, while the primal problem, i.e., Problem 9.3, has the dimension $|\mathcal{Y}|$, and that in most practical applications $|\mathcal{Y}| > J$.

Theorem 9.2 states that the relaxed MRE measure is an exponential distribution, the parameters of which are estimated by means of an ℓ_κ-regularized maximum-likelihood method, i.e., are given by the ℓ_κ-regularized maximum-likelihood estimator from Definition 9.3. Hence, we can use the results from Section 9.2 to analyze the properties of the relaxed-MRE measure.

As it is the case for the standard MRE problem, the parameters of the exponential distribution $\hat{q}^{(\beta^*)}$ are related to the sensitivities of this distribution with respect to the right hand sides of the equality constraints, i.e., Corollary 9.1 holds.

The most popular choices for the norm parameter ω are $\omega = \infty$, which corresponds to $\kappa = 1$ (see, for example, Goodman (2003), Perkins et al. (2003), Riezler and Vasserman (2004), Dudik et al. (2004), or Kazama and Tsujii (2003)), and $\omega = 2$, which corresponds to $\kappa = 2$ (see, for example, Chen and Rosenfeld (1999), Lebanon and Lafferty (2001), Skilling (1991), or Wu (1997)). In the first case, we perform an ℓ_1-regularized maximum-likelihood estimation, in which regularization is combined with feature selection. The second choice does not lead to any feature selection, but the resulting measure is regularized. As we have discussed in Section 9.2.2, both of these choices can perform well in practical applications; which of them performs better depends on various factors such as the number of irrelevant features and the sample size.

Feature transformation

The standard MRE problem, Problem 9.1, is invariant with respect to a linear feature transformation of the form

$$f'(y) = Af(y) + a \ , \ \forall y \in \mathcal{Y} \ , \tag{9.69}$$

where $f(y) = (f_1(y), ..., f_J(y))^T$ is the feature vector, A is an invertible $J \times J$-matrix, and a is a J-vector. That is, if we replace the constraint (9.54) in Problem 9.1, which is

$$E_q[f] = E_{\tilde{p}}[f] \tag{9.70}$$

by

$$E_q[Af + a] = E_{\tilde{p}}[Af + a] \ , \tag{9.71}$$

the problem has exactly the same solution as before. The reason for this is that, for invertible A, (9.70) implies (9.71) and vice versa. This invariance seems desirable, since there is usually no practical guidance to choosing the scale and offset of features or to how to combine them.

The relaxed MRE problem, Problem 9.3, on the other hand, is not invariant with respect to a transformation of the form (9.69). This can be seen from the fact that replacing f by f' in the constraints (9.63) and (9.64) of Problem 9.3 results in

$$E_q[Af + a] - E_{\tilde{p}}[Af + a] = c \tag{9.72}$$
$$\text{and } \ell_\omega(c) \le \alpha \ , \tag{9.73}$$

which is equivalent to

$$E_q[f] - E_{\tilde{p}}[f] = c' \tag{9.74}$$
$$\text{and } \ell_\omega \left(A^{-1}c' \right) \le \alpha \ , \tag{9.75}$$

but generally different from the original constraints. We can see from the fact that the above equation doesn't contain the constant vector a that Problem 9.3 is invariant with respect to a constant shift. However, it is not invariant with respect to rescaling the features or rotating the features vector.

In practical applications, one often assumes that the features are normalized in some sense. For example, one can assume that each feature is scaled such that its empirical variance is one (see, for example, Lebanon and Lafferty (2001), or Chen and Rosenfeld (1999)). This leaves us still with a non-uniqueness, since one can find linear combination of variance-one features that have a variance of one too. To overcome this nonuniqueness, one sometimes makes the stronger assumption that the empirical covariance matrix of the feature is the identity matrix (see, for example, Wu (1997), or Skilling (1991)). Another way to choose a scale for features is to fix all their empirical minima and maxima to given values, say zero and one.

9.4.4   Proof of Theorem 9.1

In order to prove this theorem, we apply the framework from Section 2.2.5. Based on this framework and on Definition 2.7 of the relative entropy, we find the Lagrangian of Problem 9.1:

$$\mathcal{L}(q, \beta, \mu, \xi) = \sum_{y \in \mathcal{Y}} q_y \log \left(\frac{q_y}{q_y^0} \right) - \beta^T \sum_{y \in \mathcal{Y}} q_y f(y) + \beta^T E_{\tilde{p}}[f]$$

$$+ \mu \left\{ \sum_{y \in \mathcal{Y}} q_y - 1 \right\} - \sum_{y \in \mathcal{Y}} \xi_y q_y , \tag{9.76}$$

where $\beta \in \mathbf{R}^J$, $\mu \in \mathbf{R}$, and $\xi \in (\mathbf{R}^+)^{|\mathcal{Y}|}$ are Lagrange multipliers.

We minimize w.r.t. q by solving

$$0 = \frac{\partial \mathcal{L}(q, \beta, \mu, \xi)}{\partial q_y} \tag{9.77}$$

$$= 1 + \log \left(\frac{q_y}{q_y^0} \right) - \beta^T f(y) + \mu - \xi_y . \tag{9.78}$$

The solution is

$$\hat{q}_y^{(\beta)} = q_y^0 \, e^{\beta^T f(y) - 1 - \mu + \xi_y} > 0 . \tag{9.79}$$

It follows that $\hat{q}_y^{(\beta)} > 0$ and from complementary slackness, or, equivalently, from the KKT conditions, that the optimal ξ_y is $\xi_y^* = 0$. The optimal μ has to be chosen such that $\sum_{y \in \mathcal{Y}} q_y^{(\beta)} = 1$, i.e., as

$$\mu^* = -1 - \log Z(\beta) , \tag{9.80}$$

$$\text{where } Z(\beta) = \sum_{y \in \mathcal{Y}} q_y^0 \, e^{\beta^T f(y)} . \tag{9.81}$$

So we can write

$$\hat{q}_y^{(\beta)} = \frac{1}{Z(\beta)} q_y^0 \, e^{\beta^T f(y)} . \tag{9.82}$$

It follows from (9.76) that the Lagrangian at the optimal points for q, μ, and ξ is given by

$$\mathcal{L}(\hat{q}^{(\beta)}, \beta, \mu^*, \xi^*) = -\log Z(\beta) + \beta^T E_{\tilde{p}}[f]$$

$$= E_{\tilde{p}} \left[\log \frac{\hat{q}^{(\beta)}}{q^0} \right] . \tag{9.83}$$

We note that

$$E_{\tilde{p}} \left[\log \frac{\hat{q}^{(\beta)}}{q^0} \right] = \frac{1}{N} \left[\log L_{\mathcal{D}} \left(q^{(\beta)} \right) - \log L_{\mathcal{D}} \left(q^0 \right) \right] , \tag{9.84}$$

where $\mathcal{D} = \{y_1, y_2, ..., y_N\}$ is the observed dataset and $L_{\mathcal{D}}$ denotes the likelihood from Definition 6.1. So it follows from (9.83) that

$$\mathcal{L}(\hat{q}^{(\beta)}, \beta, \mu^*, \xi^*) = \frac{1}{N}\left[\log L_{\mathcal{D}}\left(\hat{q}^{(\beta)}\right) - \log L_{\mathcal{D}}\left(q^0\right)\right]. \qquad (9.85)$$

It follows from (9.82), (9.81), (9.85), and the fact that $L_{\mathcal{D}}\left(q^0\right)$ is independent of β, that Problem 9.2 is the dual of Problem 9.1. (9.58) follows then from Theorem 2.9.

The uniqueness of the solution of Problem 9.1 follows from the fact that $D(q\|q^*)$ is a strictly convex function of q (see Lemma 2.10, *(iii)*) and Theorem 2.8. This completes the proof of the theorem.□

9.5   Exercises

1. Let $y_1, ..., y_N$ denote the observed values for Y. Show that the sample mean,

$$\bar{y}_N = \frac{1}{N}\sum_{i=1}^{N} y_i,$$

is an unbiased estimator for $E_{p^{\theta_0}}[Y]$.

2. Let $y_1, ..., y_N$ denote the observed values for Y. Show that the following variance estimator

$$\hat{\sigma}_N^2 = \frac{1}{N}\sum_{i=1}^{N}(y_i - \bar{y}_N)^2$$

is a biased estimator for the variance of Y under p^{θ_0}.

3. Let $y_1, ..., y_N$ denote the observed values for Y. Show that the following variance estimator

$$\frac{1}{N-1}\sum_{i=1}^{N}(y_i - \bar{y}_N)^2$$

is an unbiased estimator for the variance of Y under p^{θ_0}.

4. Let Y be the number of heads in N spins of a coin, whose probability of "head" is θ. We did an experiment (with n spins) in which $\bar{y}$ heads were observed. Assuming a prior distribution for θ that is uniform on $[0, 1]$,

   (a) derive the posterior distribution for θ,

   (b) derive the prior predictive distribution for Y,

   (c) derive the posterior predictive distribution for Y,

   (d) show that the posterior mean of θ lies always between the prior mean, and the observed relative frequency of heads, $\frac{\bar{y}}{N}$, and

   (e) plot the posterior predictive distribution for Y for the case $\bar{y} = 5$ and $N = 10$.

5. Suppose you want to estimate the parameter vector $\theta = (\theta_1, ..., \theta_J)$ of the probability distribution of a random variable Y in a Bayesian approach by finding the maximum of the posterior distribution. Let the resulting estimator be $\hat{\theta}^B$. Assume that the prior distribution of θ is given by

$$P(\theta) = \frac{\alpha^J}{2^J} e^{-\alpha \sum_{j=1}^{J} |\theta_j|}, \tag{9.86}$$

where $\alpha > 0$ is some parameter.

   (a) Write down, in its general form, the optimization problem you have to solve in order to find $\hat{\theta}^B$.

   (b) State two limits in which $\hat{\theta}^B = \hat{\theta}^{ML}$, where $\hat{\theta}^{ML}$ is the maximum-likelihood estimator for θ?

   (c) Assume that the model distribution is from the single-parameter $(J = 1)$ family

$$P(y|\theta) = \frac{1}{\sqrt{2\pi}} e^{-\frac{(y-\theta_1)^2}{2}} \tag{9.87}$$

   and that we have one observation with $Y = 1$. Compute

       i. $\hat{\theta}^{ML}$,

       ii. $\hat{\theta}^B$ for $\alpha = 0.1$, and

       iii. $\hat{\theta}^B$ for $\alpha = 2$.

6. A coin has the probability θ for showing heads when tossed. In an experiment, the coin was tossed N times and heads were observed $\tilde{k}$ times. Our prior for θ is a beta-distribution with parameters α and β.

   (a) What is the posterior distribution for θ?

   (b) What is the posterior distribution for θ in a normal approximation, assuming that N is large?

7. Suppose you have N observations, $D = \{y_1, ..., y_N\}$ of a random variable Y, you believe that the probability measure of Y is from the family p^θ, and your prior measure for θ is $P(\theta)$. Show that the following to Bayesian estimation procedures give the same result.

(*i*) Sequentially use the information provided by the data, i.e., use the first observation to imply a posterior measure for θ, then use this posterior measure as your prior and imply a new posterior from the second observation, and repeat this procedure till all data are used.

(*ii*) Use all the information provided in the dataset D in one step to imply a posterior measure for θ.

Chapter 10

A Utility-Based Approach to Probability Estimation

In Chapter 8, we assumed that there are a number of candidate models and that a decision maker seeks the most useful of these models. A related, but harder, question is: how can a decision maker learn a useful model from data? We address this question in this chapter. The approach that we describe in this chapter is the one outlined in Section 1.3 of the introduction to this book, specialized to the horse race setting, which was introduced in Chapter 1. Under this approach, we explicitly take into account the decision consequences of the model, measuring these decision consequences as discussed in Chapter 8. That is, we measure, on an out-of-sample dataset, the decision consequences by the success of the strategy that a rational investor (who believes the model) would choose to place bets in the horse race setting. We assume that a decision maker strives to build models that perform well with respect to this model performance measure.

Our discussion, which is based on material from Friedman and Sandow (2003a) and Grünwald and Dawid (2004),[1] occurs on two levels:

(I) *(Model estimation principles)* establishing three equivalent economically motivated model estimation principles,

  (i) the robust outperformance principle, Principle 10.1, below (which is built around the model performance measurement principle — Principle 8.1 — in Chapter 8),

  (ii) the minimum market exploitability principle, Principle 10.2, below, and

  (iii) the minimum relative $(U, \mathcal{O})$-principle, Principle 10.3, below

  and

(II) *(Tuning consistency with the data and dual problem formulation)* given one of these (equivalent) model estimation principles and a set of features, establishing a method to tune the extent to which the model is

---

[1]The setting in Friedman and Sandow (2003a) is less general than that of Grünwald and Dawid (2004), which allows for development of certain explicit results that do not seem to be available in the more general setting.

consistent with the data (depending on the set of features, a model that is too consistent with the sample-averaged feature values can overfit). We also develop dual problem formulations suited for numerical implementation.

The model estimation principles and the formulation of the primal problems (in Sections 10.1.4 and 10.2.2, below) can be developed in a more general setting than the horse race. However, our dual problem formulation (see Sections 10.1.5 and 10.2.3, below) depends on the horse race setting. To keep things as simple as possible, we shall confine the discussion in this chapter to the horse race setting.

Model estimation principles
As in the introductory chapter, Section 1.3.2, we use the robust outperformance principle in order to maximize (over all measures) the worst-case (over all measures consistent with the data-consistency constraints) outperformance over a rival investor. By employing such a robust measure, we hope to immunize ourselves against attuning our model too precisely to the individual observations in the training set.

We then show that the robust outperformance principle is equivalent to a minimum market exploitability principle, under which we select the model that minimizes (over all data-consistent measures) the maximum (over all allocation strategies) outperformance over a rival investor.

In this chapter, we are more explicit about the setting (the horse race) than we were in the introduction to this book; this allows for a more explicit formulation of the model building problem. In this setting, the minimum market exploitability principle can be interpreted as a minimum relative $(U, \mathcal{O})$-entropy principle.

Tuning consistency with the data and dual problem formulation
The above model estimation principles provide some protection from overfitting, i.e., from models that fit the training data well but perform poorly out of sample. However, in practical application, there often remains some risk of overfitting; the magnitude of this risk depends on the data and the choice of data consistency constraints. In order to mitigate this risk, we shall introduce a family of data consistency constraint sets, and estimate a model as follows:

(i) for each set of data consistency constraints, estimate a model via the aforementioned principles (stated precisely as Principle 10.1, or, equivalently, Principle 10.2 or Principle 10.3, below), and then

(ii) from the collection of models obtained in step (i), select the one with the best performance (in the sense of Principle 8.1) on a test dataset (or via k-fold cross validation). This step is stated more precisely in Principle 10.4, below.

We recast this approach as a search among models on an efficient frontier (Pareto optimal models),[2] which we shall define in terms of consistency with a prior (benchmark or rival) model, measured by means of the relative $(U, \mathcal{O})$-entropy introduced in Chapter 7, and in terms of consistency with the data, measured by means of functions of feature expectations.

The models on the efficient frontier, each of which can be obtained by solving a convex optimization problem (see Problems 10.2 and 10.8 below), form a single-parameter family. Given the equivalence of the robust outperformance principle and the minimum $(U, \mathcal{O})$-relative entropy principle (Principles 10.1 and 10.3, respectively), it follows that each Pareto optimal model is robust in the sense that, for its level of consistency with the data, the model maximizes the worst-case outperformance relative to the benchmark model.

For each level of consistency with the data, we derive the dual problem (see Problems 10.3, 10.4, and 10.9) which has a Pareto optimal measure as its solution; this dual problem amounts to the maximization of expected utility with a regularization penalty over a well-defined family of functions.[3]

Once we have obtained the models on the efficient frontier, we rank the models by estimating their expected utilities on a hold-out sample, and select the model with maximum estimated expected utility. For ease of exposition, we consider only one hold-out sample; our procedure can be modified for k-fold cross validation.

Odds ratio independent formulations
The above economic paradigm, in general, requires the specification of the payoff structure of the horse race. In situations where such a structure is not obvious, this requirement imposes an encumbrance on the model builder. However, as we shall see, the optimization problems that follow from this paradigm are independent of the payoffs if (and only if) the investor's utility function is in the generalized logarithmic family $U(z) = \gamma_1 \log(z - \gamma B) + \gamma_2$. As we have seen in Section 8.1.5, this logarithmic family is rich enough to de-

---

[2]We shall define these terms precisely below; the notion of an efficient frontier originally comes from portfolio theory, where it represents the collection of portfolios that cannot be improved upon, allowing for the tradeoff between risk and reward.

[3]It is possible to formulate these dual problems and interpret them in terms of maximization of expected utility (with a regularization penalty) over a family of functions, by virtue of the following:

  (*i*) we have used a monotone, concave utility function, rather than a more general (negative) loss function,

  (*ii*) we have adopted the appropriate compatibility conditions,

  (*iii*) the decision maker invests in a horse race setting, and

  (*iv*) the right hand side of our data consistency constraints is expressed in terms of empirical expectations,

but not in more general settings. As far as we know, the aforementioned formulation/interpretation is not possible in completely general situations.

scribe a wide range of risk aversions, and it can be used to well-approximate (under reasonable conditions) nonlogarithmic utility functions; it is therefore applicable to many practical problems. In the case of a utility function from this logarithmic family, we obtain a regularized relative entropy minimization similar to the method discussed in Section 9.4. This means that the above economic principles provide additional motivation for this regularized relative entropy minimization.

Robust absolute performance
So far, in this chapter introduction, the model estimation principles that we have discussed have been formulated, or could be formulated, in terms of performance relative to that of a benchmark investor. It is also possible to formulate a generalized MRE problem for which the solution is robust in an absolute sense, rather than the relative sense of the robust outperformance principle, Principle 10.1.

Organization of this chapter
In Section 10.1, we formulate our modeling approach in the simplest context: we seek a discrete probability model. In Section 10.2, we briefly discuss the same in a more general context: we seek a model that describes the conditional distribution of a possibly vector-valued random variable with a continuous range and discrete point masses. In Section 10.3, we discuss estimation methods geared to maximizing robust (absolute) performance. In Section 10.4, we show how the data consistency constraints can be expressed in purely economic terms. Numerical experiments based on methods consistent with the methodology described in this chapter are reported in a wide variety of sources. For particular examples, see Chapter 12.

10.1 Discrete Probability Models

In this section, we describe a decision-theoretic modeling paradigm in the simplest context:

(*i*) an investor has a utility function that satisfies Assumption 4.1 stated in Section 4.6,

(*ii*) this investor operates in the discrete horse race setting of Section 3.1,

(*iii*) the horse race and the investor's utility function are compatible, in the sense of Definition 5.2 of Section 5.1.1. This (technical) compatibility condition is imposed to insure that the investor's optimal allocation is well defined, (*iv*) the investor allocates his assets so as to maximize his expected utility according to his beliefs, i.e., the investor allocates so as to maximize the expectation of his utility under the model probability measure he believes, and

(*v*) the investor measures model performance as per Principle 8.1, in Chapter 8.

We explicitly develop,[4] in the horse race setting, three model estimation principles,

(*i*) the robust outperformance principle (which is built around the model performance measurement principle (Principle 8.1) in Chapter 8),

(*ii*) the minimum market exploitability principle, and

(*iii*) the minimum relative $(U, \mathcal{O})$-principle.

We interpret these principles and show that they are equivalent. In the sections that follow, we introduce the notion of data constraint relaxation as a way to balance consistency with the data and "prior beliefs," a dual problem, and discuss the importance of the logarithmic family.

10.1.1 The Robust Outperformance Principle

Suppose that there is an investor who wants to estimate a model that he can use to make decisions in the future. At first blush, it might seem natural for this investor to choose the model that maximizes the utility-based performance measures on the data available for building the model (the training data). However, it can be shown (see Lemma 8.1 of Section 8.1) that this course of action leads to the selection of the empirical measure — the model that we obtain by assigning the empirical frequency of each datum to its probability — which can be a very poor model indeed, if we want our model to generalize well on out-of-sample data. We illustrate this idea in the following Example, which is a slightly more detailed version of Example 1.3 from Section 1.6.

Example 10.1 *An overfit model*

Let the random variable X denote the daily return of a stock. We observe the daily stock returns $x_1, \ldots, x_{10}$, over a two week period (10 trading days).

---

[4]The approach that we take follows the same logic as the approach described in Chapter 1, but we are more explicit here, in the horse race setting, making use of results from previous chapters.

The empirical measure is then

$$prob(X = x) = \begin{cases} \frac{1}{10}, & \text{if } x \in \{x_1, \ldots, x_{10}\}, \text{ and} \\ 0, & \text{otherwise,} \end{cases} \qquad (10.1)$$

assuming that the daily returns are unique. A model builder who aims to maximize the utility-based performance measure on the observed data would pick the above empirical measure as his model. This model reflects the training data perfectly, but will fail out-of-sample, since it only attaches nonzero probability to events that have already occurred. If this model is to be believed, then it would make sense to risk all on the bet that $X \in \{x_1, \ldots, x_{10}\}$, a strategy doomed to fail when a previously unobserved return (inevitably) occurs.

In Exercise 1, we see that when the investor and market are compatible, the model performance measure of Chapter 8 yields a value of $-\infty$ if there is a datum in the test dataset that is not in the training set, $\{x_1, \ldots, x_{10}\}$.

Though it is, generally speaking, unwise to build a model that adheres too strictly to the individual outcomes that determine the empirical measure, the observed data contain valuable statistical information that can be used for the purpose of model estimation. We incorporate statistical information from the data into a model via data consistency constraints, expressed in terms of features, as described in Section 1.3.1.1 and 9.4.1.

Armed with the notions of features and data-consistency constraints, we return to the model estimation problem. The empirical measure typically does not generalize well because it is all too precisely attuned to the observed data. We seek a model that is consistent with the observed data, in the sense of conforming to the data-consistency constraints, yet is not too precisely attuned to the data. The question is, which data-consistent measure should we select? We want to select a model that will perform well (in the sense of the model performance measurement principle, Principle 8.1, in Chapter 8), no matter which data-consistent measure might govern a potential out-of-sample test set. To address this question, below, we suppose that we have a rival and consider a game against nature[5] (who sides with our rival) that occurs in the horse race setting. We recall that in this setting, our investor allocates according to

$$b^*(q) = \arg \max_{\{b : \sum_y b_y = 1\}} \sum_y q_y U(b_y \mathcal{O}_y). \qquad (10.2)$$

From Lemma 5.1 in Section 5.1.2, we know that the investor's optimal allocation exists, is unique, and is given by

$$b_y^*(q) = \frac{1}{\mathcal{O}_y} (U')^{-1} \left(\frac{\lambda}{q_y \mathcal{O}_y} \right), \qquad (10.3)$$

---

[5]This game is a special case of a game in Grünwald and Dawid (2004), which was preceded by the "log loss game" of Good (1952).

where λ is the solution of the following equation:

$$\sum_y \frac{1}{\mathcal{O}_y}(U')^{-1}\left(\frac{\lambda}{q_y\mathcal{O}_y}\right) = 1. \tag{10.4}$$

We also suppose that a rival, or benchmark investor, allocates according to the measure q^0, i.e., he allocates according to

$$b_y^*(q^0) = \frac{1}{\mathcal{O}_y}(U')^{-1}\left(\frac{\lambda^0}{q_y^0\mathcal{O}_y}\right), \tag{10.5}$$

where λ^0 is the solution of the following equation:

$$\sum_y \frac{1}{\mathcal{O}_y}(U')^{-1}\left(\frac{\lambda^0}{q_y^0\mathcal{O}_y}\right) = 1. \tag{10.6}$$

Having specified how our investor and his rival allocate, we specify the game; in the game specification we allocate according to the (to be determined) measure q, and our rival allocates according to the measure q^0.

To make headway, we will need the following assumption.

Assumption 10.1 *(Compactness and convexity of the set of data-consistent measures) The set of data-consistent measures, which we denote by K, is compact and convex.*

We shall see below that this assumption holds for the sets of data-consistent measures considered in the book.

A game against "nature" in the horse race setting *Let Q denote the set of all probability measures and let K denote the set of data-consistent probability measures.*

(i) *(Our move) We choose a model, $q \in Q$; then,*

(ii) *(Nature's move) given our choice of a model, and, as a consequence, the allocations we would make, "nature," who sides with our rival investor, cruelly inflicts on us the worst (in the sense of our outperforming the rival model with respect to the model performance measurement principle, Principle 8.1, in Chapter 8) possible data-consistent measure; that is, "nature" chooses the measure[6]*

$$p^* = \arg\min_{p \in K} E_p[U(b^*(q), \mathcal{O}) - U(b^*(q^0), \mathcal{O})], \tag{10.7}$$

---

[6]The careful reader will observe that in (10.7), by writing min, rather than inf, we have tacitly assumed that the minimum in fact exists. We could argue that this must be so, based on the convexity and compactness of K, and the convexity in p of the expectation, E_p. We shall not expend much effort on such justifications going forward, using min's or max's without explicit justification, rather than more correctly using sup's or inf's, and then providing justification. For a rigorous related treatment, the reader can consult Grünwald and Dawid (2004).

where

$$E_p[U(b^*(q), \mathcal{O}) - U(b^*(q^0), \mathcal{O})] \equiv E_p[U(b^*(q), \mathcal{O})] - E_p[U(b^*(q^0), \mathcal{O})] \tag{10.8}$$

and (using the same notation as in Chapter 8 — see Section 8.1),

$$E_p[U(b^*(q), \mathcal{O})] = \sum_y p_y U(b_y^*(q)\mathcal{O}_y)$$

and

$$E_p[U(b^*(q^0), \mathcal{O})] = \sum_y p_y U(b_y^*(q^0)\mathcal{O}_y).$$

If we want to perform as well as possible in this game we will estimate our model according to the following principle:

Principle 10.1 *(Robust Outperformance Principle) Given a set of data-consistent measures, K, we seek*

$$q^* = \arg\max_{q \in Q} \min_{p \in K} E_p[U(b^*(q), \mathcal{O}) - U(b^*(q^0), \mathcal{O})]. \tag{10.9}$$

By solving (10.9), we estimate a measure that (as we shall see later[7]) conforms to the data-consistency constraints; moreover, by construction, this measure is *robust*, in the sense that the excess (over our rival) expected utility that we can derive from it will be attained, or surpassed, no matter which data-consistent measure "nature" chooses. The resulting estimate therefore, in particular, avoids being too precisely attuned to the individual observations in the training dataset, thereby mitigating overfitting.[8]

If we allocate according to q^*, we maximize the worst-case outperformance over our competitor (who allocates according to the measure $q^0 \in Q$), in the presence of a "nature" that conforms to the data-consistency constraints and tries to minimize our outperformance (in the sense of the model performance measurement principle, Principle 8.1, in Chapter 8) over our rival.

We note that this formulation has been cast entirely in the language of utility theory. The model that is produced is therefore specifically tailored to the risk preferences of the model user with utility function U. We also note that we have not made use of the concept of a "true" measure in this formulation.

---

[7]See Theorem 10.1, below.

[8]This strategy does not guarantee a cure to overfitting, though! If there are too many data-consistency constraints, or the data consistency constraints are not chosen wisely, problems can arise. We shall discuss these issues, and countermeasures that can be taken to further protect against overfitting, at greater length later in this chapter.

10.1.2   The Minimum Market Exploitability Principle

In this section, we shall state the minimum market exploitability principle[9] and show that, for the discrete horse race, the model obtained from the robust outperformance principle is the same as the model obtained from the minimum market exploitability principle.

Principle 10.2 *(Minimum Market Exploitability Principle) Given a set of data-consistent measures, K, we seek*

$$p^* = \arg \min_{p \in K} \max_{q \in Q} E_p[U(b^*(q), \mathcal{O}) - U(b^*(q^0), \mathcal{O})]. \tag{10.10}$$

Here,

$$E_p[U(b^*(q), \mathcal{O}) - U(b^*(q^0), \mathcal{O})] \tag{10.11}$$

can be interpreted as the gain in expected utility, for an investor who allocates according to the model q, rather than q^0, when the "true" measure is p. Under the minimum market exploitability principle, we seek the data-consistent measure, p, that minimizes the maximum gain in expected utility over an investor who uses the model q^0. After a little reflection, this principle is consistent with a desire to avoid overfitting. The intuition here is that the data-consistency constraints completely reflect the characteristics of the model that we want to incorporate, and that we want to avoid introducing additional (spurious) characteristics. Any additional characteristics (beyond the data-consistency constraints) could be exploited by an investor; so, to avoid introducing additional such characteristics, we minimize the exploitability of the market by an investor, given the data-consistency constraints.

We now return to our goal of establishing the equivalence of Principles 10.1 and 10.2. To see the equivalence of the two principles, we first recall that from Lemma 5.1 from Section 5.1.2, we know that when the investor and the horse race market are compatible, given a measure, q, the investor's optimal allocation exists, is unique, and is given by

$$b_y^*(q) = \frac{1}{\mathcal{O}_y}(U')^{-1}\left(\frac{\lambda}{q_y \mathcal{O}_y}\right), \tag{10.12}$$

where λ is the solution of the following equation:

$$\sum_y \frac{1}{\mathcal{O}_y}(U')^{-1}\left(\frac{\lambda}{q_y \mathcal{O}_y}\right) = 1. \tag{10.13}$$

Suppose, on the other hand, that we are given an optimal allocation, b_y^*, and we seek the probability measure that generated it. It is easy to show[10]

---

[9] Also discussed in Section 1.3.2.2.

[10] See Exercise 2 to fill in the details.

that

$$q_y(b_y^*) = \frac{1}{\mathcal{O}_y U'(b_y^* \mathcal{O}_y)} \frac{1}{\sum_{y'} \frac{1}{\mathcal{O}_{y'} U'(b_{y'}^* \mathcal{O}_{y'})}}. \tag{10.14}$$

Given that for every q there is a b^*, and for every b^*, there is a q, we can cast our robust outperformance principle in terms of robust allocations, rather than robust measures, and seek the robust allocation

$$b^* = \arg\max_{b \in \overline{\mathcal{B}}} \min_{p \in K} E_p[U(b, \mathcal{O}) - U(b^*(q^0), \mathcal{O})], \tag{10.15}$$

where $\overline{\mathcal{B}}$ denotes the set of allocations generated (under (10.12)) by the measures $q \in Q$. From Corollary 5.1, we see that $\overline{\mathcal{B}}$ is a bounded set that is a subset of $\{b : \sum_y b_y = 1\}$ that contains all of it limit points.[11] Thus, $\overline{\mathcal{B}}$ is closed and bounded and is therefore compact. The convex combination of any two points in $\overline{\mathcal{B}}$ is also in $\overline{\mathcal{B}}$, so $\overline{\mathcal{B}}$ is a convex set.[12] For a fixed value of p, the function

$$E_p[U(b, \mathcal{O}) - U(b^*(q^0), \mathcal{O})] \tag{10.16}$$

is concave in b, by the definition of (10.16) and the concavity of the utility function, U. For a fixed b, (10.16) is linear in p and therefore convex in p. Both $\overline{\mathcal{B}}$ and K are convex and compact.[13] We can therefore apply a minimax theorem (see Section 2.2.8), obtaining

$$\max_{b \in \overline{\mathcal{B}}} \min_{p \in K} E_p[U(b, \mathcal{O}) - U(b^*(q^0), \mathcal{O})] = \min_{p \in K} \max_{b \in \overline{\mathcal{B}}} E_p[U(b, \mathcal{O}) - U(b^*(q^0), \mathcal{O})]. \tag{10.17}$$

Again, we can use that fact that for every q there is a b^*, and for every b^*, there is a q, to cast the maximization over b as a maximization over q, obtaining

$$\min_{p \in K} \max_{q \in Q} \left\{ E_p\left[U(b^*(q), \mathcal{O})\right] - E_p\left[U(b^*(q^0), \mathcal{O})\right] \right\} \tag{10.18}$$

$$= \max_{q \in Q} \min_{p \in K} \left\{ E_p\left[U(b^*(q), \mathcal{O})\right] - E_p\left[U(b^*(q^0), \mathcal{O})\right] \right\}.$$

The maxmin problem on the right attains its maximum for some pair of measures (q^*, p^*). We must have $q^* = p^*$. To see this, assume that $q^* \neq p^*$. Then with p^* fixed, it would be possible to increase the quantity

$$\left\{ E_{p^*}\left[U(b^*(q), \mathcal{O})\right] - E_{p^*}\left[U(b^*(q^0), \mathcal{O})\right] \right\} \tag{10.19}$$

by putting $q = p^*$ (by definition of the optimal betting weights under p^*), contradicting our assumption that the maximum is attained for the pair of measures (q^*, p^*).

Thus, we have obtained the following theorem:

---

[11] See Exercise 3 to fill in the details.

[12] See Exercise 3 to fill in the details.

[13] K is compact and convex by Assumption 10.1.

Theorem 10.1 *(Equivalence of the robust outperformance principle, Principle 10.1, and the minimum market exploitability principle, Principle 10.2) Let Q denote the set of all possible probability measures, and let $K \subset Q$ be compact and convex. Then*

$$p^* = \arg \min_{p \in K} \max_{q \in Q} \; \left\{ E_p\left[U(b^*(q), \mathcal{O})\right] - E_p\left[U(b^*(q^0), \mathcal{O})\right] \right\}$$

$$= \arg \max_{q \in Q} \min_{p \in K} \left\{ E_p\left[U(b^*(q), \mathcal{O})\right] - E_p\left[U(b^*(q^0), \mathcal{O})\right] \right\} = q^*.$$

By solving the resulting minimax problem, we obtain the solution to the maxmin problem (10.9) arising from the robust outperformance principle. Thus, the robust outperformance principle is equivalent to the minimum market exploitability principle.

It is interesting to note that under the robust outperformance principle, we seek a measure with robust outperformance over all measures in Q; that is, we do not require that the measure with robust outperformance be data-consistent (i.e., in K). The fact that $q^* \in K$ follows as a logical consequence of the equivalence that we have just established.

10.1.3 Minimum Relative $(U, \mathcal{O})$-Entropy Modeling

By starting with the minimum market exploitability principle, Principle 10.2, which, as we have seen is equivalent to the robust outperformance principle, Principle 10.1, we are led to the minimum relative $(U, \mathcal{O})$-entropy principle, Principle 10.3, below.

To see this, we start with the definition of relative $(U, \mathcal{O})$-entropy (Definition 7.2):

$$D_{U,\mathcal{O}}(p\|p^0) = E_p\left[U(b^*(p), \mathcal{O})\right] - E_p\left[U(b^*(p^0), \mathcal{O})\right] .$$

By Definition 5.3 of the optimal betting weights, b^*,

$$E_p\left[U(b^*(p), \mathcal{O})\right] \geq E_p\left[U(b^*(q), \mathcal{O})\right] , \tag{10.20}$$

for any measure $q \in Q$. Therefore, we have

$$D_{U,\mathcal{O}}(p\|p^0) = \max_{q \in Q} \left\{ E_p\left[U(b^*(q), \mathcal{O})\right] - E_p\left[U(b^*(p^0), \mathcal{O})\right] \right\};$$

so the minimum market exploitabilty principle is equivalent to minimum relative $(U, \mathcal{O})$-entropy minimization in the discrete unconditional horse race setting:

$$\min_{p \in K} D_{U,\mathcal{O}}(p\|p^0) = \min_{p \in K} \max_{q \in Q} \left\{ E_p\left[U(b^*(q), \mathcal{O})\right] - E_p\left[U(b^*(p^0), \mathcal{O})\right] \right\}. \tag{10.21}$$

This motivates the following principle as well as the following theorem.

Principle 10.3 *(Minimum Relative $(U, \mathcal{O})$-Entropy) Given a set of data-consistent measures, K, we seek*

$$p^* = \arg\min_{p \in K} D_{U,\mathcal{O}}(p||p^0). \qquad (10.22)$$

We summarize the relations among Principles 10.1, 10.2, and 10.3 in the following theorem:

Theorem 10.2 *(Equivalence of the robust outperformance principle, Principle 10.1, the minimum market exploitability principle, Principle 10.2, and the minimum relative $(U, \mathcal{O})$-entropy principle, Principle 10.3) Let Q denote the set of all possible probability measures, and let $K \subset Q$ be compact and convex. Then*

$$\arg\min_{p \in K} \max_{q \in Q} \quad \left\{ E_p\left[U(b^*(q), \mathcal{O})\right] - E_p\left[U(b^*(q^0), \mathcal{O})\right] \right\}$$

$$= \arg\max_{q \in Q} \min_{p \in K} \left\{ E_p\left[U(b^*(q), \mathcal{O})\right] - E_p\left[U(b^*(q^0), \mathcal{O})\right] \right\}$$

$$= \arg\min_{p \in K} D_{U,\mathcal{O}}(p||p^0).$$

We note that this theorem can be proved in a more general setting.[14]

Fortunately, as we shall see later in this chapter (making use of the properties of relative $(U, \mathcal{O})$-entropy developed in Chapter 7), the minimum relative $(U, \mathcal{O})$-entropy principle leads to a convex optimization problem with an associated dual problem that in many cases can be solved robustly via efficient numerical techniques. Moreover, as we shall also see later in this chapter, this dual problem can be interpreted as a utility maximization problem over a parametric family, and can be solved robustly via efficient numerical techniques.

Thus, given the equivalence of

(*i*) the robust outperformance principle,

(*ii*) the minimum market exploitability principle, and

(*iii*) the minimum relative $(U, \mathcal{O})$-entropy principle,

all of these principles lead us down the same path — to a tractable approach to estimate statistical models tailor-made to the risk preferences of the end user.

Henceforth, we estimate models by invoking the relative $(U, \mathcal{O})$-entropy principle, making use of the properties of relative $(U, \mathcal{O})$-entropy developed in Chapter 7, knowing that, by virtue of the equivalence of Principles 10.1, 10.2, and 10.3, the robust outperformance and minimum market exploitability properties are baked into the solution.

---

[14]See Grünwald and Dawid (2004), who adopt a more general approach; their nomenclature is different from ours: in their game against nature, rather than maximize the worst case performance in terms of expected utility, they minimize the greatest expected loss, etc.

10.1.4   An Efficient Frontier Formulation

Under Principle 10.3, given a set of data consistency constraints, K, we seek the measure with minimum relative $(U, \mathcal{O})$-entropy. But how, given a set of features, should we determine K? Should we require that the expectation, under the model that we seek, agrees exactly with the sample-average values of the features, or should we adopt a more relaxed requirement, and require only that the feature expectations under the model that we seek be "close enough" to the sample average values of the features? How can we decide what is "close enough?"

In this section, we answer these questions and build models accordingly; to do so, we consider the tradeoff between consistency with the data and consistency with our investor's prior beliefs. Given a set of feature functions, we shall consider a family of sets of data-consistency constraints, with the goal of finding the set K in this family, for which an application of the minimum relative $(U, \mathcal{O})$-entropy minimization principle, Principle 10.3, leads to good out-of-sample performance, in the sense of the model performance measurement principle, Principle 8.1. We accomplish these goals as follows:

(i) for each set of data-consistency constraints, estimate a model via the aforementioned principles (states precisely as Principle 10.1, or, equivalently, Principle 10.2 or Principle 10.3, below),[15] and then

(ii) from the collection of models obtained in step (i), select the one with the best performance (in the sense of Principle 8.1) on a test dataset (or via k-fold cross validation). This step is stated more precisely in Principle 10.4, below.

We recast this approach as a search among models on an efficient frontier (Pareto optimal models), which we shall define in terms of consistency with a prior (benchmark or rival) model, measured by means of the relative $(U, \mathcal{O})$-entropy introduced in Chapter 7, and in terms of consistency with the data, measured by means of functions of feature expectations.

Given models equally consistent with the investor's prior beliefs, we assume that the investor prefers a model that is more consistent with the data; given models equally consistent with the data, we assume that the investor prefers a model that is more consistent with the investor's prior beliefs. We make all of this precise below.

10.1.4.1   Consistency with the Data

For a model measure $q \in Q$, let $\mu^{data}(q)$ denote the investor's measure of the consistency of q with the data; this consistency is expressed in terms of expectations of the *feature vector*, $f(y) = (f_1(y), \ldots, f_J(y))^T \in \mathbf{R}^J$ where

---

[15] In practice, we estimate a finite number of models associated with a finite number of data-consistency constraints; we will describe this in greater detail below.

each feature, $f_j(y) = f_j(y_1, \ldots, y_m)$, is a mapping from the state space $\mathcal{Y}$ of Y to $\mathbf{R}$. We make the following assumption:

Assumption 10.2 *The investor measures the consistency,[16] $\mu^{data}(q)$, of the model $q \in Q$ with the data via the nonnegative convex function ψ, with*

$$\mu^{data}(q) = \psi(c), \tag{10.23}$$

where

$$c = (c_1, \ldots, c_J)^T,$$

with

$$c_j = E_q[f_j] - E_{\tilde{p}}[f_j], \tag{10.24}$$

and $\psi(0) = 0$.

We can think of the function $\psi(c)$ as a measure of the discrepancy between the expected feature values under the model that we seek and under the empirical measure.

We shall see below that Assumption 10.2 leads to data-consistency constraints for which Assumption 10.1 (the measures satisfying the data-consistency constraints form a convex compact set) holds.[17] Recall that we used Assumption 10.1 to establish Theorem 10.1, which stated the equivalence between the robust outperformance principle, Principle 10.1, and the minimum market exploitability principle, Principle 10.2.

We now motivate and describe a few possible concrete choices for the discrepancy function $\psi(c)$:

1. *Mahalanobis distance:* Given a positive definite matrix $\Sigma \in R^{n \times n}$, and two vectors, x_1 and x_2 in R^n, the Mahalanobis distance between x_1 and x_2 is given by[18]
   $$(x_2 - x_1)'\Sigma^{-1}(x_2 - x_1). \tag{10.25}$$

   This distance arises naturally in connection with the multivariate Gaussian distribution, and, as we shall see, can be used to measure the distance from a model measure to an empirical measure in terms of the feature covariance structure.

   Consider the large sample probability density of the sample feature means, evaluated at the model q feature expectations, $E_q[f]$. To elaborate, for a fixed measure $q \in Q$, the model feature mean, $E_q[f_j]$, is a deterministic quantity depending on q. The sample mean of f_j, however,

---

[16]More precisely, low μ values are associated with highly consistent models and high μ values are associated with less consistent models.

[17]See Exercise 9.

[18]See, for example, Kullback (1997), p. 190.

depends on the sample set and is therefore a random variable, ϕ_j. The quantity $E_{\tilde{p}}[f_j]$ is therefore an observation of the random variable ϕ_j. By the Central Limit Theorem, for a large number of observations, N, the random vector $\phi = (\phi_1, \dots, \phi_J)^T$ is approximately normally distributed with mean $E_{\tilde{p}}[f]$ and covariance matrix $\frac{1}{N}\Sigma$, where Σ is the empirical feature covariance matrix. Therefore, for a given measure $q \in Q$, the probability density for the random variable ϕ, evaluated at $E_q[f]$, is (approximately) given by

$$p^c(c) \equiv pdf(\phi)|_{\phi=E_q[f_j]} = (2\pi)^{-\frac{1}{2}J} N^{\frac{1}{2}} |\Sigma|^{-\frac{1}{2}} e^{-\frac{N}{2}c^T \Sigma^{-1} c}, \qquad (10.26)$$

where

$$c = (c_1, \dots, c_J)^T$$

and

$$c_j = E_q[f_j] - E_{\tilde{p}}[f_j]. \qquad (10.27)$$

The level sets of this unimodal probability density form a set of nested regions around the mean point $E_q[f_j]$, suggesting the following measure of consistency with the data:

$$\mu^{data}(q) = \psi(c) \equiv N c^T \Sigma^{-1} c \geq 0. \qquad (10.28)$$

This measure (the Mahalanobis metric) has been used to measure consistency of a model measure with the data; see, for example, Wu (1997), and Gull and Daniell (1978).

We note that though we have used the Central Limit Theorem to *motivate* this measure, we have not made any assumption on the probability distribution of the measures $q \in Q$.

Equally consistent measures, q, lie on the same level set of the function $\mu^{data}(q)$. We parameterize the nested family of sets, consisting of points $q \in Q$ that are equally consistent with the data. We note that since the feature covariance matrix Σ^{-1} is a nonnegative definite matrix, $\mu^{data}(q)$ is, indeed, a convex function of q.

Note that, by construction, in this case, $\mu^{data}(q)$ is invariant with respect to translations and rotations of the feature vectors.[19]

In order to estimate the number of elements in the feature covariance matrix, we would hope that the number of observations, N, far exceeds the number of elements in the feature covariance matrix, $\frac{J(J+1)}{2}$; that is, we would hope that $\frac{J(J+1)}{2} << N$. However, for many applications, this is not the case. In such cases (and others as well), it is often possible to obtain better numerical results by putting the off-diagonal elements of Σ to zero.

---

[19] James Huang (2003) first pointed this out to the authors.

2. ℓ_2-*norm:* The ℓ_2-norm is suggested by the large sample density of feature means, with the off-diagonal elements of Σ put to zero.

3. ℓ_∞-*norm:* As indicated in Sections 9.2.2 and Section 9.4.3, the ℓ_∞-norm allows us to combine regularization with feature selection.

10.1.4.2   Consistency with the Prior Measure

To quantify consistency of the model, $q \in Q$, with the investor's prior beliefs,[20] we make use of the relative $(U, \mathcal{O})$-entropy $D_{U,\mathcal{O}}(q||q^0)$, where U is the investor's utility function and $\mathcal{O}$ is the set of odds ratios.

Assumption 10.3 *The investor measures the consistency, $\mu^{prior}(q)$, of the model $q \in Q$ with the prior, q^0, by the relative $(U, \mathcal{O})$-entropy, $D_{U,\mathcal{O}}(q||q^0)$.*

More precisely, low μ values are associated with highly consistent models and high μ values are associated with less consistent models.

10.1.4.3   Pareto Optimal Measures

To characterize the measures $q^* \in Q$ which are optimal (in a sense to be made precise), we define *dominance, Pareto optimal* probability measures, the set of *achievable measures*, and the *efficient frontier*. These notions are from vector optimization theory (see, for example, Boyd and Vandenberghe (2004)) and portfolio theory (see, for example, Luenberger (1998)).

Definition 10.1 $q^1 \in Q$ *dominates* $q^2 \in Q$ *with respect to* μ^{data} *and* μ^{prior} *if*

(i)

$$(\mu^{data}(q^1), \mu^{prior}(q^1)) \neq (\mu^{data}(q^2), \mu^{prior}(q^2))$$

   and

(ii)

$$\mu^{data}(q^1) \leq \mu^{data}(q^2)$$

   and

$$\mu^{prior}(q^1) \leq \mu^{prior}(q^2).$$

Definition 10.2 *A model, $q^* \in Q$, is Pareto optimal if and only if no measure $q \in Q$ dominates q^* with respect to $\mu^{data}(q)$ and $\mu^{prior}(q) = D_{U,\mathcal{O}}(q||q^0)$. The efficient frontier is the set of Pareto optimal measures.*

---

[20] Alternatively, the notion of bounding the closeness to a prior measure can be interpreted as bounding the extent to which a market can be exploited.

We note that for any Pareto optimal measure $q^* \in Q$,

$$\mu^{data}(q) \leq \mu^{data}(q^*) \text{ implies that } D_{U,\mathcal{O}}(q||q^0) \geq D_{U,\mathcal{O}}(q^*||q^0) \qquad (10.29)$$

for all $q \in Q$.

The Pareto optimal measures are contained in the *achievable set, $\mathcal{A}$*, which is defined as follows:

Definition 10.3 *The achievable set, $\mathcal{A}$, is given by*

$$\mathcal{A} = \{(\alpha, D)|\mu^{data}(q) \leq \alpha \text{ and } D_{U,\mathcal{O}}(q||q^0) \leq D \text{ for some } q \in Q\} \subset \mathbf{R}^2.$$

By (10.27) and (10.28), measures q that are equally consistent with the data lie on the same level set of the function $\mu^{data}(q)$. In specific cases, for example, the Mahalanobis metric,

$$\mu^{data}(q) = \psi(c) = N(E_q[f] - E_{\tilde{p}}[f])^T \Sigma^{-1}(E_q[f] - E_{\tilde{p}}[f]), \qquad (10.30)$$

we would parameterize the nested family of sets, consisting of points $q \in Q$ that are equally consistent with the data, by (10.30).

The achievable set, $\mathcal{A}$, is convex. $\mathcal{A}$ is convex by the convexity of $\mu^{data}(q)$ and $D_{U,\mathcal{O}}(q||q^0)$ (see, for example, Boyd and Vandenberghe (2004), Section 4.7). The convexity of the achievable set follows from the particular choice $\mu^{data}(q)$ and $\mu^{prior}(q) = D_{U,\mathcal{O}}(q||q^0)$.

We may visualize the achievable set, $\mathcal{A}$, and the efficient frontier as displayed in Figure 10.1, which also incorporates the following lemma.

Lemma 10.1 *If q^* is a Pareto optimal measure, then*

(i) $\mu^{data}(q^*) \leq \alpha_{max}$*, where*

$$\alpha_{max} = \mu^{data}(q^0) \qquad (10.31)$$

(ii) $(\mu^{data}(q^*), D_{U,\mathcal{O}}(q^*||q^0))$ *lies on the lower D-boundary of $\mathcal{A}$.*

Proof: (*i*) For the measure q^0, we have $\mu^{data}(q^0) = \alpha_{max}$ and $D = D_{U,\mathcal{O}}(q^0||q^0) = 0$. If $\mu^{data}(q) > \alpha_{max}$, then q cannot be identical to q^0, and so $D_{U,\mathcal{O}}(q||q^0) > 0$; so q is dominated by q^0 and cannot be efficient. (*ii*) follows directly from (10.29). $\square$

We shall make use of the preceding lemma when we formulate our optimization problem.

We make the following assumption, which is equivalent to adopting the minimum relative $(U, \mathcal{O})$-entropy principle, Principle 10.3 (which is itself equivalent to the robust outperformance principle and the minimum market exploitability principle — principles 10.1 and 10.2, respectively).

Assumption 10.4 *The investor selects a measure on the efficient frontier.*

Thus, given a set of measures equally consistent with the prior, our investor prefers measures that are more consistent with the data, and, given a set of measures equally consistent with the data, he prefers measures that are more consistent with the prior. He makes no assumptions about the precedence of these two preferences.

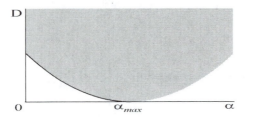

FIGURE 10.1:   Achievable set, $\mathcal{A}$: shaded region above curve; Efficient Frontier: points on bold curve with $0 \leq \alpha \leq \alpha_{max}$. (This figure originally appeared in Friedman and Sandow (2003a).)

10.1.4.4   Convex Optimization Problem

We seek the set of Pareto optimal measures. That is, motivated by Lemma 10.1, for all $q \in Q$ with $\mu^{data}(q) = \alpha$, we seek all solutions of the following problem, as α ranges from 0 to α_{max}, where α_{max} is defined in (10.31).

Problem 10.1 *(Initial Problem, Given* $\alpha, 0 \leq \alpha \leq \alpha_{max}$*)*

$$Find \quad \arg \inf_{q \in (R^+)^m, c \in R^J} D_{U,\mathcal{O}}(q\|q^0) \quad (10.32)$$

$$under\ the\ constraints\ 1 = \sum_y q_y \quad (10.33)$$

$$and \quad \psi(c) = \alpha \quad (10.34)$$

$$where \quad c_j = E_q[f_j] - E_{\tilde{p}}[f_j] \ . \quad (10.35)$$

Problem 10.1 is not the standard convex optimization problem discussed in Section 2.2, since (10.34) is a nonaffine equality constraint. However, we formulate a different (convex optimization) problem, which, as we shall show, has the same solutions:

Problem 10.2 *(Initial Convex Problem, Given* $\alpha, 0 \leq \alpha \leq \alpha_{max}$*)*

$$Find \quad \arg \min_{q \in (R^+)^m, c \in R^J} D_{U,\mathcal{O}}(q\|q^0) \quad (10.36)$$

$$under\ the\ constraints\ 1 = \sum_y q_y \quad (10.37)$$

$$and \quad \psi(c) \leq \alpha \quad (10.38)$$

$$where \quad c_j = E_q[f_j] - E_{\tilde{p}}[f_j] \ . \quad (10.39)$$

Lemma 10.2 *Problem 10.2 is a convex optimization problem and Problems 10.1 and 10.2 have the same unique solution.*

Proof: See Section 10.5.1.

In order to visualize the solutions to Problem 10.2, we define

$$S_\alpha = \{q : \mu^{data}(q) = \alpha, q \in Q_c\}, \qquad (10.40)$$

where

$$Q_c = \{q : q \geq 0, \sum_y q_y = 1, \sum_y q_y f_j(y) = c_j + E_{\tilde{p}}[f_j], j = 1, \ldots, J\}.$$

By solving Problem 10.2, for each α, we generate a one-parameter family of candidate models, $q^*(\alpha)$, indexed by α. We can visualize these models as the points of tangency (on the probability simplex, Q) of the nested surfaces of the families S_α and the level sets of $D_{U,O}(q||q^0)$ (see Figure 10.2). Each candidate model, $q^*(\alpha)$, is a solution of Problem 10.2; accordingly, each point $(\alpha, D_{U,O}(q^*(\alpha)||q^0))$ is a point on the efficient frontier (see Figure 10.1), and the efficient frontier consists of all points of the form $(\alpha, D_{U,O}(q^*(\alpha)||q^0))$, as α ranges from $(0, \alpha_{max})$.

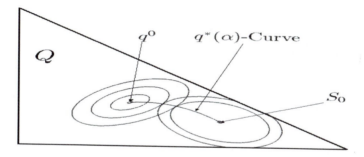

FIGURE 10.2:   The sets S_α (see (10.40)), centered at S_0, the $q^*(\alpha)$-curve, and the level sets of $D_{U,O}(q||q^0)$, centered at q^0, on the probability simplex Q. (This figure originally appeared in Friedman and Sandow (2003a).)

10.1.4.5   Choosing a Measure on the Efficient Frontier

According to the above paradigm, the best candidate model lies on a one-parameter efficient frontier. In order to choose the best candidate model from this one-parameter family, we make the following assumption.

Assumption 10.5 *The investor chooses α so as to maximize his expected utility on an out-of-sample dataset.*

Thus, given a compatible utility function, U, and odds ratios, $\mathcal{O}$, as well as a prior belief, q^0, and Assumptions 4.1 to 10.5, we are led to a method for finding a probability measure, which we state in the following principle:

Principle 10.4 *(Data Consistency Tuning Principle) Given a family of data constraint sets indexed by the parameter α, we tune the level of data consistency with respect to out-of-sample performance by selecting*

$$q^{**} = q^*(\alpha^*) \,,$$
$$\text{with } \alpha^* = \arg\max_{\alpha} E_{\tilde{p}}[U(b^*(q^*(\alpha)), \mathcal{O})] \,,$$

where $\tilde{p}$ is the empirical measure of the test set and $q^*(\alpha)$ is the minimum relative $(U, \mathcal{O})$-entropy measure associated with the parameter α.

Thus, according to Principle 10.4, the relative importance of the data and the prior is determined by the out-of-sample performance (expected utility) of the model. Principle 10.4 affords a mechanism to tune the regularization of the model (produced according to the minimum relative $(U, \mathcal{O})$-entropy. Principle 10.3 or, equivalently, Principle 10.1 or 10.2) in the hope that the model will generalize well.[21]

10.1.5 Dual Problem

We have shown in Section 10.1.4 that, in order to find the Pareto optimal model, q^*, for a given α, we have to solve Problem 10.2. As we have seen, this problem is strictly convex. We know from Section 2.2, that we can formulate a corresponding dual problem.

We show in Section 10.5.2 that the dual of Problem 10.2 can be formulated as:

---

[21] If $\alpha^* = 0$, no regularization is necessary, which suggests that it is possible that the collection of features should be enriched so that the model can learn more from the data, i.e., it might be worthwhile to add additional features.

Problem 10.3 *(More Easily Interpreted Version of Dual Problem, Given α)*

$$Find \ \beta^* = \arg\max_\beta h(\beta) \tag{10.41}$$

$$with \quad h(\beta) = \sum_y \tilde{p}_y U(b_y^*(q^*)\mathcal{O}_y)$$

$$- \inf_{\xi \geq 0} \left\{ \xi \Psi^* \left(\frac{\beta}{\xi} \right) + \alpha \xi \right\}, \tag{10.42}$$

$$where \ b_y^*(q^*) = \frac{1}{\mathcal{O}_y}(U')^{-1} \left(\frac{\lambda^*}{q_y^* \mathcal{O}_y} \right) \tag{10.43}$$

$$and \quad q_y^* = \frac{\lambda^*}{\mathcal{O}_y U'\left(U^{-1}(G_y(q^0, \beta, \mu^*))\right)}, \tag{10.44}$$

$$with \quad G_y(q^0, \beta, \mu^*) = U(b_y^*(q^0)\mathcal{O}_y) + \beta^T f(y) - \mu^*, \tag{10.45}$$

$$where \quad \mu^* \ solves \quad 1 = \sum_y \frac{1}{\mathcal{O}_y} U^{-1}\left(G_y(q^0, \beta, \mu^*)\right), \tag{10.46}$$

and

$$\lambda^* = \left\{ \sum_y \frac{1}{\mathcal{O}_y U'\left(U^{-1}(G_y(q^0, \beta, \mu^*))\right)} \right\}^{-1}. \tag{10.47}$$

Here, the β terms arise as Lagrange multipliers (see Section 10.5.2 for details) and (10.44) is the connecting equation.

We also show in Section 10.5.2 that an alternative formulation of the dual problem is the following:

Problem 10.4 *(More Easily Implemented Version of Dual Problem, Given α)*

$$Find \ \beta^* = \arg\max_\beta h(\beta) \tag{10.48}$$

$$with \quad h(\beta) = \beta^T E_{\tilde{p}}[f] - \mu^*$$

$$- \inf_{\xi \geq 0} \left\{ \xi \Psi^* \left(\frac{\beta}{\xi} \right) + \alpha \xi \right\}, \tag{10.49}$$

$$where \quad \mu^* \ solves \quad 1 = \sum_y \frac{1}{\mathcal{O}_y} U^{-1}\left(G_y(q^0, \beta, \mu^*)\right) \tag{10.50}$$

$$with \quad G_y(q^0, \beta, \mu^*) = U(b_y^*(q^0)\mathcal{O}_y) + \beta^T f(y) - \mu^*. \tag{10.51}$$

The optimal probability distribution is then

$$q_y^* = \frac{\lambda^*}{\mathcal{O}_y U'\left(U^{-1}(G_y(q^0, \beta^*, \mu^*))\right)}, \tag{10.52}$$

$$with \quad \lambda^* = \left\{ \sum_y \frac{1}{\mathcal{O}_y U'\left(U^{-1}(G_y(q^0, \beta^*, \mu^*))\right)} \right\}^{-1}. \tag{10.53}$$

Next, we examine some important special cases for the primal relaxation term and associated dual relaxations.

1. $\Psi(c) = \frac{1}{\omega}\ell_\omega^\omega(c)$, $1 < \omega < \infty$. Recall that by Corollary 2.1, if the primal relaxation term is given by

$$\Psi(c) = \frac{1}{\omega}\ell_\omega^\omega(c)\,,\ 1 < \omega < \infty\,, \tag{10.54}$$

then the dual relaxation term

$$\inf_{\xi \geq 0}\left\{\xi\Psi^*\left(\frac{\beta}{\xi}\right) + \alpha\xi\right\} \tag{10.55}$$

is given by

$$\alpha^{\frac{1}{\omega}}\kappa^{\frac{1}{\omega}}(\kappa - 1)^{-\frac{1}{\omega}}\ell_\kappa(\beta)\,, \tag{10.56}$$

where $\ell_\kappa(\beta) = \left(\sum_{j=1}^J |\beta_j|^\kappa\right)^{\frac{1}{\kappa}}$ denotes the ℓ_κ-norm, and $\kappa = \frac{\omega}{\omega-1}$.

Fortunately, when considering the α-parameterized family of solutions, it is possible to recast this dual relaxation term in a less cumbersome equivalent form, as indicated by Corollary 2.2 of Section 2.2.

2. $\Psi(c) = \ell_\infty(c)$. Also, by Corollary 2.1 of Section 2.2, if the primal relaxation term is given by

$$\Psi(c) = \ell_\infty(c)\,, \tag{10.57}$$

then the dual relaxation term

$$\inf_{\xi \geq 0}\left\{\xi\Psi^*\left(\frac{\beta}{\xi}\right) + \alpha\xi\right\} \tag{10.58}$$

is given by

$$\alpha\ell_1(\beta)\,, \tag{10.59}$$

where $\ell_1 = \sum_{j=1}^J |\beta_j|$ denotes the ℓ_1-norm.

3. Mahalanobis metric. The dual relaxation term for the case where $\Psi(c)$ is given by the Mahalanobis metric

$$Nc^T\Sigma^{-1}c \tag{10.60}$$

can be obtained from the $\frac{1}{2}\ell_2^2$ primal regularization by the transforming coordinates. We obtain the dual directly:

$$\Psi^*(y) = \sup_c \left\{c^T y - Nc^T\Sigma^{-1}c\right\}. \tag{10.61}$$

This is an unconstrained concave maximization problem. Let

$$c^* = \arg\sup_c \left\{c^T y - Nc^T\Sigma^{-1}c\right\}. \tag{10.62}$$

Putting the gradient with respect to c to zero, we obtain

$$y - 2N\Sigma^{-1}c^* = 0, \tag{10.63}$$

so

$$c^* = \frac{1}{2N}\Sigma y \tag{10.64}$$

and

$$\Psi^*(y) = \frac{1}{4N}y^T\Sigma y. \tag{10.65}$$

Applying Corollary 2.1, we see that the dual relaxation term for the Mahalanobis relaxation is given by

$$\inf_\xi \frac{1}{4N\xi}\beta^T\Sigma\beta + \alpha\xi. \tag{10.66}$$

This is an unconstrained convex optimization problem. Putting the derivative with respect to ξ to zero, we obtain

$$\xi^* = \sqrt{\frac{1}{4\alpha N}\beta^T\Sigma\beta}. \tag{10.67}$$

Substituting into (10.66), we obtain the dual relaxation term

$$\sqrt{\frac{\alpha}{N}\beta^T\Sigma\beta}. \tag{10.68}$$

We note that for the Mahalanobis metric, we can obtain the same α-parameterized family of solutions to Problems 10.3 and 10.4, if we allow α to vary over $[0,\infty)$, by dropping the square roots in the above relaxation term; this follows from the material in Section 2.2.9.

We state the following theorem:

Theorem 10.3 *Problems 10.2, 10.3, and 10.4 have the same unique solution, q_y^*.*

Proof: see Section 10.5.2.

Problems 10.3 and 10.4 are equivalent. Problem 10.4 is easier to implement and Problem 10.3 is easier to interpret. The first term in the objective function of Problem 10.3 is the utility (of the utility maximizing investor) averaged over the training sample. Thus, our dual problem is a regularized maximization of the training sample-averaged utility, where the utility function, U, is the utility function on which the relative $(U,\mathcal{O})$-entropy $D_{U,\mathcal{O}}(q||q^0)$ depends.

We know from Section 2.2 that the dual problems, Problems 10.3 and 10.4, are J-dimensional (J is the number of features), unconstrained, concave maximization problems that can be recast in the form of standard convex minimization problems. The primal problem, Problem 10.2, on the other hand, is an m-dimensional (m is the number of states) convex minimization with convex constraints. The dual problem, Problem 10.4, may be easier to solve than the primal problem, Problem 10.2, if $m > J$. In the more general context that we shall discuss in Section 10.2, the dual problem will always be easier to solve than the primal problem.

10.1.5.1   Example: The Generalized Logarithmic Family of Utilities

We consider a utility of the form

$$U(z) = \gamma_1 log(z - \gamma B) + \gamma_2 , \tag{10.69}$$

where $\gamma < 1$ to ensure compatibility with the horse race and $\gamma_1 > 0$ to ensure the proper monotonicity for the utility function.

In Section 10.5.3, we show that the dual problem is given by:

Problem 10.5 (*Dual Problem for the Generalized Logarithmic Family of Utilities*)

$$Find \; \beta^* = \arg\,\max_{\beta} h(\beta)$$

$$with \quad h(\beta) = \sum_y \tilde{p}_y \log q_y^* - \inf_{\xi \geq 0} \left\{ \xi \Psi^* \left(\frac{\beta}{\xi} \right) + \alpha \xi \right\},$$

$$where \quad q_y^* = \frac{1}{\sum_y q_y^0 e^{\beta^T f(y)}} \, q_y^0 \, e^{\beta^T f(y)} .$$

This problem is equivalent to a regularized maximum-likelihood search, which is independent of the odds ratios, $\mathcal{O}$; this is consistent with Section 10.1.6, where we show that the odds ratios drop out of the primal problem for this logarithmic family of utility functions.

10.1.5.2   Example: Power Utility

We consider a utility of the form

$$U(z) = \frac{z^{1-\kappa} - 1}{1 - \kappa} , \tag{10.70}$$

discussed in Section 4.4. In order to specify the dual problem for this utility, note that

$$U'(z) = z^{-\kappa} , \tag{10.71}$$

$$U^{-1}(z) = [1 + (1 - \kappa)z]^{\frac{1}{1-\kappa}} \tag{10.72}$$

$$\text{and } U'(U^{-1}(z)) = [1 + (1 - \kappa)z]^{\frac{-\kappa}{1-\kappa}} . \tag{10.73}$$

Recall from Section 5.1.6 that

$$b^*(q) = \frac{(q_y \mathcal{O}_y)^{\frac{1}{\kappa}}}{\mathcal{O}_y B(q, \mathcal{O})} \tag{10.74}$$

$$\text{with } B(q, \mathcal{O}) = \sum_y \frac{1}{\mathcal{O}_y} (q_y \mathcal{O}_y)^{\frac{1}{\kappa}} \;. \tag{10.75}$$

Using this equation, we can write $G_y(q^0, \beta, \mu^*)$ from (10.45) as

$$G_y(q^0, \beta, \mu^*) = \frac{1}{1-\kappa} \left[\left(\frac{(q_y^0 \mathcal{O}_y)^{\frac{1-\kappa}{\kappa}}}{(B(q^0, \mathcal{O}))^{1-\kappa}} \right) - 1 \right] + \beta^T f(y) - \mu^* \;. \tag{10.76}$$

Inserting (10.72) and (10.76) into (10.46) gives

$$1 = \sum_y \frac{1}{\mathcal{O}_y} \left[\frac{(q_y^0 \mathcal{O}_y)^{\frac{1-\kappa}{\kappa}}}{(B(q^0, \mathcal{O}))^{1-\kappa}} + (1-\kappa)[\beta^T f(y) - \mu^*] \right]^{\frac{1}{1-\kappa}} \;, \tag{10.77}$$

which is our condition for μ^*. Next, we specify the condition (10.47) for λ^*. We use (10.73) and (10.76) to write (10.47) as

$$\lambda^* = \left\{ \sum_y \frac{1}{\mathcal{O}_y} \left[\frac{(q_y^0 \mathcal{O}_y)^{\frac{1-\kappa}{\kappa}}}{(B(q^0, \mathcal{O}))^{1-\kappa}} + (1-\kappa)[\beta^T f(y) - \mu^*] \right]^{\frac{\kappa}{1-\kappa}} \right\}^{-1} \;. \tag{10.78}$$

By means of (10.44), (10.73), and (10.76) we obtain for the optimal probability distribution

$$q_y^* = \frac{1}{\mathcal{O}_y} \left[\frac{(q_y^0 \mathcal{O}_y)^{\frac{1-\kappa}{\kappa}}}{(B(q^0, \mathcal{O}))^{1-\kappa}} + (1-\kappa)[\beta^T f(y) - \mu^*] \right]^{\frac{\kappa}{1-\kappa}} \;. \tag{10.79}$$

We note from this connecting equation that q_y^* is reminiscent of the Student t-distribution. Collecting (10.77), (10.75), (10.78), and (10.79), we obtain:

Problem 10.6 *(Dual Problem for Power Utility)*

$$\text{Find } \beta^* = \arg \max_{\beta} h(\beta)$$

$$\text{with } h(\beta) = \beta^T E_{\tilde{p}}[f] - \mu^* - \inf_{\xi \geq 0} \left\{ \xi \Psi^* \left(\frac{\beta}{\xi} \right) + \alpha \xi \right\} \tag{10.80}$$

$$\text{where } \mu^* \text{ solves } 1 = \sum_y \frac{1}{\mathcal{O}_y} \left[\frac{(q_y^0 \mathcal{O}_y)^{\frac{1-\kappa}{\kappa}}}{(B(q^0, \mathcal{O}))^{1-\kappa}} + (1-\kappa)[\beta^T f(y) - \mu^*] \right]^{\frac{1}{1-\kappa}}$$

$$\text{with } B(q^0, \mathcal{O}) = \sum_y \frac{1}{\mathcal{O}_y} (q_y^0 \mathcal{O}_y)^{\frac{1}{\kappa}} \;.$$

The optimal probability distribution is then

$$q_y^* = \frac{\lambda^*}{\mathcal{O}_y} \left[\frac{(q_y^0 \mathcal{O}_y)^{\frac{1-\kappa}{\kappa}}}{(B(q^0, \mathcal{O}))^{1-\kappa}} + (1-\kappa)[\beta^{*T} f(y) - \mu^*] \right]^{\frac{\kappa}{1-\kappa}}$$

$$\text{with} \quad \lambda^* = \left\{ \sum_y \frac{1}{\mathcal{O}_y} \left[\frac{(q_y^0 \mathcal{O}_y)^{\frac{1-\kappa}{\kappa}}}{(B(q^0, \mathcal{O}))^{1-\kappa}} + (1-\kappa)[\beta^{*T} f(y) - \mu^*] \right]^{\frac{\kappa}{1-\kappa}} \right\}^{-1} .$$

10.1.5.3   Summary of Dual Problem Solution Method (Suitable for Numerical Implementation)

The modeling approach described in Sections 10.1.4 and 10.1.5 was formulated in terms of an investor who selects a Pareto optimal model, i.e., a model on an efficient frontier, which we have defined in terms of consistency with the training data and consistency with a prior distribution. We measured the former by means of the large-sample distribution of a vector of sample-averaged features, and the latter by means of a relative $(U, \mathcal{O})$-entropy. We have seen that the measures on the efficient frontier form a family which is parameterized by the single parameter $\alpha \in (0, \alpha_{max})$, and that, for a given α, the Pareto optimal measure is the unique solution of Problem 10.2, which is a strictly convex optimization problem. For a given α, the Pareto optimal measure can be found by solving the dual (concave maximization) problem in the form of Problem 10.16 or in the form of Problem 10.4. Solving this dual problem amounts to a regularized expected utility maximization (over the training sample) over a certain family of measures; for many practical examples, solving the dual problem can be easier than solving the primal problem. Having thus computed an α-parameterized family of Pareto optimal measures, we pick the measure with highest expected utility on a hold-out sample.

We note that the procedure to select α is, by virtue of the fact that α is one-dimensional, both tractable and barely susceptible to overfitting on the hold-out sample set.

The approach that we have discussed boils down to the following procedure:

1. Break the data into a training set and a hold-out sample, for example, taking 75% or 80% of the data, selected randomly, to train the model.

2. Choose a discrete set $A = \{\alpha_k \in (0, \alpha_{max}), k = 1, \ldots, K\}$.

3. For $k = 1, \ldots, K$,

   - Solve Problem 10.4 for $\beta^*(\alpha_k)$, based on the training set,

   - Compute the out-of-sample performance $P_k = E_{\tilde{p}}[U(b^*(q^*(\alpha_k)), \mathcal{O})]$ on the out-of-sample test set, where $\tilde{p}$ is the empirical measure on this test set, and q^* is determined from (10.52) with parameters $\beta^*(\alpha_k)$, and b^* is determined from (10.3).

4. Put $k^* = \arg\max_k P_k$.

5. Our model, q^{**}, is determined from (10.52) with parameters $\beta^*(\alpha_{k^*})$.

A refinement: repeat on different data partitions in order to find an optimal α^{**}, then retrain the model using α^{**} on the complete set of available training data.

10.1.6 Utilities Admitting Odds Ratio Independent Problems: A Logarithmic Family

In the setting of this chapter, model builders who use probabilistic models make decisions (bets) which result in well defined benefits or ill effects (payoffs) in the presence of risk. In principle, the payoffs associated with the various outcomes can be assigned precise values; in practice, it may be difficult to assign such values. Outside the financial modeling context, for example, there may be no "market makers" who set odds ratios. Even in the financial modeling context, the data for the payoffs (or, equivalently, market prices or odds ratios) may not exist or be of poor quality. In this context, given market prices on instruments which have nonzero payoffs for more than one state, we would need a complete market in order to calculate the odds ratios (see, for example, Duffie (1996), for a definition of complete markets). In the absence of high-quality data, one might consider modeling the odds ratios, but that introduces additional complexity; moreover, the resulting model, under a general utility function, will be sensitive to the odds ratio model.

For these reasons, we seek the most general family of utility functions for which our problem formulation is independent of the odds ratios. We recall Theorem 7.1, restated below, which specifies this family.

Theorem 7.1 *The relative $(U, \mathcal{O})$-entropy, $D_{U,\mathcal{O}}(q\|q^0)$, is independent of the odds ratios, $\mathcal{O}$, for any candidate model q and prior measure, q^0, if and only if the utility function, U, is a member of the logarithmic family*

$$U(W) = \gamma_1 \log(W - \gamma B) + \gamma_2 \ , \tag{10.81}$$

where $\gamma_1 > 0$ and $\gamma < 1$. In this case,

$$D_{U,\mathcal{O}}(q\|q^0) = \gamma_1 E_q \left[\log \left(\frac{q}{q^0} \right) \right] \ . \tag{10.82}$$

From this theorem and Problem 10.2, we obtain

Corollary 10.1 *For utility functions of the form (10.81), Problem 10.2 reduces to Problem 10.7.*

Problem 10.7 *(Initial Strictly Convex Problem for U in the Generalized Logarithmic Family, Given $\alpha, 0 \leq \alpha \leq \alpha_{max}$)*

$$Find \quad \arg\min_{q \in (R^+)^m, c \in R^J} \gamma_1 E_q \left[\log \left(\frac{q}{q^0} \right) \right]$$

$$under\ the\ constraints\ 1 = \sum_y q_y$$

$$and \quad \psi(c) = \alpha \tag{10.83}$$

$$where \quad c_j = E_q[f_j] - E_{\tilde{p}}[f_j] \ . \tag{10.84}$$

We have already explicitly derived the dual problem for utility functions of the form (10.81) in Section 10.1.5.1. We note that this is the same problem that we encountered in Section 9.4.

10.1.7 A Summary Diagram

We display some of the relationships discussed above, in the horse race setting of this chapter, in Figure 10.3.[22]

10.2 Conditional Density Models

In this section we briefly discuss the above approach in the context of a conditional density model which may include point masses, i.e., for the case where the random variable Y has the continuous conditional probability density $q(y|x)$ on the finite set $\mathcal{Y} \subset \mathbf{R}^n$ and the finite conditional point probabilities $q_{\rho|x}$ on the set of points $\{y_\rho \in R^n, \rho = 1, 2, ..., m\}$, where x denotes a value of the vector X of explanatory variables which can take any of the values $x_1, ..., x_M, x_i \in \mathbf{R}^d$.

We have only assumed that the explanatory variable vector can take a finite number of values (those that are observed in our training dataset). However, as we shall see, the models that we derive based on this assumption will allow us, potentially, to compute conditional probabilities for all $x \in \mathbf{R}^d$. This setting has interesting applications such as the modeling of recovery values of defaulted debt, which we shall discuss in Section 12.1.2.

As in Section 10.1, we make assumptions; we assume that

(*i*) an investor has a utility function that satisfies Assumption 4.1 stated in Section 4.6,

---

[22]We shall also consider settings more general than the horse race; see Section 11.2.

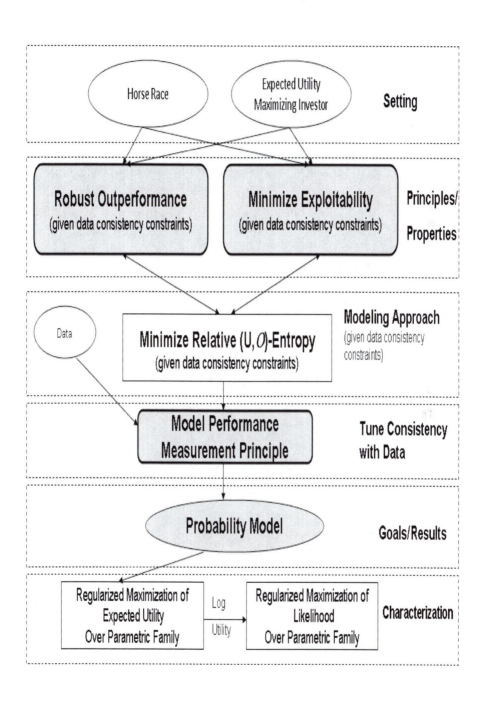

FIGURE 10.3:   Model estimation approach.

(*ii*) this investor operates in the conditional horse race setting described in Section 5.4,

(*iii*) for every value of x, the conditional horse race conditioned on x, the investor's utility function, and any conditional density $q(y|x)$ under consideration are compatible, in the sense of Definition 5.8 of Section 5.4. These compatibility conditions are imposed to ensure that the investor's optimal allocation is well defined.

(*iv*) the investor allocates his assets so as to maximize his expected utility according to his beliefs, i.e., the investor allocates so as to maximize the expectation of his utility under the model probability measure.

(*v*) the investor measures model performance as per Principle 8.1, in Chapter 8.

10.2.1   Preliminaries

In order to generalize the results and definitions from Section 10.1, we let $\tilde{p}_x$ denote the empirical probability of the vector, X, of explanatory variables, and define the following conditional probability measures:

Definition 10.4

$$\tilde{p} = \{(\tilde{p}(y|x), \tilde{p}_{\rho|x}), y \in \mathcal{Y}, \rho = 1, 2, ..., m, x = x_1, ..., x_M\}$$
$$= \textit{empirical conditional probability measure}$$
$$q = \{(q(y|x), q_{\rho|x}), y \in \mathcal{Y}, \rho = 1, 2, ..., m, x = x_1, ..., x_M\}$$
$$= \textit{model conditional probability measure}$$

Assumption 10.6 (*$p_x = \tilde{p}_x$ and some consequences*) *We assume that the following relations between conditional and joint probabilities hold:*

$$q_x = \tilde{p}_x,$$
$$q(y, x) = \tilde{p}_x q(y|x), \textit{ and}$$
$$q_{\rho,x} = \tilde{p}_x q_{\rho|x}.$$

Next, we identify the probabilistic problem with the conditional horse race of Section 5.4, and consider an investor who places bets. We assume that our investor has allocation density $b(y|x)$ and allocates $b_{\rho|x}$ to the event $Y = y_\rho$, if $X = x$ was observed, where

$$1 = \int_{\mathcal{Y}} b(y|x)dy + \sum_{\rho=1}^{m} b_{\rho|x} . \tag{10.85}$$

This means that, assuming that q, U, and $\mathcal{O}$ are compatible, such an investor who believes the model q will allocate according to

$$b^*[q] = \arg\max_{\{b \in \mathcal{B}\}} \left[\int_{\mathcal{Y}} q(y|x) U(b(y|x)\mathcal{O}(y|x)) dy + \sum_y q_{\rho|x} U(b_{\rho|x}\mathcal{O}_{\rho|x}) \right],$$

$$(10.86)$$

where

$$\mathcal{B} = \{(b(y|x), b_{\rho|x}) : \int_{\mathcal{Y}} b(y|x) dy + \sum_{\rho=1}^m b_{\rho|x} = 1\}$$

denotes the set of betting weights consistent with (10.85). From Lemma 5.4, the optimal betting density and betting weights are given by

$$b^*[q](y|x) = \frac{1}{\mathcal{O}(y|x)} (U')^{-1} \left(\frac{\lambda_x^*}{q(y|x)\mathcal{O}(y|x)} \right), \qquad (10.87)$$

$$b^*_{\rho|x}[q] = \frac{1}{\mathcal{O}_{\rho|x}} (U')^{-1} \left(\frac{\lambda_x^*}{q_{\rho|x}\mathcal{O}_{\rho|x}} \right), \qquad (10.88)$$

respectively, where λ_x^*, is the solution (which is known to exist) of the following equation:

$$1 = \int_{\mathcal{Y}} \frac{1}{\mathcal{O}(y|x)} (U')^{-1} \left(\frac{\lambda_x^*}{q(y|x)\mathcal{O}(y|x)} \right) dy + \sum_\rho \frac{1}{\mathcal{O}_{\rho|x}} (U')^{-1} \left(\frac{\lambda_x^*}{q_{\rho|x}\mathcal{O}_{\rho|x}} \right).$$

$$(10.89)$$

Recall the definition of relative $(U, \mathcal{O})$-entropy from the discrete setup of Section 7.2,

$$D_{U,\mathcal{O}}(q||q^0) = E_q[U(b^*[q], \mathcal{O})] - E_q[U(b^*[q^0], \mathcal{O})]. \qquad (10.90)$$

We note that this definition, which is expressed in terms of expectations, utility, and allocation to a horse race, can be easily generalized to our current setting. In our setting, we have

$$E_{q^1}[U(b^*[q^2], \mathcal{O})] = \sum_x \tilde{p}_x \int_{\mathcal{Y}} q^1(y|x) U(b^*[q^2](y|x)\mathcal{O}(y|x)) dy$$

$$+ \sum_{x,\rho} \tilde{p}_x q^1_{\rho|x} U(b^*_{\rho|x}[q^2]\mathcal{O}_{\rho|x}). \qquad (10.91)$$

This suggests the following definition for the conditional relative $(U, \mathcal{O})$-entropy for (conditional) probability densities with point masses.

Definition 10.5 *Given a utility function, U, and a compatible system of market prices, $\mathcal{O}$, and probability measures, q and q^0, the conditional relative $(U, \mathcal{O})$-entropy is given by:*

$$D_{U,\mathcal{O}}(q||q^0) = E_q[U(b^*[q], \mathcal{O})] - E_q[U(b^*[q^0], \mathcal{O})], \qquad (10.92)$$

where the optimal allocations $b^*(q)$ and $b^*(q^0)$ are with respect to the utility function, U, and where the expectation of a function $g_x(y)$ is defined as

$$E_q[g] = \sum_x \tilde{p}_x E_q[g|x] \qquad (10.93)$$

with $E_q[g|x] = \left\{ \int_y q(y|x)g_x(y)dy + \sum_\rho q_{\rho|x}g_x(y) \right\}$.

10.2.2 Modeling Approach

In this section, we generalize the modeling approach from Section 10.1.4 to the case of a conditional probability density with point masses. To this end, let us define the spaces

$$\mathcal{Q} = \{(q(y|x), q_{\rho|x})\} , \qquad (10.94)$$
$$\text{and } \mathcal{Q}^+ = \{q : q \in \mathcal{Q}, q(y|x) \geq 0, q_{\rho|x} \geq 0\}.$$

We further assume that Assumptions 10.2-10.6 hold; that is, we assume that the investor

(*i*) measures the consistency of the model with the data in terms of a convex function of the difference between expectations of the feature vector under the model and the empirical measures,

(*ii*) measures the consistency of the model with the prior via the conditional relative $(U, \mathcal{O})$-entropy,

(*iii*) selects a measure on the efficient frontier, and

(*iv*) selects the measure on the efficient frontier that maximizes his expected utility on an out-of-sample test set.

This leads to

Problem 10.8 (*Minimum Relative $(U, \mathcal{O})$-Entropy Problem: Conditional, Given* α)

$$\textit{Find} \quad \arg\min_{q \in \mathcal{Q}^+, c \in \mathbf{R}^J} D_{U,\mathcal{O}}(q||q^0) \qquad (10.95)$$
$$\textit{under the constraints } 1 = E_q[1|x] \qquad (10.96)$$
$$\textit{and } \psi(c) \leq \alpha \qquad (10.97)$$
$$\textit{where } c_j = E_q[f_j] - E_{\tilde{p}}[f_j] . \qquad (10.98)$$

Here, as in the context of discrete probabilities, f_j denotes a feature; there are J features, each of which is a real-valued function of x and y.

According to Assumption 10.5, i.e., by applying the data consistency tuning principle, Principle 10.4, our investor will choose the measure that maximizes his expected utility among the measures that are the family (parameterized by α) of solutions to Problem 10.8.

We note that we haven't proved that the solution to Problem 10.8 exists. From a practical point of view, we can solve Problem 10.8 and then check the solution for compatibility (in the sense of Definition 5.8).

We now argue heuristically[23] that the minimum conditional relative $(U, \mathcal{O})$-entropy problem, Problem 10.8, is equivalent to the minimum market exploitability principle. To see this, we shall cast the definition of conditional relative entropy, Definition 10.5, as a maximization over measures, rather than an expectation over optimal allocations.

To do this, we first need to show that Definition 10.5 can be cast as a maximization over allocations. Note that the quantities $b^*[q]$ in Definition 10.5, which are given by (10.87) and (10.88), result from maximizations of expected utility, given the particular individual value x. On the other hand, the maximization

$$\tilde{b}[q] = \max_{(b(y|x),b_{\rho|x})\in\mathcal{B}} \left(E_q[U(b,\mathcal{O})] - E_q[U(b^*[q^0],\mathcal{O})]\right) \qquad (10.99)$$

is (see the definition of E_q in (10.93)) a maximization over weighted sums of terms involving concave increasing functions of $b(y|x)$ or $b_{\rho|x}$, where $(b(y|x), b_{\rho|x}) \in \mathcal{B}$. This difference notwithstanding, we must have $\tilde{b}[q] = b^*[q]$. To see this, note that the right hand side of the preceding equation consists of finite nonnegatively-weighted sums of terms of the form of the objective function in (10.86); such sums can be maximized by maximizing the individual terms that comprise the sum. The solution that results coincides with that given by (10.87) and (10.88).

Therefore we can write

$$D_{U,\mathcal{O}}(q||q^0) = \max_{(b(y|x),b_{\rho|x})\in\mathcal{B}} \left(E_q[U(b,\mathcal{O})] - E_q[U(b^*[q^0],\mathcal{O})]\right). \qquad (10.100)$$

From (10.87) and (10.88), for every b there is a q and conversely; thus, the conditional relative entropy can also be viewed as a maximization over measures, rather than allocations; that is,

$$D_{U,\mathcal{O}}(q||q^0) = \max_{(q(y|x),q_{\rho|x})\in\mathcal{Q}^+} \left(E_q[U(b^*[q],\mathcal{O})] - E_q[U(b^*[q^0],\mathcal{O})]\right). \qquad (10.101)$$

Thus, the minimum (conditional) relative entropy problem, Problem 10.8, can be written as the minimum market exploitability problem

$$\text{Find} \quad \arg\min_{q\in K} \max_{(q(y|x),q_{\rho|x})\in\mathcal{Q}^+} \left(E_q[U(b^*[q],\mathcal{O})] - E_q[U(b^*[q^0],\mathcal{O})]\right)$$

---

[23] For a more rigorous argument, see Grünwald and Dawid (2004).

where we have used K to denote the subset of $\mathcal{Q}^+$ that satisfies the constraints (10.96) to (10.98).

Rigorously proving that the Pareto-optimal, i.e., conditional relative $(U, \mathcal{O})$-entropy minimizing, measure is robust is a considerably more difficult undertaking than proving the analogous statement in the discrete setting. We refer the reader to Grünwald and Dawid (2004) for a discussion. For our purposes, in the continuous market setting, we take the minimum market exploitability principle as our starting point.

10.2.3   Dual Problem

Like Problem 10.2, Problem 10.8 has a dual. We indicate how this dual is obtained in Section 10.5.4.

Problem 10.9 *(Dual Problem: Conditional Probability Density, Given α)*

$$Find \quad \beta^* = \arg \max_{\beta} h(\beta) \tag{10.102}$$

$$with \quad h(\beta) = E_{\tilde{p}}[U(b^*[q^*], \mathcal{O})]$$

$$- \inf_{\xi \geq 0} \left\{ \xi \Psi^* \left(\frac{\beta}{\xi} \right) + \alpha \xi \right\} \tag{10.103}$$

$$where \quad b^*[q^*](y|x) = \frac{1}{\mathcal{O}(y|x)} (U')^{-1} \left(\frac{\lambda_x^*}{q^*(y|x)\mathcal{O}(y|x)} \right) , \tag{10.104}$$

$$b^*_{\rho|x}[q^*] = \frac{1}{\mathcal{O}_{\rho|x}} (U')^{-1} \left(\frac{\lambda_x^*}{q^*_{\rho|x}\mathcal{O}_{\rho|x}} \right) , \tag{10.105}$$

$$and \quad q^*(y|x) = \frac{\lambda_x^*}{\mathcal{O}(y|x)U'\left(U^{-1}(G(x, y, q^0, \beta, \mu_x^*))\right)} , \tag{10.106}$$

$$q^*_{\rho|x} = \frac{\lambda_x^*}{\mathcal{O}_{\rho|x}U'\left(U^{-1}(G_{\rho|x}(q^0, \beta, \mu_x^*))\right)} , \tag{10.107}$$

with

$$G(x, y, q^0, \beta, \mu_x^*) = U(b^*[q^0](y|x)\mathcal{O}(y|x)) + \beta^T f(x, y) - \mu_x^*, \tag{10.108}$$

$$G_{\rho|x}(q^0, \beta, \mu_x^*) = U(b^*_{\rho|x}[q^0]\mathcal{O}_{\rho|x}) + \beta^T f(x, y_\rho) - \mu_x^* , \tag{10.109}$$

$$where \quad \mu_x^* \ solves \quad 1 = \int_y \frac{1}{\mathcal{O}(y|x)} U^{-1}\left(G(x, y, q^0, \beta, \mu_x^*)\right) dy \tag{10.110}$$

$$+ \sum_\rho \frac{1}{\mathcal{O}_{\rho|x}} U^{-1}\left(G_{\rho|x}(q^0, \beta, \mu_x^*)\right) , \tag{10.111}$$

$$and \quad (\lambda_x^*)^{-1} = \int_y \frac{1}{\mathcal{O}(y|x)U'\left(U^{-1}(G(x, y, q^0, \beta, \mu_x^*))\right)} dy$$

$$+ \sum_\rho \frac{1}{\mathcal{O}_{\rho|x}U'\left(U^{-1}(G_{\rho|x}(q^0, \beta, \mu_x^*))\right)} . \tag{10.112}$$

This dual problem is a generalization of the dual problem for discrete probabilities, Problem 10.3. In general, it is easier to solve the dual problem than the primal problem, since the primal problem is an infinite dimensional optimization problem, and the dual problem is a J dimensional optimization problem, where J is the length of the feature function vector. Even so, the dual problem need not be well posed. For example, the integrals that must be evaluated depend on the prior measure, $q^0(y|x)$, the feature vector, $f(x,y)$, and the odds ratios, $\mathcal{O}(y|x)$. In order to solve this problem, these functions of y must be compatible in the sense that the integrals that involve them converge.

10.2.3.1   Example: The Generalized Logarithmic Family and Mahalanobis Distance

Because of its practical relevance, we state the above dual problem for the case of a generalized logarithmic utility and the Mahalanobis distance as ψ. We assume that

$$\psi(c) = N c^T \Sigma^{-1} c. \tag{10.113}$$

It is easy to see that, in this case, the (10.87)-(10.89) for the optimal betting weights give:

$$b^*_{\rho|x}[q] = q_{\rho|x}\left[1 - \gamma B \sum_{\rho'} \frac{1}{\mathcal{O}_{x,\rho'}}\right] + \frac{\gamma B}{\mathcal{O}_{\rho|x}} \tag{10.114}$$

$$b^*[q](y|x) = q(y|x)\left[1 - \gamma B \sum_{y'} \frac{1}{\mathcal{O}(x,y')}\right] + \frac{\gamma B}{\mathcal{O}(y|x)}. \tag{10.115}$$

The conditional relative $(U, \mathcal{O})$-entropy, which enters Problem 10.8, is then

$$D_{U,\mathcal{O}}(q\|q^0) = \gamma_1 E_q\left[\log\left(\frac{q}{q^0}\right)\right]. \tag{10.116}$$

Inserting (10.69), (10.114), and (10.115) into Problem 10.9, with $\psi(c)$ put to the Mahalanobis distance, we derive the dual problem as:

Problem 10.10 *(Dual Problem for Probability Densities and the Generalized Logarithmic Family of Utilities)*

$$\text{Find } \beta^* = \arg\max_{\beta} h(\beta)$$

$$\text{with} \quad h(\beta) = \frac{\gamma_1}{N}\sum_i \log q^{(\beta)}(y_i|x_i) - \sqrt{\frac{\alpha}{N}\beta^T \Sigma \beta}, \tag{10.117}$$

$$where \quad q^{(\beta)}(y|x) = Z_x^{-1} e^{\beta^T f(x,y)} \times \begin{cases} q^0_{\rho|x} & if\ y = y_\rho\ for\ some\ \rho \\ q^0(y|x) & otherwise, \end{cases}$$

$$and\ Z_x = \int_{\mathcal{Y}} q^0(y|x) e^{\beta^T f(x,y)} dy + \sum_{\rho} q^0_{\rho|x} e^{\beta^T f(x,y_\rho)},$$

where the (x_i, y_i) are the observed values and N is the number of observations. The measure on the efficient frontier is then

$$q^* = \{(q^*(y|x), q^*_{\rho|x}), y \in \mathcal{Y}, \rho = 1, 2, ..., m, x = x_1, ..., x_M\}$$

with $q^*(y|x) = q^{(\beta^*)}(y|x)$

and $q^*_{\rho|x} = q^{(\beta^*)}(y_\rho|x)$.

Recall (from Section 2.2.9) that we can obtain the same α-parameterized family of solutions to Problem 10.10, if we allow α to vary over $[0, \infty)$, by dropping the square root in (10.117).

10.2.3.2   Example: Conditional MRE in the Discrete Setting

We note that in the special case where $U(W)$ is in the generalized logarithmic family (10.69), $\mathcal{Y} = \emptyset$, and $\alpha = 0$, the minimum conditional relative $(U, \mathcal{O})$-entropy problem becomes:

Problem 10.11 *(MRE When Seeking a Conditional Discrete Distribution)*
Find

$$q^*_{\rho|x} = \arg \min_{q_{\rho|x} \geq 0, \sum_{\rho} q_{\rho|x} = 1} D(q\|q^0), \tag{10.118}$$

subject to

$$E_q[f] = E_{\tilde{p}}[f], \tag{10.119}$$

where the conditional relative entropy, $D(q\|q^0)$, is given by

$$D(q\|q^0) = \sum_x \tilde{p}_x \sum_{\rho} q_{\rho|x} \log \frac{q_{\rho|x}}{q^0_{\rho|x}}, \tag{10.120}$$

$$E_q[f] = \sum_x \tilde{p}_x \sum_{\rho} q_{\rho|x} f(x, \rho), \tag{10.121}$$

and

$$E_{\tilde{p}}[f] = \sum_x \tilde{p}_x \sum_{\rho} \tilde{p}_{\rho|x} f(x, \rho). \tag{10.122}$$

This problem has the associated dual problem:

Problem 10.12 *(Dual Problem Associated with MRE When Seeking a Conditional Discrete Distribution)*
Find

$$\beta^* = \arg \max_{\beta} \sum_{x,\rho} \tilde{p}_x \tilde{p}_{\rho|x} \log q^{(\beta)}_{\rho|x}, \tag{10.123}$$

where

$$q_{\rho|x}^{(\beta)} = \frac{e^{\beta^T f(x,\rho)}}{Z(\beta, x)} q_{\rho|x}^0, \tag{10.124}$$

and

$$Z(\beta, x) = \sum_{\rho} e^{\beta^T f(x,\rho)} q_{\rho|x}^0. \tag{10.125}$$

The solution of Problem 10.11 is then given by

$$q_{\rho|x}^* = q_{\rho|x}^{(\beta^*)}, \tag{10.126}$$

where β^* is the solution of Problem 10.12.

10.2.3.3 Example: Conditional ME in the Discrete Setting

We note that in the special case where $U(W)$ is in the generalized logarithmic family (10.69), $\mathcal{Y} = \emptyset$, the prior is flat, and $\alpha = 0$, the minimum conditional relative $(U, \mathcal{O})$-entropy problem becomes:

Problem 10.13 *(ME When Seeking a Conditional Distribution)*
Find

$$q_{\rho|x}^* = \arg \max_{q_{\rho|x} \geq 0, \sum_{\rho} q_{\rho|x}=1} H(q), \tag{10.127}$$

subject to

$$E_q[f] = E_{\tilde{p}}[f], \tag{10.128}$$

where

$$H(q) = -\sum_x \tilde{p}_x \sum_{\rho} q_{\rho|x} \log q_{\rho|x}, \tag{10.129}$$

$$E_q[f] = \sum_x \tilde{p}_x \sum_{\rho} q_{\rho|x} f(x, \rho), \tag{10.130}$$

and

$$E_{\tilde{p}}[f] = \sum_x \tilde{p}_x \sum_{\rho} \tilde{p}_{\rho|x} f(x, \rho). \tag{10.131}$$

This problem has the associated dual problem:

Problem 10.14 *(Dual Problem Associated with ME When Seeking a Conditional Discrete Distribution)*
Find

$$\beta^* = \arg \max_{\beta} \sum_{x,\rho} \tilde{p}_x \tilde{p}_{\rho|x} \log q_{\rho|x}^{(\beta)}, \tag{10.132}$$

where

$$q_{\rho|x}^{(\beta)} = \frac{e^{\beta^T f(x,\rho)}}{Z(\beta, x)}, \tag{10.133}$$

and

$$Z(\beta, x) = \sum_{\rho} e^{\beta^T f(x,\rho)}. \tag{10.134}$$

The solution of Problem 10.13 is then given by

$$q^*_{\rho|x} = q^{(\beta^*)}_{\rho|x},$$

$$(10.135)$$

where β^* is the solution of Problem 10.14.

10.2.3.4   Example: Conditional MMI in the Discrete Setting

Recall (from Definition 2.10) the definition of mutual information:

$$I(X;\rho) = \sum_{x,\rho} q_{x,\rho} log \frac{q_{x,\rho}}{q_x q_\rho},$$

$$(10.136)$$

which can be rewritten as

$$I(X;\rho) = \sum_x q_x \sum_\rho q_{\rho|x} log \frac{q_{\rho|x} q_x}{q_x q_\rho} = \sum_x q_x \sum_\rho q_{\rho|x} log \frac{q_{\rho|x}}{q_\rho},$$

$$(10.137)$$

So, under Assumption 10.6,[24] the mutual information can be written as

$$I(X;\rho) = \sum_x \tilde{p}_x \sum_\rho q_{\rho|x} log \frac{q_{\rho|x}}{q_\rho}.$$

$$(10.138)$$

and the MMI principle is a special case of the MRE principle where the prior distribution is given by q_ρ.

10.2.3.5   Example: Logistic Regression

We note that in the special case where $U(W)$ is in the generalized logarithmic family (10.69), $\mathcal{Y} = \emptyset$, $m = 2$, the prior is flat, and $\alpha = 0$, and the jth feature, f_j, is given by

$$f_j(x,y) = \left(y - \frac{1}{2}\right) x_j, j = 1, \ldots, J,$$

$$(10.139)$$

the dual problem, Problem 10.10, has a solution of the form

$$q(\rho = 1|x) = \frac{1}{1 + e^{-\beta^T x}}.$$

$$(10.140)$$

The optimal parameters, β^*, are found by maximizing likelihood. This procedure (calibrating β in (10.140)) is known as logistic regression.

---

[24] We note that $\tilde{p}_x$ enters the sum on the right hand side of (10.138) only as a sample average weight term. For those who are unwilling to make Assumption 10.6, there are (significantly more complicated) methods to minimize the mutual information while simultaneously estimating q_x (see Globerson and Tishby (2004) and Globerson et al. (2009)).

10.2.4 Summary of Modeling Approach

The logic of our modeling approach in this section's more general context is similar to the logic described in Section 10.1.5.3. We have the following procedure:

1. Break the data into a training set and a hold-out sample, for example, taking 75% or 80% of the data, selected randomly, to train the model.

2. Choose a discrete set $A = \{\alpha_k \in (0, \alpha_{max}), k = 1, \ldots, K\}$.

3. For $k = 1, \ldots, K$,

   - Solve Problem 10.9 for $\beta^*(\alpha_k)$,

   - Compute the out-of-sample performance

   $$P_k = E_{\tilde{p}}[U(b^*(q^*(\alpha_k)), \mathcal{O})]$$

   on the out-of-sample test set, where $\tilde{p}$ is the empirical measure on this test set, and q^* is determined from (10.106) and (10.107) with parameters $\beta^*(\alpha_k)$, and b^* is determined from (10.104) and (10.105).

4. Put $k^* = \arg \max_k P_k$.

5. Our model, q^{**}, is determined from (10.106) and (10.107) with parameters $\beta^*(\alpha_{k^*})$.

A refinement: repeat on different data partitions in order to find an optimal α^{**}, then retrain the model using α^{**} on the complete set of available training data.

10.3 Probability Estimation via Relative U-Entropy Minimization

In this section, we address, in the discrete horse race setting, the following question: is it possible to formulate a generalized MRE problem for which the solution is robust in an absolute sense, rather than the relative sense of the robust outperformance principle, Principle 10.1?[25]

This is indeed possible, as the following Corollary, which follows directly from the robust outperformance principle, Principle 10.1, indicates:

---

[25]Much of the material in this section can be found in Friedman et al. (2007).

Corollary 10.2 *If*

$$q_y^0 = \frac{B}{\mathcal{O}_y} = \frac{1}{\mathcal{O}_y}, \forall y \in \mathcal{Y}, \tag{10.141}$$

then

$$q^* = \arg\min_{q \in K} D_U(q\|q^0) = \arg\max_{q \in Q} \min_{q' \in K} E_{q'}[U(b^*(q), \mathcal{O})], \tag{10.142}$$

where Q is the set of all possible probability measures and K is the set of all $q \in Q$ that satisfy the feature constraints.

This Corollary states that, by choosing q^*, a rational (expected utility-optimizing) investor maximizes his (model-based optimal) expected utility in the most adverse environment consistent with the feature constraints. Therefore, an investor with a general utility function, who bets on this horse race and wants to maximize his worst case (in the sense of (10.142)) expected utility can set q^0 to the homogeneous expected return measure determined by the odds ratios and solve the following problem:[26]

Problem 10.15 *(MRUE Problem)*
Find

$$q^* = \arg\min_{q \in Q} D_U(q\|q^0) \tag{10.143}$$

$$s.t. \ E_q[f^j] - E_{\tilde{p}}[f^j] = 0 \ , \ j = 1, \ldots, J \ , \tag{10.144}$$

where each f^j represents a feature (i.e., a function that maps $\mathcal{X}$ to $\mathbf{R}$), q^0 represents the prior measure, $\tilde{p}$ represents the empirical measure, and Q is the set of all probability measures.

As before, q^0, here, has a specific and nontraditional interpretation. It no longer represents prior beliefs about the real world measure that we estimate with q^*. Rather, it represents the homogeneous expected return measure. In the horse race setting, the homogeneous expected return measure is equivalent to a prominent measure from finance: the risk neutral pricing measure consistent with the odds ratios. If an investor wants the measure that he estimates to have the optimality property (10.142), he is not free to choose q^0 to represent his prior beliefs, in general. To attain the optimality property (10.142), he can set q^0 to represent the risk neutral pricing measure determined by the odds ratios. The MRUE measure, q^*, is a function of the odds ratios (which are incorporated in the measure q^0).

In order to obtain the dual to Problem 10.15, we specialize, in Problem 10.3, relative $(U, \mathcal{O})$-entropy to relative U-entropy by setting the "prior" measure to the risk neutral pricing measure generated by the odds ratios, $q_y^0 = \frac{1}{\mathcal{O}_y}$, with $\alpha = 0$. We obtain

---

[26]As noted above, we keep the context and notation as simple as possible by confining our discussion to unconditional estimation without regularization. Extensions are straightforward.

Problem 10.16 (*Dual of MRUE Problem*)
Find

$$\beta^* = \arg \max_\beta \sum_y \tilde{p}_y U \left(b^*(\hat{q}_y^\beta) \mathcal{O}_y \right), \qquad (10.145)$$

$$\text{where } b^*(\hat{q}_y^\beta) = \frac{1}{\mathcal{O}_y} (U')^{-1} \left(\frac{\lambda^*}{\hat{q}_y^\beta \mathcal{O}_y} \right) \qquad (10.146)$$

$$\text{and } \hat{q}_y^\beta = \frac{\lambda^*}{\mathcal{O}_y U' \left(U^{-1}(\beta^T f_y - \mu^*) \right)}, \qquad (10.147)$$

$$\text{where } \mu^* \text{ solves } 1 = \sum_x \frac{U^{-1} \left(\beta^T f_y - \mu^* \right)}{\mathcal{O}_y}, \qquad (10.148)$$

and

$$\lambda^* = \left\{ \sum_x \frac{1}{\mathcal{O}_y U' \left(U^{-1}(\beta^T f_y - \mu^*) \right)} \right\}^{-1}. \qquad (10.149)$$

Problem 10.16 is easy to interpret.[27] The objective function of Problem 10.16 is the utility (of the expected utility maximizing investor) averaged over the training sample. Thus, our dual problem is a maximization of the training sample-averaged utility, where the utility function, U, is the utility function on which the U-entropy depends.

We note that the primal problem (Problem 10.15) and the dual problem (Problem 10.16) have the same solution, in the sense that $q^* = \hat{q}_{\beta^*}$.

This problem is a J-dimensional (J is the number of features), unconstrained, concave maximization problem. The primal problem, Problem 10.15, on the other hand, is an m-dimensional (m is the number of states) convex minimization with linear constraints. The dual problem, Problem 10.16, may be easier to solve than the primal problem, Problem 10.15, if $m > J$. For conditional probability models, the dual problem will always be easier to solve than the primal problem.

As we have seen above, in cases where odds ratios are available, the MRUE problem yields a solution with the optimality property (10.142). However, in real statistical learning applications, as mentioned above, it is often the case that odds ratios are not observable. In this case, the builder of a statistical learning model can use assumed odds ratios, on which the model will depend. Given the relation

$$q_y^0 = \frac{B}{\mathcal{O}_y} = \frac{1}{\mathcal{O}_y}, \forall y \in \mathcal{Y}, \qquad (10.150)$$

as a perhaps more convenient alternative, the model builder can directly specify a risk neutral pricing measure consistent with the assumed odds ratios.

---

[27] A version that is more easily implemented is given in Friedman and Sandow (2003a).

Either way, the model will possess the optimality property (10.142) under the odds ratios consistent with the assumption. The necessity of providing a risk neutral pricing measure, perhaps, imposes an onus on the MRUE modeler comparable to the onus of finding a prior for the MRE modeler. However, we note that, as for MRE models, the importance of p^0 will diminish as the number of feature constraints grows.

The benefits of solving the MRUE problem, rather than the MRE problem, extend beyond being able to tailor probabilistic models to the risk preferences of investors. By solving the MRUE problem, it is possible to discover compact, elegant representations of flexible, yet fat-tailed models. At first blush, it might appear that this is possible with MRE methods, since, subject to technical regularity conditions, MRE methods can, *in principle*, be used to generate *any* probability distribution function, *if* we are prescient in our selection of a prior distribution or feature functions. To see this, consider the standard MRE problem, Problem 9.1 and its dual problem, Problem 9.2, with the connecting equation

$$\hat{q}_y^{(\beta)} = \frac{1}{Z(\beta)} \, q_y^0 \, e^{\beta^T f(y)}. \tag{10.151}$$

Given a "target" distribution function q_y, suppose that we take $q_y^0 = q_y$; in this case, if $\beta = 0$, $\hat{q}_y^{(\beta)}$ from (10.151) will reproduce our target distribution function. Alternatively, by choosing a feature of the form $\log q_y$ together with a flat prior distribution, we can recover q_y.

Of course, in practice, we are not prescient in our selection of a prior distribution and features. If we restrict ourselves to polynomial features or fractional-power features, as we can see by a quick inspection of (10.151), MRE methods result in thin-tailed distributions, regardless of the prior distribution. We are not aware of a "spanning" set of features that lend themselves to fitting, via convex programming methods, a rich set of fat-tailed flexible distributions via MRE methods.

Given the recent roiling of financial markets often attributed to reliance on models that do not adequately capture the likelihood of extreme events, fat-tailed distributions seem to be of particular interest (see, for example, the following recent New York Times articles: Nocera (2009), Bookstaber (2009), and Safire (2009)).

MRUE methods provide a way of building compact, flexible, parametric, fat-tailed models, using a small set of well-chosen features. Such methods are employed by Friedman et al. (2010b), who consider the MRUE problem in the particularly tractable and important case where the utility function is the power utility,

$$U(W) = \frac{W^{1-\kappa} - 1}{1 - \kappa}, \tag{10.152}$$

where κ denotes the investor's (constant) relative risk aversion.[28] They note

---

[28] As we have mentioned, power utility functions are used widely in industry (see, for ex-

that, given a seat of features, increased relative risk aversion results in thicker tailed distributions. They also note that a number of well-known power-law distributions, including the student-t, generalized Pareto, and exponential distributions, can be obtained as special cases of the connecting equation associated with MRUE approach with power utility and linear or quadratic features; the skewed generalized-t distribution is a special case with power features. We briefly review their approach in Section 12.4.

10.4   Expressing the Data Constraints in Purely Economic Terms

The primal problems described earlier in this chapter were expressed in terms of feature expectation constraints. We shall see that we can express these problems in purely economic terms by making use of the following quantity, defined in (7.32),

$$G_{U,\mathcal{O}}(p^2, p^1; p) = \sum_y p_y [U(b_y^*(p^2)\mathcal{O}_y) - U(b_y^*(p^1)\mathcal{O}_y)], \qquad (10.153)$$

which is the gain in expected utility, under the measure p, from allocating according to p^2, rather than p^1.

Consistency of the (model) probability measure q with the data is often measured in terms of quantities of the form

$$c^j(q) = E_q[f^j] - E_{\tilde{p}}[f^j], j = 1, \ldots, J, \qquad (10.154)$$

where the f^j are functions $f^j : \mathbf{R} \rightarrow \mathbf{R}$, which are usually referred to as *features*, and $\tilde{p}$ refers to the empirical measure on the training set.

Here we provide a decision-theoretic interpretation of c_j. To do so, we introduce the *data-probing measures* q^j, which can be used to generate features f^j via

$$f^j(y) = U(b_y^*(q^j)\mathcal{O}_y) - U(b_y^*(q^0)\mathcal{O}_y). \qquad (10.155)$$

That is, the feature values can be interpreted as utility gains for an investor who allocates according to the data-probing measures rather than the benchmark measure. In the case of the generalized logarithmic family of utilities, we obtain

$$f^j(y) = \gamma_1 \log \frac{q_y^j}{q_y^0}. \qquad (10.156)$$

---

ample, Morningstar (2002)). Moreover, power utility functions have constant relative risk aversion and important optimality properties (see, for example, Stutzer (2003)).

By (10.153), (10.154), and (10.155), we see that

$$c^j(q) = G_{U,\mathcal{O}}(q^j, q^0; q) - G_{U,\mathcal{O}}(q^j, q^0; \tilde{p}). \qquad (10.157)$$

Thus, for each j, c_j is a measure of the consistency of the candidate model with the data that is defined in terms of relative performance (as measured by expected utility) for an investor who allocates according to the data-probing measure q^j.

For a general logarithmic utility, we note that the c_j do not depend on the odds ratios and that

$$c^j(q) = \gamma_1 \sum_y \left(q_y log\frac{q_y^j}{q_y^0} - \tilde{p}_y log\frac{q_y^j}{q_y^0} \right). \qquad (10.158)$$

We now reformulate our decision-theoretic, robust modeling approach.

Problem 10.17 *Find*

$$q^* = \max_{q \in Q} \min_{w \in Q} G_{U,\mathcal{O}}(q, q^0; w) \qquad (10.159)$$

$$s.t. \ \psi(w) \le \alpha \ , \ j = 1, \dots, J, \qquad (10.160)$$

where

$$c^j(w) = G_{U,\mathcal{O}}(q^j, q^0; w) - G_{U,\mathcal{O}}(q^j, q^0; \tilde{p}) \ , \ j = 1, \dots, J . \qquad (10.161)$$

Here, $\alpha \ge 0$ is a hyperparameter that controls consistency with the data and Q is the probability simplex. That is, for each candidate model q, we consider the most adverse environment, as described by $w^*(q)$, where

$$w^*(q) = \arg\min_{w \in Q} G_{U,\mathcal{O}}(q, q^0; w)$$

$$s.t. \ \psi(w) \le \alpha \ , \ j = 1, \dots, J,$$
$$\text{where } c^j(w) = G_{U,\mathcal{O}}(q^j, q^0; w) - G_{U,\mathcal{O}}(q^j, q^0; \tilde{p}) \ , \ j = 1, \dots, J.$$

We select the measure $q \in Q$ with the greatest outperformance, over the benchmark model, in its own most adverse environment. Models that are completely consistent with the data ($\alpha = 0$, which implies $c^j = 0$ for all j) may generalize poorly on out-of-sample sets (overfitting). We allow imperfect consistency with the data by allowing α to be positive.

In light of (10.157), the constraints (10.160) and (10.161) require that some convex function of the difference between the gains from

(*i*) allocating according to q^j rather than q^0 under q, and

(*ii*) allocating according to q^j rather than q^0 under $\tilde{p}$

is less than or equal to α. That is, model consistency with the data is enforced by regulating (via the tolerance, α) the consistency of the gains $G_{U,\mathcal{O}}(q^j, q^0; \cdot)$ under q and $\tilde{p}$.

We note that $c^j(w)$ is linear in w and that it is possible to generalize the constraints (10.160) to any convex function of the vector $c(w)$.

10.5   Some Proofs

10.5.1   Proof of Lemma 10.2

We restate Lemma 10.2:

Lemma 10.2 *Problem 10.2 is a strictly convex optimization problem and Problems 10.1 and 10.2 have the same unique solution.*

Proof: We note that the objective function, $D_{U,\mathcal{O}}(q||q^0)$, is strictly convex (see Theorem 7.2 in Section 7.2.2). The inequality constraint, (10.38), of Problem 10.2 is also convex; this follows from the fact that Σ is a covariance matrix and therefore nonnegative definite. The equality constraints, (10.37) and (10.39), are both affine. Therefore, Problem 10.2 is a strictly convex programming problem (see Problem 2.1).

We now show that Problems 10.1 and 10.2 have the same unique solution.

We first *assume that* $\alpha < \alpha_{max}$, and show that, in this case, the solution to Problem 10.2 satisfies

$$\psi(c) = \alpha.$$

To this end, recall from Theorem 7.2 that $D_{U,\mathcal{O}}(q||q^0)$ is strictly convex in q, for q in the simplex Q, and that the global minimum of the function $D_{U,\mathcal{O}}(q||q^0)$ occurs at $q = q^0$, which occurs only if $\alpha = \alpha_{max}$; therefore,

$$\nabla_q D_{U,\mathcal{O}}(q||q^0) \neq 0 \qquad (10.162)$$

for $q \neq q^0$. Suppose that q^* is such that $\psi(c(q^*)) < \alpha$ where

$$c(q) = E_q[f] - E_{\tilde{p}}[f].$$

Then there exists a neighborhood of q^* on the simplex Q, such that for all q in the neighborhood, $\psi(c(q)) \leq \alpha$. From (10.162), we see that there is a direction of decrease of the objective function $D_{U,\mathcal{O}}(q||q^0)$ on the simplex Q, so q^* cannot be the optimal solution. Therefore, we cannot have $\psi(c(q^*)) < \alpha$. It follows that $\psi(c(q^*)) = \alpha$, so the solution to Problem 10.2 is the solution to 10.1 for the case $\alpha < \alpha_{max}$.

In the case $\alpha = \alpha_{max}$, it is obvious that both problems have the unique solution $q^* = q^0$.

The objective function, $D_{U,\mathcal{O}}(q||q^0)$, is strictly convex in q, so the solution of Problem 10.2 is unique (see, for example, Rockafellar (1970), Section 27). It follows that the solution to Problem 10.1 is also unique. $\square$

10.5.2   Proof of Theorem 10.3

We will show that Problem 10.2, which we restate below for convenience, has the (equivalent) dual formulations Problems 10.3 and 10.4.

Problem 10.2 *(Initial Convex Problem, Given $\alpha, 0 \le \alpha \le \alpha_{max}$)*

$$Find \quad \min_{q \in (R^+)^m, c \in R^J} D_{U,O}(q \| q^0) \quad (10.163)$$

$$under \ the \ constraints \ 1 = \sum_y q_y \quad (10.164)$$

$$and \quad \psi(c) \le \alpha \quad (10.165)$$

$$where \quad c_j = E_q[f_j] - E_{\tilde{p}}[f_j] \ . \quad (10.166)$$

We will derive the dual of Problem 10.2 now. To do so, we shall first derive the dual in the special case where the feature expectation constraints are satisfied exactly. We will then apply Theorem 2.13 to obtain the dual in the general case. Note that the Lagrangian, when the feature expectation constraints are satisfied exactly, is given by

$$\mathcal{L}(q, c, \beta, \mu, \nu) = D_{U,O}(q \| q^0) + \beta^T \left\{ E_{\tilde{p}}[f] - E_q[f] \right\}$$

$$+ \mu \left\{ \sum_y q_y - 1 \right\} - \nu^T q, \quad (10.167)$$

where $\beta = (\beta_1, ..., \beta_J)^T$, μ, and $\nu^T = (\nu_1, \ldots, \nu_m) \ge 0$ are Lagrange multipliers and q varies over $\mathbf{R}^m$.

10.5.2.1   The Connecting Equation

In order to derive the connecting equation, we have to solve

$$0 = \frac{\partial \mathcal{L}(q, c, \beta, \xi, \mu, \nu)}{\partial q_y} \ . \quad (10.168)$$

In order to solve (10.168), we insert (10.167) and the equation (see Exercise 5)

$$\frac{\partial D_{U,O}(q \| q^0)}{\partial q_y} = U(b_y^*(q)O_y) - U(b_y^*(q^0)O_y) \ , \quad (10.169)$$

into (10.168), and obtain

$$0 = U(b_y^*(q)O_y) - U(b_y^*(q^0)O_y) - \beta^T f(y) + \mu - \nu_y \ . \quad (10.170)$$

We rewrite this equation as

$$U(b_y^*(q)O_y) = G_y(q^0, \beta, \mu, \nu) \quad (10.171)$$

$$with \ G_y(q^0, \beta, \mu, \nu) = U(b_y^*(q^0)O_y) + \beta^T f(y) - \mu + \nu_y \ , \quad (10.172)$$

where $G_y(q^0, \beta, \mu, \nu)$ does not depend on q. In order to solve for q, we substitute (10.3) into (10.171), to obtain

$$U\left(U'^{-1}\left(\frac{\lambda}{q_y \mathcal{O}_y}\right)\right) = G_y(q^0, \beta, \mu, \nu) . \tag{10.173}$$

Solving for q_y, we obtain the connecting equation

$$q_y^* \equiv \frac{\lambda}{\mathcal{O}_y U'\left(U^{-1}(G_y(q^0, \beta, \mu, \nu))\right)} . \tag{10.174}$$

From (10.174), by the positivity of the $\mathcal{O}_y$ and the fact the U is a monotone increasing function, we conclude that all of the q_y^* and λ have the same sign. We note, from (10.164), that the q_y^* and λ must be positive. From the Karush-Kuhn-Tucker conditions, we must have $\nu_y q_y^* = 0$; it follows that $\nu_y^* = 0$ for all y. Accordingly, we may suppress the dependence of G and $\mathcal{L}$ on ν.

The connecting equation, (10.174), depends on β, λ, and μ. We now show how to calculate λ and μ in terms of β. Solving (10.173) for $U'^{-1}\left(\frac{\lambda}{q_y \mathcal{O}_y}\right)$ and substituting into (10.4), we obtain a condition for μ^*:

$$\sum_y \frac{1}{\mathcal{O}_y} U^{-1}\left(G_y(q^0, \beta, \mu^*)\right) = 1 . \tag{10.175}$$

This equation is easy to solve numerically for μ^*, by the following lemma.

Lemma 10.3 *There exists a unique solution, μ^*, to (10.175). The left hand side of (10.175) is a strictly monotone decreasing function of μ^*.*

Proof: First, we note that since U is a strictly increasing function,

$$(U^{-1})' = \frac{1}{\frac{dU}{dW}} > 0,$$

so U^{-1} is a strictly increasing function and the left hand side of (10.175) is a strictly decreasing function of μ^*.
Letting

$$\bar{\mu} = \max_y \beta^T f(y),$$

we see that

$$\beta^T f(y) - \bar{\mu} \le 0 \text{ for all } y.$$

In this case, it follows from (10.172) that

$$G_y(q^0, \beta, \bar{\mu}) \le U(b_y^*(q^0)\mathcal{O}_y) \text{ for all } y;$$

so, by the monotonicity of U^{-1},

$$\sum_y \frac{1}{\mathcal{O}_y} U^{-1} \left(G_y(q^0, \beta, \bar{\mu}) \right) \leq \sum_y \frac{1}{\mathcal{O}_y} U^{-1} \left(U(b_y^*(q^0)\mathcal{O}_y) \right) \quad (10.176)$$

$$= \sum_y b_y^*(q^0) = 1.$$

Note that $G_y(q^0, \beta, \bar{\mu}) \in dom(U^{-1})$ for all y, by (10.171). Similarly, by letting

$$\underline{\mu} = \min_y \beta^T f(y),$$

we can guarantee that

$$\sum_y \frac{1}{\mathcal{O}_y} U^{-1} \left(G_y(q^0, \beta, \underline{\mu}) \right) \geq 1.$$

By the Intermediate Value Theorem and the monotonicity and continuity of the left hand side of (10.175), there exists a unique solution to (10.175). □

We now show how to calculate λ in terms of β and μ^*. We insert (10.174) into (10.164), and obtain:

$$1 = \lambda \sum_y \frac{1}{\mathcal{O}_y U' \left(U^{-1}(G_y(q^0, \beta, \mu^*)) \right)} \quad ;$$

solving for λ, we obtain

$$\lambda^* \equiv \left\{ \sum_y \frac{1}{\mathcal{O}_y U' \left(U^{-1}(G_y(q^0, \beta, \mu^*)) \right)} \right\}^{-1}. \quad (10.177)$$

Summarizing the result of this subsection:

The connecting equation, which describes q^* as a member of a parametric family (in β), is given by

$$q_y^* = \frac{\lambda^*}{\mathcal{O}_y U' \left(U^{-1}(G_y(q^0, \beta, \mu^*)) \right)}, \quad (10.178)$$

where we determine μ^* from (10.175) via Lemma 10.3 and λ^* from (10.177).

10.5.2.2   Dual Problems

We now show that

Lemma 10.4 *Problem 10.4 is the dual of Problem 10.2.*

Proof: (10.174), together with the (10.175) and (10.177), gives the probabilities q_y^* and the Lagrange multipliers μ^*, ν^* for which the Lagrangian is at its minimum for given multipliers β. This allows us to formulate the dual problem as an optimization with respect to β. To this end, we have to compute $\mathcal{L}(q^*, c^*, \beta, \mu^*)$.

Substituting $D_{U,\mathcal{O}}(q^*\|q^0)$ (from Definition 7.2) into (10.167), we obtain:

$$\mathcal{L}(q^*, c^*, \beta, \mu^*) = \sum_y q_y^* U(b_y^*(q^*)\mathcal{O}_y) - \sum_y q_y^* U(b_y^*(q^0)\mathcal{O}_y)$$

$$+\beta^T \left\{ E_{\tilde{p}}[f] - \sum_y q_y^* f(y) \right\}$$

$$+\mu^* \left\{ \sum_y q_y^* - 1 \right\};$$

so

$$\mathcal{L}(q^*, c^*, \beta, \mu^*) = \sum_y q_y^* \left\{ U(b_y^*(q^*)\mathcal{O}_y) - U(b_y^*(q^0)\mathcal{O}_y) - \beta^T f(y) + \mu^* \right\}$$

$$+\beta^T E_{\tilde{p}}[f] - \mu^* .$$

Because of (10.170), the first line on the right hand side of above equation is zero, i.e., we obtain

$$\mathcal{L}(q^*, c^*, \beta, \mu^*) = \beta^T E_{\tilde{p}}[f] - \mu^* . \tag{10.179}$$

The dual problem is to maximize the function $h(\beta) = \mathcal{L}(q^*, c^*, \beta, \mu^*)$ with respect to β. Now we are ready to formulate the dual problem: maximize $h(\beta) = \mathcal{L}(q^*, c^*, \beta, \mu^*)$ with respect to β. From (10.179), (10.172), (10.178), (10.177), (10.175), and an application of Theorem 2.13 we obtain Problem 10.4, which completes the proof of the equivalence of the solutions to Problems 10.4 and 10.2. $\square$

In the following lemma, we show that we can express the dual problem objective function in a more easily interpreted form.

Lemma 10.5 *Problem 10.4 can be restated as Problem 10.3.*

Proof: Using (10.170) to replace $\beta^T E_{\tilde{p}}[f] - \mu^*$ in (10.49), and noticing that $U(b_y^*(q^0)\mathcal{O}_y)$ does not depend on β, we obtain:

$$h(\beta) = \sum_y \tilde{p}_y U(b_y^*(q^*)\mathcal{O}_y) - \inf_{\xi \geq 0} \left\{ \xi \Psi^* \left(\frac{\beta}{\xi} \right) + \alpha \xi \right\}, \tag{10.180}$$

up to an unimportant constant. This means that the dual problem can be restated as in Problem 10.3. $\square$

The proof of Theorem 10.3 is a direct consequence of Lemmas 10.4 and 10.5 and the fact that the primal problem satisfies the Slater condition and therefore there is no duality gap (see, for example, Section V, Theorem 4.2 in Berkovitz (2002)). The primal problem is strictly convex and therefore has a unique solution (see, for example, Rockafellar (1970), Section 27).

10.5.3 Dual Problem for the Generalized Logarithmic Utility

In order to specify the dual problem for the generalized logarithmic utility (10.69), we first notice that

$$U'(z) = \frac{\gamma_1}{z - \gamma B} \, , \tag{10.181}$$

$$U^{-1}(z) = e^{\frac{z - \gamma_2}{\gamma_1}} - \gamma \tag{10.182}$$

$$\text{and } U'(U^{-1}(z)) = \gamma_1 e^{-\frac{z - \gamma_2}{\gamma_1}} \, . \tag{10.183}$$

Using the relation

$$b_y^*(q) = q_y \left[1 - \gamma \right] - \frac{\gamma}{\mathcal{O}_y} \tag{10.184}$$

(see (5.56)), we can write $G_y(q^0, \beta, \mu^*)$ from (10.45) as

$$G_y(q^0, \beta, \mu^*) = \gamma_1 \left[\log \left(q_y^0 \mathcal{O}_y \left[1 - \gamma \right] \right) + \beta^T f(y) - \mu^* \right] + \gamma_2 \, . \tag{10.185}$$

Inserting (10.182) and (10.185) into (10.46) gives

$$1 = \sum_y \frac{1}{\mathcal{O}_y} \left\{ q_y^0 \mathcal{O}_y \left[1 - \gamma \right] e^{\beta^T f(y) - \mu^*} - \gamma \right\} \tag{10.186}$$

$$= e^{-\mu^*} \left[1 - \gamma \right] \sum_y \left[q_y^0 e^{\beta^T f(y)} \right] - \gamma \sum_{y'} \frac{1}{\mathcal{O}_{y'}} \, , \tag{10.187}$$

which can be solved for μ^*:

$$\mu^* = \log \left(\sum_y q_y^0 e^{\beta^T f(y)} \right) \, . \tag{10.188}$$

Next, we solve (10.47) for λ^*. We use (10.183) and (10.185) to write (10.47) as

$$1 = \lambda^* \sum_y \left\{ \frac{1}{\mathcal{O}_y} \left(q_y^0 \mathcal{O}_y \left[1 - \gamma \right] e^{\beta^T f(y) - \mu^*} \right) \right\}$$

$$= \lambda^* \left[1 - \gamma \right] \sum_y \left\{ q_y^0 e^{\beta^T f(y) - \mu^*} \right\} \, .$$

After inserting (10.188) we can solve for λ^* and get:

$$\lambda^* = \frac{1}{1-\gamma} \; . \tag{10.189}$$

By means of (10.44), (10.183), (10.189), and (10.185) we obtain for the optimal probability distribution

$$
\begin{aligned}
q_y^* &= \frac{1}{1-\gamma} \frac{1}{\mathcal{O}_y} \left(q_y^0 \mathcal{O}_y \left[1-\gamma\right] e^{\beta^T f(y) - \mu^*} \right) \\
&= q_y^0 \, e^{\beta^T f(y) - \mu^*} \tag{10.190} \\
&= \frac{1}{\sum_y q_y^0 e^{\beta^T f(y)}} \, q_y^0 \, e^{\beta^T f(y)} \quad \text{(by (10.188))} \; . \tag{10.191}
\end{aligned}
$$

We can now compute the objective function $h(\beta)$. Based on (10.42) and (10.190), we obtain

$$h(\beta) = \sum_y \tilde{p}_y \log q_y^* - \inf_{\xi \geq 0} \left\{ \xi \Psi^* \left(\frac{\beta}{\xi} \right) + \alpha\xi \right\} \; , \tag{10.192}$$

up to the constants $E_{\tilde{p}}[\log q_y^0]$ and γ_2 and the factor γ_1.

Collecting (10.191) and (10.192), we obtain Problem 10.5.

10.5.4   Dual Problem for the Conditional Density Model

In order to derive this dual problem (Problem 10.9), we note that $\mathcal{Q} \times \mathbf{R}^J$ is a convex subset of a vector space, the constraints expressed by (10.96)-(10.98) can be rewritten in terms of convex mappings into a normed space, and the equality constraints expressed by (10.96) and (10.98) are linear. By Theorem 1 of Section 8.6 in Luenberger (1969), the dual problem is the maximization over $\xi \geq 0$, $\beta = (\beta_1, ..., \beta_J)^T$, $\mu = \{\mu_x, x = x_1, ..., x_M\}$, and $\nu = \{(\nu(y|x) \geq 0, \nu_{\rho|x} \geq 0), y \in \mathcal{Y}, \rho = 1, 2, ..., m, x = x_1, ..., x_M\}$ of $\inf_{q \in \mathcal{Q}, c \in \mathbf{R}^J} \mathcal{L}(q, c, \beta, \xi, \mu, \nu)$, where

$$
\begin{aligned}
\mathcal{L}(q, c, \beta, \xi, \mu, \nu) = \; & D_{U, \mathcal{O}}(q||q^0) + \beta^T \left\{ c - E_q[f] + E_{\tilde{p}}[f] \right\} + \xi \left\{ \psi(c) - \alpha \right\} \\
& + \sum_x \mu_x \tilde{p}_x \left\{ E_q[1|x] - 1 \right\} - E_q[\nu]
\end{aligned}
$$

is a generalization of the Lagrangian (10.167) for the case of discrete probabilities. One can find $\inf_{q \in \mathcal{Q}, c \in \mathbf{R}^J} \mathcal{L}(q, c, \beta, \xi, \mu, \nu)$ the same way as we do in Section 10.5.2 for discrete probabilities; the only difference is that we have to use Fréchet derivatives instead of ordinary ones. As a result, we obtain the analog of the connecting equation described in Section 10.5.2. We can then continue along the lines from Section 10.5.2, showing that $\nu = 0$ and finding ξ^* and μ^*. This leads to the dual of Problem 10.8: Problem 10.9.

10.6   Exercises

1. Show that in the setting of Example 10.1, when the investor and market are compatible, the model performance measure of Chapter 8 yields a value of $-\infty$ whenever there exists a test set datum that is not in the training dataset.

2. Show that by solving the following equation for q_y:

$$b_y^* = \frac{1}{\mathcal{O}_y}(U')^{-1}\left(\frac{\lambda}{q_y\mathcal{O}_y}\right), \qquad (10.193)$$

   we obtain

$$q_y(b_y^*) = \frac{1}{\mathcal{O}_y U'(b_y^*\mathcal{O}_y)}\frac{1}{\sum_{y'}\frac{1}{\mathcal{O}_{y'}U'(b_{y'}^*,\mathcal{O}_{y'})}}. \qquad (10.194)$$

3. Let $\overline{\mathcal{B}}$ denote the set of all allocations generated by

$$b_y^*(q) = \frac{1}{\mathcal{O}_y}(U')^{-1}\left(\frac{\lambda}{q_y\mathcal{O}_y}\right), \qquad (10.195)$$

   as q ranges over the probability simplex Q. Prove that

   (*i*) $\overline{\mathcal{B}}$ contains all of it limit points, and

   (*ii*) the convex combination of any two points in $\overline{\mathcal{B}}$ is also in $\overline{\mathcal{B}}$.

4. Prove the statement in Section 10.2.3.5; in particular, show explicitly that the connecting equation for the logistic regression problem is indeed of the form

$$p(\rho = 1|x) = \frac{1}{1+e^{-\beta^T x}}. \qquad (10.196)$$

5. Show that

$$\frac{\partial D_{U,\mathcal{O}}(q\|q^0)}{\partial q_y} = U(b_y^*(q)\mathcal{O}_y) - U(b_y^*(q^0)\mathcal{O}_y). \qquad (10.197)$$

6. A more general form of logistic regression includes a constant term β_0, i.e., is given by (10.196). What feature function needs to be added to the problem formulation from Section 10.2.3.5, in order to obtain this more general form?

7. Derive the dual to Problem 10.2 for an exponential utility function (see Section 4.4).

8. Derive the minimum relative $(U, \mathcal{O})$-measure under the following assumptions:

    (*i*) discrete unconditional horse race,

    (*ii*) U is a generalized logarithmic utility, and

    (*iii*) $q_y^0 = I_{y=y'}$ where I is the indicator function and y' is some value.

9. Let $\psi(c)$ denote a nonnegative convex function of

$$c = (c_1, \ldots, c_J)^T,$$

with

$$c_j = E_q[f_j] - E_{\tilde{p}}[f_j], \tag{10.198}$$

and $\psi(0) = 0$. Prove that the set of measures for which $\psi(c) \leq \alpha$ is convex and compact.

10. Suppose that $p_{x,\rho}$ and $p_{x,\rho}^0$ share the same marginal distribution in x, i.e., $p_x = p_x^0$. Show that, under this assumption, the relative entropy $D(p_{x,\rho}\|p_{x,\rho}^0)$ is the same as the conditional relative entropy, $D(p_{\rho|x}\|p_{\rho|x}^0)$.

11. Recall (from Definition 2.10) the definition of mutual information:

$$I(X; \rho) = \sum_{x,\rho} p_{x,\rho} \log \frac{p_{x,\rho}}{p_x p_\rho}, \tag{10.199}$$

which can be rewritten as

$$I(X; \rho) = \sum_x p_x \sum_\rho p_{\rho|x} \log \frac{p_{\rho|x} p_x}{p_x p_\rho} = \sum_x p_x \sum_\rho p_{\rho|x} \log \frac{p_{\rho|x}}{p_\rho}. \tag{10.200}$$

    (*i*) Interpret $I(X; \rho)$ from (10.199) in terms of the outperformance under $p_{x,\rho}$ of an investor with a generalized logarithmic utility, who allocates according to $p_{x,\rho}$ in a discrete unconditional horse race on (X, ρ) over an investor who allocates under the assumption of independence of X and ρ.

    (*ii*) Interpret $I(X; \rho)$ from (10.200) in terms of the outperformance under $p_{\rho|x}$ of an investor with a generalized logarithmic utility, who allocates according to $p_{\rho|x}$ in a discrete conditional horse race on $\rho|x$ over an investor who does not make use of the side information, X.

    (*iii*) For the generalized logarithmic utility investor, both of the preceding interpretations are consistent with the same definition of $I(X; \rho)$. Is the same true for investors with other utility functions? Formulate two analogs to mutual information, based on the interpretations in (*i*) and (*ii*) for investors with general utility functions.

Chapter 11

Extensions

In this chapter we generalize the utility-based model performance measures and model building approaches from Chapters 8 and 10 to performance measures for leveraged investors in a horse race and for investors in incomplete markets. We also introduce a utility-based performance measure for regression models.[1]

11.1 Model Performance Measures and MRE for Leveraged Investors

In Chapter 8, we have measured the performance of a probabilistic model from the point of view of a gambler who invests a fixed amount of money in a horse race. A perhaps more realistic idealization of a general decision maker, at least in a financial context, is an investor who can withhold or borrow cash according to the opportunities the horse race offers. In this section, we shall, following Friedman and Sandow (2004), consider such a leveraged investor, and we shall see that most of the interesting results from Chapter 8 can be easily generalized. We shall also briefly discuss the $(U, \mathcal{O})$-entropy and MRE model building from the viewpoint of a leveraged investor.

For ease of exposition, we restrict ourselves to evaluating probabilistic models of a random variable Y with values, y, in a finite set $\mathcal{Y}$. However, the results in this section can be generalized to conditional probability and probability density models.

11.1.1 The Leveraged Investor in a Horse Race

As in Chapter 8, we consider the horse race from Definition 3.1. We generalize, however, the definition of an investor, Definition 3.4, as follows.

---

[1]Another extension of the utility-based model building approaches from Chapter 10, which is not discussed in this book, is the semi-supervised learning approach from Sandow and Zhou (2007).

Definition 11.1 *(Leveraged investor)* *An investor is a gambler with $1 initial capital, who allocates b_y to the event $Y = y$, with the constraint*

$$\underline{B} \leq \hat{b} \leq \overline{B} , \tag{11.1}$$

$$\text{where } \hat{b} = \sum_{y \in \mathcal{Y}} b_y - 1 , \tag{11.2}$$

and $\underline{B} \leq 0, \overline{B} \geq 0$ are constants. We denote the investor's allocation by

$$b = \{b_y, \ y \in \mathcal{Y}\} . \tag{11.3}$$

As in Definition 3.4, we have made the assumption of $1 initial capital for the sake of convenience, but without loss of generality; we may view this $1 as the investor's total wealth in some appropriate currency. We have not required b_y to be nonnegative. The investor may choose to "short" a particular horse.

Since we did not require that $\sum_{y \in \mathcal{Y}} b_y = 1$, we have included the possibility that the investor does not bet all his capital or borrows additional money to leverage his bet. The investor keeps the amount

$$\hat{b} = \sum_{y \in \mathcal{Y}} b_y - 1 \tag{11.4}$$

in cash. We have defined $\hat{b}$ such that $\hat{b} < 0$ describes an investor who borrows the amount $-\hat{b}$, i.e., a leveraged investor. The constraint (11.1) limits the size of the investor's bets. Such limits exist in most practical problems; they certainly do in financial markets.

For the sake of simplicity, we assume that the investor can borrow cash or lend withheld cash at an interest rate of 0. It is easy to generalize our discussion to nonzero interest rates.

Based on the above assumptions, the investor's wealth after the bet is

$$W = b_{y'} \mathcal{O}_{y'} - \hat{b} , \tag{11.5}$$

where y' denotes the winning horse.

11.1.2 Optimal Betting Weights

As in Chapter 8, we assume that our investor is rational and has a utility function, U, for which Assumption 4.1 holds. According to Utility Theory, a rational investor who believes the measure q allocates so as to maximize the expectation, under q, of his utility function (as applied to his post-bet wealth). In conjunction with (11.5), this means that our investor chooses the following allocation.

Definition 11.2 *(Optimal allocation for a leveraged investor)*

$$b^*(q) = \arg \max_{b \in \mathcal{B}} \hat{E}_q[U(b, \mathcal{O})] , \tag{11.6}$$

where

$$\hat{E}_q[U(b,\mathcal{O})] = \sum_{y\in\mathcal{Y}} q_y U(b_y\mathcal{O}_y - \hat{b}) , \tag{11.7}$$

$$\hat{b} = \sum_{y\in\mathcal{Y}} b_y - 1 , \tag{11.8}$$

and

$$\mathcal{B} = \{b : \underline{B} \le \hat{b} \le \overline{B}\} . \tag{11.9}$$

The following lemma shows how the optimal betting weights can be computed.

Lemma 11.1

(i) *If the equation*

$$1 + \left(1 - B^{-1}\right)\hat{b}^* = \sum_{y\in\mathcal{Y}} \frac{1}{\mathcal{O}_y}(U')^{-1}\left(\frac{\lambda}{q_y\mathcal{O}_y}\right) , \tag{11.10}$$

where B is given by Definition 3.2 and

$$\hat{b}^* = \begin{cases} \underline{B} & \text{if } B < 1 \\ \in [\underline{B}, \overline{B}] & \text{if } B = 1 \\ \overline{B} & \text{if } B > 1 , \end{cases} \tag{11.11}$$

has a solution for λ, then $\hat{b}^*$ is an optimal borrowed cash amount (which is unique for $B \ne 1$) and the optimal betting weights are given by

$$b_y^*(q) = \frac{1}{\mathcal{O}_y}(U')^{-1}\left(\frac{\lambda}{q_y\mathcal{O}_y}\right) + \frac{\hat{b}^*}{\mathcal{O}_y} . \tag{11.12}$$

(ii) *If*

$$\frac{1}{(U')^{-1}(0) - \hat{b}^*(B-1)} < B^{-1} < \frac{1}{max\{0, (U')^{-1}(\infty) - \hat{b}^*(B-1)\}} , \tag{11.13}$$

then (11.10) has a solution for λ.

Proof: The proof of statement (i) is a straightforward generalization of the proof of Lemma 5.1. It can be found in Friedman and Sandow (2004). Statement (ii) follows from Lemma 8.2 by setting $V = \hat{b}^*(B-1)$. □

In the case of fair odds, i.e., for $B = 1$, the optimal allocation (11.12) is not unique, since the cash amount, $\hat{b}^*$, can be chosen arbitrarily in $[\underline{B}, \overline{B}]$. The reason for this is that allocating $\frac{\hat{b}'}{\mathcal{O}_y}$ on each horse y (which requires a total allocation of $\hat{b}'\sum_{y\in\mathcal{Y}}\frac{1}{\mathcal{O}_y} = \frac{\hat{b}'}{B} = \hat{b}'$) always results in the payoff $\hat{b}'$, no matter which horse wins. Therefore, borrowing an additional (feasible) $\hat{b}'$ and adding it to the bet according to (11.12) has no effect.

11.1.3   Performance Measure

We generalize Definition 8.1 as follows.

Definition 11.3 *For a leveraged investor with utility function U, the utility-based model performance measure for the model q is*

$$\hat{E}_{\tilde{p}}[U(b^*(q), \mathcal{O})] , \tag{11.14}$$

where

$$\hat{E}_{\tilde{p}}[U(b, \mathcal{O})] = \sum_{y \in \mathcal{Y}} \tilde{p}_y U(b_y \mathcal{O}_y - \hat{b}). \tag{11.15}$$

We note that, because of (11.6), for any measure q,

$$\hat{E}_{\tilde{p}}[U(b^*(q), \mathcal{O})] \le \hat{E}_{\tilde{p}}[U(b^*(\tilde{p}), \mathcal{O})] \text{ , with equality if } q = \tilde{p} . \tag{11.16}$$

This means that the best performance is achieved by a model that accurately predicts the frequency distribution of the test set. This statement holds true for an investor with an arbitrary utility function. All investors agree on what is the perfect probability measure; they may disagree only (if they have different utility functions) on the ranking of imperfect probability measures.

We also note that, in general, our performance measure depends on the odds ratios, $\mathcal{O}$; we will get back to this point later.

Next, we generalize Definition 8.2 as follows.

Definition 11.4 *For a leveraged investor with utility function U, the utility-based relative model-performance measure for the models $q^{(1)}$ and $q^{(2)}$ is*

$$\Delta_U\left(q^{(1)}, q^{(2)}, \mathcal{O}\right) = \hat{E}_{\tilde{p}}\left[U\left(b^*\left(q^{(1)}\right), \mathcal{O}\right)\right] - \hat{E}_{\tilde{p}}\left[U\left(b^*\left(q^{(1)}\right), \mathcal{O}\right)\right].$$

In the special case $B = 1$, if the assumptions of Lemma 11.1 hold, the above performance measures reduce to the ones from Section 8.1. This follows from Lemmas 5.1 and 11.1.

Example: the power utility

The power utility is given by

$$U(W) = \frac{W^{1-\kappa} - 1}{1 - \kappa} , \tag{11.17}$$

where $\kappa \ge 0$. It follows from Lemma 11.1, (11.10), that

$$\lambda = \left\{ \frac{\sum_{y \in \mathcal{Y}} \frac{1}{\mathcal{O}_y}(q_y \mathcal{O}_y)^{\frac{1}{\kappa}}}{1 + (1 - B^{-1})\hat{b}^*} \right\}^{\kappa} ; \tag{11.18}$$

and from (11.12), we see that

$$b_y^*(q) = \{1 + (1 - B^{-1})\,\hat{b}^*\} \frac{\frac{1}{\mathcal{O}_y}(q_y\mathcal{O}_y)^{\frac{1}{\kappa}}}{\sum_{y'\in\mathcal{Y}} \frac{1}{\mathcal{O}_{y'}}(q_{y'}\mathcal{O}_{y'})^{\frac{1}{\kappa}}} + \hat{b}^* . \tag{11.19}$$

Hence,

$$U(b_y^*(q)\mathcal{O}_y - \hat{b}^*)$$

$$= \frac{1}{1-\kappa}\left(\left[\{1 + (1 - B^{-1})\,\hat{b}^*\}\frac{(q_y\mathcal{O}_y)^{\frac{1}{\kappa}}}{\sum_{y'\in\mathcal{Y}} \frac{1}{\mathcal{O}_{y'}}(q_{y'}\mathcal{O}_{y'})^{\frac{1}{\kappa}}}\right]^{1-\kappa} - 1\right)$$

$$\equiv \hat{\Phi}_y^\kappa(q) ,$$

and

$$\Delta\left(q^{(1)}, q^{(2)}, \mathcal{O}\right) = \sum_{y\in\mathcal{Y}} \tilde{p}_y \left\{\hat{\Phi}_y^\kappa(q^{(2)}) - \hat{\Phi}_y^\kappa(q^{(1)})\right\} . \tag{11.20}$$

11.1.4 Generalized Logarithmic Utility Functions: Likelihood Ratio as Performance Measure

In this section, we discuss the generalized logarithmic family of utility functions, for which the relative performance measure from Definition 8.2 is odds-ratio independent and essentially reduces to the likelihood ratio of the two models that we compare. We begin by stating the following theorem:

Theorem 11.1 *For a utility function of the form*

$$U(W) = \alpha\log(W - \gamma B) + \beta \tag{11.21}$$

with $\alpha > 0$ and

$$\gamma < 1 + (1 - B^{-1})\,\hat{b}^* , \tag{11.22}$$

where B and $\hat{b}^*$ are given by Definition 3.2 and (11.11), respectively, the relative performance measure, $\Delta_U\left(q^{(1)}, q^{(2)}, \mathcal{O}\right))$, is given by

$$\Delta_U\left(q^{(1)}, q^{(2)}, \mathcal{O}\right) = \alpha\,\Delta_{\log}\left(q^{(1)}, q^{(2)}\right) \tag{11.23}$$

$$= \alpha\,l\left(q^{(1)}, q^{(2)}\right) . \tag{11.24}$$

Proof: See Section 11.1.7.

It follows from Theorem 11.1 that for a generalized logarithmic utility function of the form (11.21) the performance measure, $\Delta_U\left(q^{(1)}, q^{(2)}, \mathcal{O}\right))$, does not depend — apart from a trivial factor — on the parameters α, β, and γ of the utility function, and does not depend on the odds ratios, $\mathcal{O}$. This means that all investors with utility functions from the family (11.21) have the same model performance measure. This performance measure is, up to a positive multiplicative constant, the likelihood ratio, no matter what the investor's specific utility function is, and that this performance measure does not depend on the odds ratios.

11.1.5 All Utilities That Lead to Odds-Ratio Independent Relative Performance Measures

We have seen in Section 11.1.4 that generalized logarithmic utility functions lead to relative model performance measures that are independent of the odds ratios. In our next theorem, we answer the following question: are there other utility functions that lead to odds-ratio independent relative performance measures?

Theorem 11.2 *If, for any empirical measure, $\tilde{p}$, any cash limits $\underline{B} \leq 0$ and $\overline{B} \geq 0$, and any candidate model measures, $q^{(1)}$, $q^{(2)}$, the relative model performance measure, $\Delta_U(q^{(1)}, q^{(2)}, \mathcal{O})$, is independent of the odds ratios, $\mathcal{O}$, then the utility function, U, must have the form*

$$U(W) = \alpha \log(W - \gamma B) + \beta \ , \qquad (11.25)$$

where α, β, and γ are constants.

Proof: Since we demand independence of the odds ratios for any cash limits $\underline{B} \leq 0$ and $\overline{B} \geq 0$, we demand it, in particular, for $\underline{B} = \overline{B} = 0$. Theorem 8.4, which considers the case $\underline{B} = \overline{B} = 0$, therefore implies Theorem 11.2. $\square$

11.1.6 Relative $(U, \mathcal{O})$-Entropy and Model Learning

In order to generalize the relative $(U, \mathcal{O})$-entropy learning approach from Chapter 10, we replace the expectations, $E[\cdot]$ in Definition 7.2, by the expectations $\hat{E}[\cdot]$ from Definition 11.3. This provides a definition of the relative $(U, \mathcal{O})$-entropy for leveraged investors. The learning approach from Chapter 10 can then be generalized in a straightforward manner. In particular, μ^{data} remains unchanged, and μ^{prior} can be defined in terms of the more general relative $(U, \mathcal{O})$-entropy

11.1.7 Proof of Theorem 11.1

Let $\gamma' = -\gamma B$. We note that, for a utility function of the form (11.21),

$$U'(z) = \frac{\alpha}{z + \gamma'} \ , \text{ and} \qquad (11.26)$$

$$(U')^{-1}(z) = \frac{\alpha}{z} - \gamma \ . \qquad (11.27)$$

Next, we solve (11.10), i.e.,

$$1 + (1 - B^{-1})\hat{b}^* = \sum_{y\in\mathcal{Y}} \frac{1}{\mathcal{O}_y}\left[\frac{\alpha q_y \mathcal{O}_y}{\lambda} - \gamma\right] \tag{11.28}$$

$$= \frac{\alpha}{\lambda} - \gamma' B^{-1}, \tag{11.29}$$

for λ (we have inserted (11.27) into (11.10)). The solution is

$$\lambda = \frac{\alpha}{1 + (1 - B^{-1})\hat{b}^* + \gamma' B^{-1}}. \tag{11.30}$$

By Lemma 11.1, (11.12) we obtain

$$b_y^* = \frac{1}{\mathcal{O}_y}\left[\frac{\alpha q_y \mathcal{O}_y}{\lambda} - \gamma'\right] + \frac{\hat{b}^*}{\mathcal{O}_y}. \tag{11.31}$$

From (11.31) we obtain

$$U(b_y^* \mathcal{O}_y - \hat{b}^*) = U\left(\frac{\alpha q_y \mathcal{O}_y}{\lambda} - \gamma'\right) \tag{11.32}$$

$$= \alpha \log\left(\frac{\alpha q_y \mathcal{O}_y}{\lambda}\right) + \beta \tag{11.33}$$

$$= \alpha \log\left(q_y \mathcal{O}_y[1 + (1 - B^{-1})\hat{b}^* + \gamma' B^{-1}]\right) + \beta \tag{11.34}$$

(by (11.30)).

The condition (11.22) in Theorem 11.1 ensures the existence of the logarithm.
By means of Definition 11.4, we now proceed to compute the relative performance measure

$$\Delta_U\left(q^{(1)}, q^{(2)}, \mathcal{O}\right) = E_{\tilde{p}}\left[U\left(b^*\left(q^{(2)}\right), \mathcal{O}\right)\right] - E_{\tilde{p}}\left[U\left(b^*\left(q^{(1)}\right), \mathcal{O}\right)\right]$$

$$= \sum_{y\in\mathcal{Y}} \tilde{p}_y\left[U\left(b_y^*\left(q^{(2)}\right)\mathcal{O}_y - \hat{b}^*\right) - U\left(b_y^*\left(q^{(1)}\right)\mathcal{O}_y - \hat{b}^*\right)\right]$$

$$= \alpha \sum_{y\in\mathcal{Y}} \tilde{p}_y \log\left(\frac{q_y^{(2)}}{q_y^{(1)}}\right), \tag{11.35}$$

where we have used (11.35) and the fact that $\hat{b}^*$, B^{-1}, and $\mathcal{O}$ are independent of the model measure. Theorem 11.1 follows then from Definition 6.2 and Theorem 8.1.□

11.2   Model Performance Measures and MRE for Investors in Incomplete Markets

In the previous chapters, we have evaluated and built probabilistic models for a gambler in a horse race. This horse race gambler, which we have defined in Chapter 3, is an idealization of a decision maker in an uncertain environment. In this idealized picture, the decision maker can place a bet on each possible state that the environment might take, i.e., for each possible state there is a bet that pays off only if the state occurs. It is obvious that this assumption doesn't always hold in practice. For example, financial markets are often incomplete, i.e., they don't offer bets on all the possible market states, but rather a set of trading instruments that corresponds to bets that pay off for more than a single state. In this section, following Huang et al. (2006), we briefly discuss a generalization of the approach from prior chapters for investors in such an incomplete market. Assuming that the market is arbitrage free, i.e., doesn't offer any opportunities for riskless return in excess over the bank account, we generalize our model performance measures, the relative U-entropy and the MRE-principle. We will restrict our attention to discrete unconditional probability measures and to one-period markets, but the ideas in this section can be generalized.

In Figure 11.1, below, we depict the modeling approach that we shall describe in the incomplete market setting of this section; Figure 11.1 is the analog of Figure 10.3, which depicts the modeling approach in the horse-race-setting of Chapter 10. As we shall see, the minimum market exploitability principle, Principle 10.2, is closely related to the least favorable market completion principle discussed below. Also, Figure 11.1 reflects the fact that we do not formulate a dual problem in this more general setting.

In this section, we assume that the reader has some familiarity with contingent claim pricing (see, for example, Duffie (1996)).

11.2.1   Investors in Incomplete Markets

We consider models (probability measures) that assign positive probabilities to the states $y \in \mathcal{Y}$ of the random variable Y, where $\mathcal{Y}$ is a finite set representing states that occur over a single trading period. If q is such a probability measure, q_y denotes the probability that $Y = y$. We denote the set of all such (positive) probability measures on $\mathcal{Y}$ by Q.

Incomplete market

We define an incomplete market as follows

Definition 11.5 *(Incomplete market) An incomplete market is characterized by a random variable Y with finite state space $\mathcal{Y}$ and a set $\mathcal{Z}$ of instruments*

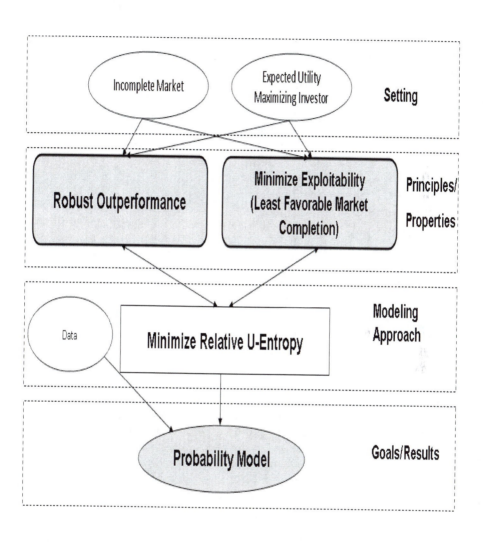

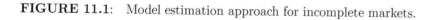

FIGURE 11.1: Model estimation approach for incomplete markets.

(contingent claims) on Y. The payoff for one dollar invested in instrument z is $\Omega_{z,y} \geq 0$ if the state y occurs. The market is arbitrage-free and frictionless, and there exists a bank account (risk-free investment) with the payoff B on a $\$1$-investment.

The above definition refers to the concept of arbitrage, which is fundamental to modern financial theory. An arbitrage is a "free lunch," i.e., an investment strategy that requires no initial capital, leads to a positive probability of a gain, and zero probability of a loss. We assume that our market is arbitrage-free, i.e., offers no arbitrage opportunities. This is a fairly reasonable assumption in financial markets, since arbitrage opportunities are rare, and, when they occur, they get exploited by those market participants who notice them and, as a consequence, disappear quickly.

We have also assumed that our market is frictionless, i.e., that there are no costs or restraints associated with transactions.

The absence of arbitrage opportunities in the market implies that there exists a set, $\Pi \neq \emptyset$, of pricing measures (see, for example, Duffie (1996), Luenberger (1998), or Bingham and Kiesel (2004)), which we can define as follows.

Definition 11.6 *(Pricing measures)*

$$\Pi = \left\{ \pi \ , \pi_y > 0 \ , \ y \in \mathcal{Y} \ , \ \frac{1}{B} \sum_{y \in \mathcal{Y}} \pi_y \Omega_{z,y} = 1 \ , \ \forall z \in \mathcal{Z} \right\} . \quad (11.36)$$

The pricing measures $\pi \in \Pi$ are normalized, i.e., $\sum_{y \in \mathcal{Y}} \pi_y = 1$, which follows from the above definition for the special case where $z = z'$ is the bank account, i.e., for $\Omega_{z',y} = B$, $\forall y \in \mathcal{Y}$. Although the pricing measures have the same mathematical form as probability measures, they should not be interpreted as probabilities. Only in a market where the market makers are risk-neutral (which, in practice, they are not), i.e., don't expect any reward for the risk they are taking, the market makers would set the payoffs to Ω if they believed in some $\pi \in \Pi$ as a probability measure for the random variable Y (see, for example, Duffie (1996), Luenberger (1998), or Bingham and Kiesel (2004)). This explains why the π_y are often referred to as risk-neutral probabilities. Another commonly used name for the π_y is state prices.

It follows from Definition 11.6 that all information about the market is contained in Π, so we can view the market as being specified by Π.

In what follows we will use the set of all contingent claims with price $\$1$, which can be expressed in terms of Π as follows.

Definition 11.7 *($\$1$ Claims)*

$$W_\Pi = \left\{ w \ , \ \sum_{y \in \mathcal{Y}} w_y \pi_y = 1, \forall \pi \in \Pi \right\} . \quad (11.37)$$

Here, w_y is the discounted (divided by B) payoff of the contingent claim w, which costs \$1, in the state y. Alternatively, we could have considered more general claims with the constant price V; in this case, the above definition would still be meaningful with w_y representing the contingent claim's payoff divided by VB.

The interpretation of the preceding definition is reinforced by the following lemma, in which we express W_Π in terms of the payoffs.

Lemma 11.2

$$W_\Pi = \left\{ w \,\middle|\, w_y = \frac{1}{B} \sum_{z \in \mathcal{Z}} \beta_z \Omega_{z,y}, \ \sum_{z \in \mathcal{Z}} \beta_z = 1 \right\}. \tag{11.38}$$

Proof: See Lemma 1 in Huang et al. (2006).

Here, β plays the role of the allocation in the horse race.

The incomplete-market investor

We make the following definition.

Definition 11.8 *(Incomplete-market investor) An incomplete-market investor is somebody who invests \$1 in an incomplete market, i.e., who invest in a contingent claim from W_Π. The investor has a utility function, $U : R \to R \cup \{-\infty\}$, that*

 (i) *is strictly concave on $\{t | U(t) > -\infty\}$,*

 (ii) *is twice differentiable on $\{t | U(t) > -\infty\}$,*

 (iii) *is strictly monotone increasing on $\{t | U(t) > -\infty\}$,*

 (iv) $\lim_{t \to \infty} U'(t) = 0$,

 (v) *satisfies*

   (a) $\lim_{t \to a+} U'(t) = \infty$ *if $dom(U) = (\alpha, \infty)$, $\alpha < 1$, and*
   (b) $\lim_{t \to -\infty} U'(t) = \infty$ *if $dom(U) = R$,*

   where $dom(U)$ is the interior of the set $\{t | U(t) > -\infty\}$, and

 (vi) *has the property $U(B) = 0$ (this doesn't cause any loss of generality since, as is well known, adding a constant to an investor's utility function does not affect his behavior).*

We have made slightly different assumptions about the investor's utility functions here than we have made for the horse race investors in the previous chapters. The reason for this is of a technical nature. As it was the case for our previous assumptions, most popular utility functions (for example, the logarithmic, exponential, and power utilities (see Luenberger (1998))) are consistent with the assumptions we made in this section.

By Lemma 11.2, the assumption that our investor invests in a contingent claim from W_Π is equivalent to assuming that our investor allocates his \$1 to the traded instruments $Z \in \mathcal{Z}$.

For the sake of simplicity, we assume from now on that $B = 1$, i.e., that the bank account pays off exactly the invested amount.

In the financial context, the setting of this section corresponds to a discrete market with a single trading period and zero interest rate. One can generalize the results from this section to multiple trading periods and to the more general state spaces considered by Kallsen (2002), Karatzas et al. (1991), and Schachermayer (2004). One can also generalize the approach to conditional probabilities and probability densities. These generalizations, however, are beyond the scope of this book.

11.2.2 Relative U-Entropy

We generalize the relative U-entropy from Definition 7.6 to the incomplete market setting.

Definition 11.9 *The relative $U-$entropy is the maximum expected utility for an incomplete-market investor with \$1 endowment, i.e., the relative U-entropy is given by*

$$D_U(q\|\Pi) = \sup_{w \in W_\Pi} \sum_{y \in \mathcal{Y}} q_y U(w_y). \tag{11.39}$$

In the special case of a complete market, where Π has only the single element π, the above expression reduces to the relative U-entropy $D_U(q\|\pi)$ from Definition 7.6. Specializing further to a generalized logarithmic utility function we obtain the Kullback-Leibler relative entropy.

We shall justify below why we call $D_U(q\|\Pi)$ a relative entropy; we shall show that $D_U(q\|\Pi)$ indeed has many properties one would expect a relative entropy to have. In particular, we shall see that $D_U(q\|\Pi)$ is

(*i*) a measure of the discrepancy between the probability measure, q, and the set of probability measures, Π,

(*ii*) consistent with the second law of thermodynamics.

The following theorems, which are proved in Huang et al. (2006), make these statements precise and list some additional properties of the relative U-entropy in an incomplete market.

Theorem 11.3

(i) $D_U(q\|\Pi)$ *is convex in q.*

(ii) $D_U(q\|\Pi) \geq 0$, with equality if and only if $q \in \Pi$.

(iii) If $\Pi^{(1)} \subset \Pi^{(2)}$, then $D_U(q\|\Pi^{(1)}) \geq D_U(q\|\Pi^{(2)})$.

Like Kullback-Leibler relative entropy, our relative U-entropy is consistent with the Second Law of Thermodynamics.

Theorem 11.4 *(Second law of thermodynamics in incomplete markets) Let $q^{(n)}$ be a probability distribution over time-n states of a stationary Markov chain with transition matrix $r_{\tilde{y}|y}$, i.e., let*

$$q_{\tilde{y}}^{(n+1)} = \sum_{y \in \mathcal{Y}} q_y^{(n)} r_{\tilde{y}|y} , \qquad (11.40)$$

and let $\Pi^{(n)}$ be the set of pricing measures corresponding to time n. Then $D_U(q^{(n)}\|\Pi^{(n)})$ decreases with n.

Next, we relate the relative U-entropy to the concept of least favorable market completion. Since each $\pi \in \Pi$ is a pricing measure, which can be used to price all securities on $\mathcal{Y}$, it specifies a complete market. In general, an expected utility maximizing investor will prefer such a complete market to the incomplete market specified by Π, because (by Theorem 11.3, (iii)) $D_U(q\|\pi) \geq D_U(q\|\Pi)$, i.e., his expected utility is greater for the complete market. However, if there exists a $\pi^* \in \Pi$, such that $D_U(q\|\pi^*) = D_U(q\|\Pi)$, the investor will derive the same optimal expected utility in both the incomplete market and the completed market, so he will not be motivated to make use of or invest in any of the new, fictitious securities used to complete the market. In this case, it makes sense to use π^* as the pricing measure for all securities, since the completed market is equivalent to the original market from the point of view of an expected utility maximizer.

By Theorem 11.3, (iii), if such a measure, π^* exists, we can make the following definition.

Definition 11.10 *(Least favorable market completion) The least favorable market completion is given by*

$$\pi^* = \arg\min_{\pi \in \Pi} D_U(q\|\pi) . \qquad (11.41)$$

The following theorem shows that such a measure exists.

Theorem 11.5 *(Relative U-entropy characterization of the least favorable market completion) Let Π be the set of all pricing probabilities of some market, then*

(i) the least favorable market completion,

$$\pi^* = \arg\min_{\pi \in \Pi} D_U(q\|\pi), \qquad (11.42)$$

exists and is unique, and

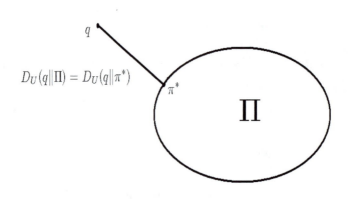

FIGURE 11.2: The discrepancy between a probability measure, q, and a set of probability measures, Π, is the minimum of all discrepancies between q and each of the elements of Π. (This figure originally appeared in Huang et al. (2006).)

(ii)

$$D_U(q\|\Pi) = D_U(q\|\pi^*). \tag{11.43}$$

Proof: See Karatzas et al. (1991), Schachermayer (2004), or Theorem 5 from Huang et al. (2006).

It follows from Theorem 11.5 that, if we measure discrepancy by the relative U-entropy, the discrepancy between a probability measure, q, and a set of probability measures, Π, is the minimum of all discrepancies between q and each of the elements of Π. This property is depicted in Figure 11.2.

One can easily derive some explicit expressions for the optimal contingent claim (the claim that affords the greatest expected utility)

$$w^* = \arg \max_{w \in W_\Pi} \sum_y q_y U(w_y) \tag{11.44}$$

and the least favorable market completion pricing measure. For that purpose, we pick a finite subset $\{\pi^{(j)}\}_{j=1,2,\ldots J} \subset \Pi$ which spans Π, i.e.,

$$\Pi \subset \text{aff}(\{\pi^{(j)}\}_{j=1,2,\ldots J}), \tag{11.45}$$

where $\text{aff}(S)$ denotes the affine hull of S

$$\text{aff}(S) = \{\lambda s_1 + (1-\lambda)s_2 | s_1, s_2 \in S; \lambda \in R\}. \tag{11.46}$$

We have

(i) The optimal contingent claim can be represented as:

$$w_y^* = (U')^{-1} \left(\frac{\sum_j \lambda_j \pi_y^{(j)}}{q_y} \right), \tag{11.47}$$

where the λ_j's are picked such that $\sum_y w_y^* \pi_y^{(j)} = 1$. (This follows from Lagrangian optimization.)

(ii) The least favorable completion is given by the Davis Formula (see, for example, Bingham and Kiesel (2004), p. 291):

$$\pi_y^* = \frac{q_y \cdot U'(w_y^*)}{\sum_{\tilde{y}} q_{\tilde{y}} U'(w_{\tilde{y}}^*)} \tag{11.48}$$

$$= \frac{\sum_j \lambda_j \pi_y^{(j)}}{\sum_{\tilde{y}} \sum_j \lambda_j \pi_{\tilde{y}}^{(j)}} \tag{11.49}$$

(this follows from Lagrangian optimization).

11.2.3 Model Performance Measure

In this section we measure the performance of probabilistic models for investors in incomplete markets. To accomplish this, we utilize the same ideas that we have used in Section 8.1 for investors in a complete market, i.e., we assume that there is a rational investor who believes the candidate model and invests so as to maximize his expected utility, and we measure model performance by means of an (out-of-sample) estimate of the expected utility attained by this investor. In the incomplete market setting considered in this section, we make the following definition.

Definition 11.11 *(Relative model performance measure in incomplete market) Let $\tilde{p}$ denote the empirical measure on our test set. The relative performance, Δ, between the models $q^{(1)}$ and $q^{(2)}$ is*

$$\Delta_U \left(q^{(1)}, q^{(2)}, \Omega \right) = E_{\tilde{p}} \left[U \left(\sum_{z \in \mathcal{Z}} \beta_z^*(q^{(2)}) \Omega_{z,\cdot} \right) \right] - E_{\tilde{p}} \left[U \left(\sum_{z \in \mathcal{Z}} \beta_z^*(q^{(1)}) \Omega_{z,\cdot} \right) \right], \tag{11.50}$$

where

$$\beta^*(q) = \arg \max_{\beta: \sum_{z \in \mathcal{Z}} \beta_z = 1} \sum_{y \in \mathcal{Y}} q_y U \left(\sum_{z \in \mathcal{Z}} \beta_z \Omega_{z,y} \right) \tag{11.51}$$

is an optimal betting strategy corresponding to the model q.

From Lemma 11.2, it follows that we can express this relative model performance measure in terms of the optimal contingent claims as follows.

$$\Delta_U\left(q^{(1)}, q^{(2)}, \Omega\right) = E_{\tilde{p}}\left[U\left(w^*(q^{(2)})\right)\right] - E_{\tilde{p}}\left[U\left(w^*(q^{(1)})\right)\right], \qquad (11.52)$$

where

$$w^*(q) = \arg\max_{w \in W_\Pi} E_q[U(w)] \qquad (11.53)$$

is the optimal \$1 contingent claim the investor chooses based on the model q, and W_Π is related to Ω via Lemma 11.2. The optimal claim w^* is known to exist and to be unique (see Theorem 2.18 of Schachermayer (2004)).

Below we shall compute the above performance measure for a particular utility function and discuss a numerical example.

Model performance measure for an investor with a generalized logarithmic utility

The theorem below expresses our relative model performance measure for an investor with a generalized logarithmic utility in terms of the differences between two log-likelihood ratios.

Theorem 11.6 *For an investor with the utility function*

$$U(W) = \alpha \log(W - \gamma B) + \beta , \qquad (11.54)$$

the relative performance, Δ, between the models $q^{(1)}$ and $q^{(2)}$ is

$$\Delta_U\left(q^{(1)}, q^{(2)}, \Omega\right) = \alpha\left[l\left(q^{(1)}, q^{(2)}\right) - l\left(\pi^*\left(q^{(1)}\right), \pi^*\left(q^{(1)}\right)\right)\right], \qquad (11.55)$$

where

$$\pi^*(q) = \arg\min_{\pi \in \Pi} E_q\left[\log\left(\frac{q}{\pi}\right)\right] , \qquad (11.56)$$

Π *is related to Ω via Definition 11.6, and l denotes the log-likelihood ratio from Definition 6.2.*

Proof: See Section 11.2.6.

Theorem 11.6 states that for a generalized-logarithmic utility investor, the relative performance measure between two models is — up to a constant positive factor — the difference between the log-likelihood ratio of the two models and the log-likelihood ratio of the least favorable pricing measures corresponding to the two models.

As an example, let us consider Theorem 11.6 in an extreme case: a complete market, i.e., a market that allows an investor to place bets on every possible state; such a complete market is the same as the horse race from Chapter 3. In a complete market with odds ratios $\mathcal{O}$, it follows from Definition 11.6

that $\Pi = \{\pi : \pi_y = BO_y^{-1}\}$, i.e., that the set Π contains only a single pricing measure. It follows then from (11.56) that $\pi_y^*(q) = BO_y^{-1}$. Therefore, by (11.55), $\Delta_U\left(q^{(1)}, q^{(2)}, \Omega\right) = \alpha E_{\tilde{p}}\left[\log\left(\frac{q^{(2)}}{q^{(1)}}\right)\right] = \alpha l\left(q^{(1)}, q^{(2)}\right)$, where l denotes the log-likelihood ratio from Definition 6.2. That is, in a complete market, the model performance measure (for an investor with a generalized logarithmic utility function) is, up to a constant positive factor, the log-likelihood ratio. This, of course, we have shown already in Chapter 8, where we considered investors in a horse race, i.e., investors in a complete market (see Theorem 8.2). Hence Theorem 11.6 is consistent with the results from Chapter 8.

As a second example, let us consider the other extreme: a market that offers the bank account as its only traded instrument. In this case, it follows from Definition 11.6 that $\Pi = \{\pi : \sum_{y \in \mathcal{Y}} \pi_y = 1\}$, i.e., that Π is the set of all measures, and it follows from (11.56) that $\pi_y^*(q) = q_y$. Therefore, by (11.55), we have $\Delta_U\left(q^{(1)}, q^{(2)}, \Omega\right) = 0$. That is, in a market that offers only the bank account, all models perform the same. This is consistent with our intuition: if all the investor can do is put his money in a bank account, he has no opportunity to exploit the information (we use the word in loose sense here) contained in a model; therefore, this information is useless from a decision-theoretic perspective and doesn't contribute to the model's performance.

Example: Loss models for defaultable debt

As an example, we measure the relative performance of loss models for default-able debt from the viewpoint of an investor with the generalized logarithmic utility. We consider probabilistic models for the random variable Y, which is the loss relative to the par amount of a defaultable debt security. Let us assume that the state space for Y is $\mathcal{Y} = \{0, \frac{1}{m-1}, \frac{2}{m-1}, ..., 1\}$ where m is some large integer. Here, the event of no default is captured by the state $Y = 0$. We further assume that there are the following three traded instruments.

(*i*) a bank account with the payoff

$$\Omega_{1,y} = B , \ \forall y , \tag{11.57}$$

(*ii*) a defaultable zero-coupon bond with the payoff

$$\Omega_{2,y} = a(1 - y) , \tag{11.58}$$

and

(*iii*) a digital default swap with the payoff

$$\Omega_{3,y} = c\left(1 - \delta_{y,0}\right) . \tag{11.59}$$

(For the sake of simplicity, we assume that $y > 0$ whenever there is a default, i.e., that there is no complete recovery in the case of default.)

In order to compute our relative model performance measure, we first compute the worst-case pricing measure. From (11.56) and Definition 11.6 we obtain

$$\pi^*(q) = \arg\min_{\pi \in \Pi} E_q \left[\log \left(\frac{q}{\pi} \right) \right] , \tag{11.60}$$

where

$$\Pi = \left\{ \pi, \sum_{y \in \mathcal{Y}} \pi_y = 1, \sum_{y \in \mathcal{Y}} \pi_y y = \frac{a - B}{b}, \pi_0 = \frac{c - B}{c}, \pi_y > 0 , y \in \mathcal{Y} \right\}. \tag{11.61}$$

It follows from Lagrangian duality (see Huang et al. (2006) for details) that the solution of the problem posed by (11.60) is

$$\pi_y^*(q) = \frac{q_y}{\lambda^* + \eta^* y + \nu^* \delta_{y,0}} , \tag{11.62}$$

where

$$(\lambda^*, \eta^*, \nu^*) = \arg \max_{\lambda,\eta,\nu:\lambda+\eta y+\nu\delta_{y,0}>0 , \forall y \in \mathcal{Y}} h(\lambda, \eta, \nu) \tag{11.63}$$

and

$$h(\lambda, \eta, \nu) = \sum_{y \in \mathcal{Y}} q_y \log (\lambda + \eta y + \nu \delta_{y,0}) - \lambda - \eta \frac{a - B}{a} - \nu \frac{c - B}{c}. \tag{11.64}$$

Based on these equations and Theorem 11.6, we can compute the relative performance measure for the models $q^{(1)}$ and $q^{(2)}$; we obtain

$$\Delta_U \left(q^{(1)}, q^{(2)}, \Omega \right) = \alpha E_{\tilde{p}} \left[\log \left(\frac{\lambda_2^* + \eta_2^* y + \nu_2^* \delta_{y,0}}{\lambda_1^* + \eta_1^* y + \nu_1^* \delta_{y,0}} \right) \right] , \tag{11.65}$$

where the $\lambda_i^*, \eta_i^*, \nu_i^*$ are given by (11.63) with q replaced by $q^{(i)}$.

We note that this performance measure depends only on three parameters per model, which encode all the model information necessary to evaluate model performance.

We illustrate the above logic via a numerical example. Suppose that we have a defaultable loan with $a = 1.2$, a bank account with $B = 1.05$, and a digital default swap with $c = 1.3$. We assume that the empirical loss distribution is the one shown in Fig. 11.3, and that we want to evaluate the relative performance of the probability measures $q^{(1)}$, $q^{(2)}$, and $q^{(3)}$ from Fig. 11.3 with respect to the reference measure $q^{(0)}$ shown in Fig. 11.3. The probability measures $q^{(0)}, q^{(1)}, q^{(2)}$, and $\tilde{p}$ all assign the probability 0.95 to the state $Y = 0$, i.e., to the event of no default; measure $q^{(3)}$ assigns the probability 0.9 to the state $Y = 0$. The measures $\tilde{p}$, $q^{(0)}$, $q^{(1)}$, and $q^{(2)}$ differ in the probabilities that they assign to loss, given that a default has occurred. The measure $q^{(3)}$ assigns the same (conditional) probabilities to loss given default as $q^{(1)}$. The default-conditioned measures $q^{(1)}$, $q^{(2)}$, and $q^{(3)}$ are discretized beta-distributions for $Y > 0$, while the measure $q^{(0)}$ is uniform for $Y > 0$.

Using (11.65), we compute the performance measures $\Delta_U \left(q^{(0)}, p, \Omega \right)$, where we have set $\alpha = 1$ and p is $\tilde{p}, q^{(1)}, q^{(2)}$, or $q^{(3)}$. For comparison, we also compute the complete-market performance measure $\Delta_U^{(c)} \left(q^{(0)}, p, \Omega \right) = l(q^{(0)}, p)$, which reflects the model performance from the viewpoint of an investor who invests in a market that allows for bets on each state of $Y \in \mathcal{Y}$. We note that $\Delta_U^{(c)} \left(q^{(0)}, p, \Omega \right)$ is generally different from $\Delta_U \left(q^{(0)}, p, \Omega \right)$ although the two expected utility terms in the latter measure can be expressed as expected utilities in certain worst-case complete markets. The reason for this difference is that the least favorable complete market depends on the model.

The results are shown in Table 11.1. In a complete market (see the third column of Table 11.1), where an investor can bet on every possible outcome, an investor who knows the empirical measure, $\tilde{p}$, achieves the best outperformance (with respect to the reference investor), an investor who believes the model $q^{(1)}$ fares almost as well as the $\tilde{p}$-investor, and an investor who believes one of the misleading models $q^{(2)}$ or $q^{(3)}$ does worse than the reference investor. In our incomplete market the model performance measures reflect a different picture: the differences between the models $\tilde{p}$, $q^{(1)}$, and $q^{(2)}$ are much smaller than for the complete market, while the model $q^{(3)}$ considerably underperforms, as is the case for the complete market. The reason for the similarity in the performance of the models $\tilde{p}$, $q^{(1)}$, and $q^{(2)}$ (and the benchmark $q^{(0)}$) is that these measures differ only in the conditional probabilities of loss given default, on which the investor can bet only in a very crude way. The reason for the significant underperformance of the model $q^{(3)}$ is that the digital default swap allows for bets on the event of default.

The above example illustrates that the relative model performance measure Δ rewards only that information from a model that can be used to improve investment strategies. On the other hand, information that cannot be exploited by an investor is not reflected in Δ.

11.2.4   Model Value

We generalize Definition 8.6 (see Section 8.6) of the monetary value of a model upgrade to the incomplete market setting as follows.

Definition 11.12 *(Monetary value of a model upgrade in incomplete market)*
The monetary value, $V_U \left(q^{(1)}, q^{(2)}, \Omega \right)$, of upgrading from model $q^{(1)}$ to model $q^{(2)}$ is the solution for V of the following equation:

$$E_{\tilde{p}} \left[U \left(\sum_{z \in \mathcal{Z}} \beta_z^* \left(q^{(2)}, 0 \right) \Omega_{z,\cdot} \right) \right] = E_{\tilde{p}} \left[U \left(\sum_{z \in \mathcal{Z}} \beta_z^* \left(q^{(1)}, V \right) \Omega_{z,\cdot} + V \right) \right],$$

(11.66)

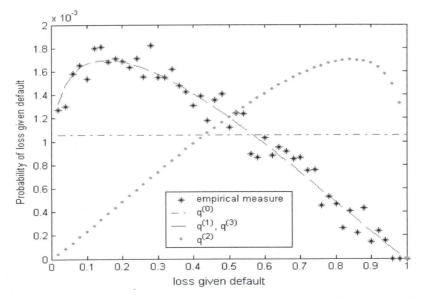

FIGURE 11.3: Conditional probabilities of loss given default (corresponding to the empirical probabilities $\tilde{p}$, the reference measure $q^{(0)}$, and the three models, $q^{(1)}$, $q^{(2)}$, and $q^{(3)}$) for a particular loan as an example. Only the probabilities for $Y > 0$ given default are shown; the probability for $Y = 0$ given default is zero (the default probabilities are 0.05 for $\tilde{p}$, $q^{(0)}$, $q^{(1)}$, and $q^{(2)}$, and 0.1 for $q^{(3)}$). (This figure originally appeared in Huang et al. (2006).)

where

$$\beta^*(q, V) = \arg\max_{\beta: \sum_{z \in \mathcal{Z}} \beta_z = 1} \sum_{y \in \mathcal{Y}} q_y U\left(\sum_{z \in \mathcal{Z}} \beta_z \Omega_{z,y} + V\right), \qquad (11.67)$$

and $\tilde{p}$ is the empirical probability measure of the test set.

It is easy to show that in the case of a generalized logarithmic utility function, i.e., for $U(W) = \alpha \log(W - \gamma B) + \beta$, the following equation holds

$$V_{U,\Omega}\left(q^{(1)}, q^{(2)}\right) = B(1 - \gamma)\left(e^{\frac{1}{\alpha}\Delta_U\left(q^{(1)}, q^{(2)}, \Omega\right)} - 1\right), \qquad (11.68)$$

as it does for a complete market (see Theorem 8.12).

11.2.5   Minimum Relative U-Entropy Modeling

We modify the MRE approach from Section 10.3 by replacing the horse-race relative U-entropy by the incomplete-market relative U-entropy from Definition 11.9. This leads to the following problem.

TABLE 11.1: Relative performance measures, $\Delta_U\left(q^{(0)}, \cdot, \Omega\right)$, for an investor in our incomplete market, and $\Delta_U^{(c)}\left(q^{(0)}, \cdot, \Omega\right)$, for an investor in a complete market, for various loss probability measures (see Fig. 11.3) with respect to the measure $q^{(0)}$.

Model	$\Delta_U\left(q^{(0)}, \cdot, \Omega\right)$	$\Delta_U^{(c)}\left(q^{(0)}, \cdot, \Omega\right)$
$\tilde{p}$	0.0003	0.0092
$q^{(1)}$	0.0002	0.0088
$q^{(2)}$	-0.0041	-0.0240
$q^{(3)}$	-0.0165	-0.0079

Problem 11.1 *Find*

$$q^* = \arg\min_{q\in K} D_U(q\|\Pi) \tag{11.69}$$

$$\text{where } K = \{q, q \in Q\,,\, E_p[f] = E_{\tilde{p}}[f]\}\,, \tag{11.70}$$

where f is a (vector-valued) features function of Y.

We note that by Theorem 11.3, Problem 11.1 is a convex problem.

From an information-theoretic viewpoint, q^* is the measure that minimizes, by means of $D_U(q\|\Pi)$, the "distance" to the set of pricing measures Π among all measures consistent with the constraint. From an expected utility point of view, on the other hand, it is not obvious why q^* is a useful probability measure. Its usefulness is derived from the property stated in the following theorem, which is based on the same argument from Topsøe (1979) that we have used in Theorem 10.2 in the horse race context.

Theorem 11.7 *Suppose that the domain of the utility function U is $(\zeta, \infty]$, $\lim_{t\to\zeta+} U(t) = -\infty$, and that there exists $p^{(0)} \in K$, such that $p_y^{(0)} > 0, \forall y \in \mathcal{Y}$, then*

$$\arg\min_{q\in K} D_U(q\|\Pi) = \arg\max_{q\in Q} \min_{p'\in K} E_{p'}[U(w^*(p))] \tag{11.71}$$

where

$$w^*(p) = \arg\max_{w\in W_\Pi} E_p[U(w)] \tag{11.72}$$

is the optimal \$1 contingent claim the investor chooses based on the model p.

Proof: See Theorem 6 from Huang et al. (2006).

This theorem states that, under additional (beyond the assumption from Definition 11.8) regularity conditions, by choosing q^* an expected utility maximizing investor maximizes his (model-based optimal) expected utility in the most adverse environment consistent with the feature constraints. In other words, the measure q^* is the worst-case optimal measure — this measure is a measure that could be used by an investor who bets in an incomplete market and wants to maximize his worst-case (in the sense of (11.71)) expected utility. Thus, Problem 11.1 specifies a statistical learning problem appropriate for this investor.

11.2.6   Proof of Theorem 11.6

Let $\gamma' = -\gamma B$. Equation (11.56) follows directly from Definition 11.10. In order to prove (11.55), we start with (11.48), i.e., with

$$\pi_y^*(q) = \frac{q_y \cdot U'(w_y^*)}{\sum_{\tilde{y}} q_{\tilde{y}} U'(w_{\tilde{y}}^*)} \; , \tag{11.73}$$

which specializes for the utility function $U(W) = \alpha \log(W + \gamma') + \beta$, to

$$\pi_y^*(q) = \frac{q_y}{(w_y^* + \gamma')S} \; , \tag{11.74}$$

where

$$S = \sum_{y \in \mathcal{Y}} \frac{q_y}{w_y^* + \gamma'} \; . \tag{11.75}$$

(11.74) can be written as

$$\pi_y^*(q)w_y^* + \gamma'\pi_y^*(q) = \frac{1}{S}q_y. \tag{11.76}$$

Summing over y and using the fact that $w \in W_{\Pi}$ together with Definition 11.7, the above equation results in

$$S = \frac{1}{1 + \gamma'} \; . \tag{11.77}$$

Combining this equation with (11.74), we obtain

$$w_y^* + \gamma' = (1 + \gamma')\frac{q_y}{\pi_y^*(q)} \; , \tag{11.78}$$

which, after inserting into (11.52) and using Definition 6.2, results in (11.55). This completes the proof of the theorem. □

11.3   Utility-Based Performance Measures for Regression Models

So far we have discussed performance measures for models that provide conditional or unconditional probability distributions. However, many models that practitioners use do not provide probabilities, but rather relate some characteristics of a certain variable, such as its conditional expectation, to a vector of explanatory variables. Let us, in a slight deviation from the usual convention, call such models regression models. In this section, following Friedman and Sandow (2006a), we introduce utility-based performance measures for such regression models that are constructed in the same spirit as the probabilistic model performance measures from Chapter 8.

As in Chapter 8, we adopt the viewpoint of an investor in a horse race who uses a model to place his bets. The investor has a regression model, but does not have a probability measure for the variable under consideration. All he knows is some expectations of this variable, and he has no way of attaching weights to all (usually infinitely many) probability measures that are compatible with these expectations. We assume that our investor is conservative and, therefore, prepares himself for the most adverse (as measured by the expected utility) scenario. This leads to the following paradigm for regression model valuation: We evaluate the strategy that maximizes the gain in expected utility under the most adverse probability measure consistent with the model, where the gain is defined with respect to the investor's prior strategy (the strategy that seems expected utility-optimal based on the investor's prior knowledge). We do so by measuring the out-of-sample gain in expected utility of the investor who invests in a horse race according to this strategy.

We shall see that there is an interesting relationship between the above model performance measure and the relative $(U, \mathcal{O})$-entropy, which allows for an alternative interpretation of the performance measure: The investor constructs a probabilistic model from the regression model by minimizing the information-theoretic distance to his prior over all measures consistent with the regression model, and then evaluates this probabilistic model; the result is the above performance measure for regression models.

If the investor's utility function is a member of the generalized logarithmic family, the problem of minimizing the relative $(U, \mathcal{O})$-entropy takes a particularly simple form and the model performance measure is independent of the odds ratios. We shall derive the dual problem of the relative $(U, \mathcal{O})$-entropy minimization in this case, which leads to a recipe for calculating model performance measures.

We shall discuss in more detail the performance measures for a few specific, commonly used, regression models, under the assumption that the investor's utility function is from the generalized logarithmic family. Under some additional assumptions, we derive approximations for these performance measures, which can be easily implemented. By means of an example, we demonstrate how the different pieces of information that are provided by the specific regression models create value for an investor.

The approach we take to measuring the performance of regression models overcomes some of the drawbacks of measuring model performance in terms of the mean squared error or some related quantity, which is arguably the most commonly used approach (see, for example, Davidson and MacKinnon (1993), or Hastie et al. (2009)). One of these drawbacks, for example, is the fact that the mean squared error does not make a distinction between an upward and a downward prediction error. Of course, there are alternatives to this approach that don't suffer from this deficiency. Such alternatives include various loss functions (see, for example, Hastie et al. (2009), or Berger (1985)), or the recently introduced deviation measures from Rockafellar et al. (2002b) and Rockafellar et al. (2002a). The latter ones are closely related to the coherent risk measures from Artzner et al. (1999). We shall not discuss any of these alternatives in this book.

11.3.1 Regression Models

We consider models that relate some characteristics of the random variable Y (with values, y, in a finite set $\mathcal{Y} \subset \mathbf{R}^n$) to the random variable X (with values, x, in a finite set $\mathcal{X} \subset \mathbf{R}^d$). Throughout Section 11.3, we assume that $\mathcal{X}$ is the set of actually observed X-values in the test set, so that $\tilde{p}_x > 0$, $\forall x \in \mathcal{X}$, where $\tilde{p}$ is the empirical measure on the test set. We have chosen finite state spaces for the sake of convenience; it is straightforward to generalize the performance measures in this section to models on infinite state spaces. Specifically, we consider models of the following type.

Definition 11.13 *(Regression model) A regression model is a collection of relations between the random variables X and Y of the following form:*

$$f_j(X, Y) = \overline{f}_j(X) + \epsilon_j \ , \ j = 1, ..., N_f \ , \tag{11.79}$$

$$g_j(X, Y) = \overline{g}_j + \zeta_j \ , \ j = 1, ..., N_g \ , \tag{11.80}$$

where the ϵ_j and the ζ_j are random variables with some unknown joint probability measure, p, such that

$$E_p[\epsilon_j | X = x] = 0 \ , \forall \, x \in \mathcal{X} \ , j = 1, ..., N_f \tag{11.81}$$

$$and \qquad E_p[\zeta_j] = 0 \ , \forall \, j = 1, ..., N_g \ . \tag{11.82}$$

We denote the regression model by the quadruple $(f, \overline{f}, g, \overline{g})$, where $f = (f_1, ..., f_{N_f})^T$, $\overline{f} = (\overline{f}_1, ..., \overline{f}_{N_f})^T$, $g = (g_1, ..., g_{N_g})^T$ and $\overline{g} = (\overline{g}_1, ..., \overline{g}_{N_g})^T$.

The above definition of a regression model is slightly more general than the standard definition (see, for example, Davidson and MacKinnon (1993), or Hastie et al. (2009)). Usually, the term "regression model" refers to a model of the form $y = \mu(x) + \delta$; our definition can include "side information," for example, about the conditional variance. Also, according to our definition, a regression model might provide the conditional variance, without providing the conditional expectation.

In Definition 11.13, we assume that the probability distribution for the error terms is unknown; in the case where the distribution is known, it is possible to apply the performance measures for conditional probability models from Chapter 8.

Below, we provide a few simple examples where $\mathcal{Y} \subset \mathbf{R}^1$.

Model 11.1 *(Point estimator)*

$$Y = \mu(X) + \epsilon_1 \, . \tag{11.83}$$

Model 11.2 *(Point estimator with unconditional variance)*

$$Y = \mu(X) + \epsilon_1 \, , \tag{11.84}$$
$$\text{and} \quad (Y - \overline{y})^2 = \sigma^2 + \zeta_1, \tag{11.85}$$
$$\text{where } Y = \overline{y} + \zeta_2 \, . \tag{11.86}$$

We assume that the model parameters are such that the model is consistent.

Model 11.3 *(Point estimator with conditional variance)*

$$Y = \mu(X) + \epsilon_1 \tag{11.87}$$
$$\text{and} \quad (Y - \mu(X))^2 = \sigma^2(X) + \epsilon_2. \tag{11.88}$$

11.3.2 Utility-Based Performance Measures

As we have done in Section 8.3 when we evaluated conditional probability models, we evaluate a regression model in terms of its usefulness for an investor who places bets in a conditional horse race (see Definition 3.7 for the conditional horse race). A regression model typically provides less information than a complete probabilistic model; it is consistent with more than one probability measure. In order to evaluate a regression model, we assume that our investor is conservative and prepares for the most adverse measure consistent with the regression model. Hence, we assume that our investor chooses a betting strategy (the so-called robust allocation) that maximizes the gain in expected utility under a worst-case measure compatible with the model; we measure the gain with respect to the prior strategy

$$b^{(0)} = b^* \left(q^{(0)} \right) \, , \tag{11.89}$$

where $q^{(0)}$ is the prior probability measure and b^* is the optimal allocation corresponding to this prior measure (see Definition 5.6). We then measure model performance in terms of the gain in expected utility under this strategy on an out-of-sample dataset. Thus, we measure model performance based on the following two definitions.

Definition 11.14 *(Robust allocation for a regression model) The robust allocation for the model $M = (f, \overline{f}, g, \overline{g})$ is given by*

$$\hat{b}(M) = \arg \max_{b \in \mathcal{B}} \min_{p \in C(M)} \left\{ E_p[U(b, \mathcal{O})] - E_p \left[U \left(b^{(0)}, \mathcal{O} \right) \right] \right\}, \tag{11.90}$$

where

$$E_p[U(b, \mathcal{O})] = \sum_{x \in \mathcal{X}} \tilde{p}_x \sum_{y \in \mathcal{Y}} p_{y|x} U(b_{y|x} \mathcal{O}_{y|x}) , \qquad (11.91)$$

$$\mathcal{B} = \left\{ b, \sum_{y \in \mathcal{Y}} b_{y|x} = 1 , \ x \in \mathcal{X} \right\} , \qquad (11.92)$$

$\tilde{p}$ is the empirical measure of an out-of-sample test set, $C(M)$ denotes the set of all measures compatible with the model M, i.e.

$$C(M) = \{p, \ 0 \leq p_{y|x} \leq 1, \sum_{y \in \mathcal{Y}} p_{y|x} = 1 , \ E_p[f|x] = \overline{f}(x) , \ \forall x \in \mathcal{X} , \ E_p[g] = \overline{g}\},$$

$$(11.93)$$

and $b^{(0)} \in \mathcal{B}$ is the prior strategy from (11.89).

Definition 11.15 *(Regression model performance measure) The performance of the model M is measured by*

$$\Delta_U \left(M, b^{(0)}, \mathcal{O} \right) = E_{\tilde{p}} \left[U \left(\hat{b}(M), \mathcal{O} \right) \right] - E_{\tilde{p}} \left[U \left(b^{(0)}, \mathcal{O} \right) \right] . \qquad (11.94)$$

According to Definition 11.14, our investor uses a minimax decision rule (see, for example, Berger (1985) for minimax decision rules). As we shall see, the robust allocation is also optimal with respect to less adverse measures.

Depending on how many relations the regression model provides, the set, $C(M)$, of probability measures consistent with the regression model can contain many or few measures. If there are relatively few relations in our regression model, $C(M)$ contains many probability measures. In this case, preparing for a worst-case measure may suggest a rather conservative allocation strategy, which leads to a rather conservative performance measure. For a regression model with relatively many relations, on the other hand, there is a relatively limited set of probability measures consistent with the model. In this case, preparing for a worst-case measure may suggest a less conservative allocation strategy, which leads to a less conservative performance measure. In the extreme case, where the regression model consists of a set of relations (for example δ-kernels for all states) that completely define a conditional probability measure, the model performance measure from Definition 11.15 is identical to the one for conditional probability models from Definition 8.4.

11.3.3   Robust Allocation and Relative $(U, \mathcal{O})$-Entropy

In Chapter 7, we have introduced the relative $(U, \mathcal{O})$-entropy, $D_{U,\mathcal{O}}(p||q^{(0)})$ (see Definition 7.2), which can be interpreted as the loss in expected utility experienced by an investor who bets according to model $q^{(0)}$ when p is the "true" probability measure, and we have discussed the minimum relative entropy (MRE) problem. The following theorem relates the robust allocation from the previous section to the MRE measure, i.e., to the measure that minimizes the relative $(U, \mathcal{O})$-entropy.

Theorem 11.8 *If $dom(U)$ is a closed set or $dom(U) = (\xi, \infty)$ and $\lim_{W \to \xi+} U(W) = -\infty$, then the robust allocation exists, is unique, and can be expressed as*

$$\hat{b}(M) = b^*(p^*), \tag{11.95}$$

$$where \; p^* = \arg \min_{p \in C(M)} D_{U,\mathcal{O}}(p||q^{(0)}), \tag{11.96}$$

$$b^*(p^*) = \arg \max_{b \in \mathcal{B}} E_{p^*}[U(b, \mathcal{O})] \tag{11.97}$$

$$= \left\{ \frac{1}{\mathcal{O}_{y|x}} (U')^{-1} \left(\frac{\lambda_x}{p^*_{y|x} \mathcal{O}_{y|x}} \right), \; x \in \mathcal{X}, \; y \in \mathcal{Y} \right\}, \tag{11.98}$$

and, for each $x \in \mathcal{X}$, λ_x is the solution of

$$\sum_{y \in \mathcal{Y}} \frac{1}{\mathcal{O}_{y|x}} (U')^{-1} \left(\frac{\lambda_x}{p^*_{y|x} \mathcal{O}_{y|x}} \right) = 1. \tag{11.99}$$

Proof. See Friedman and Sandow (2006a).

According to Theorem 11.8, the smooth MRE measure induces an allocation that is robust, in the sense that it maximizes the outperformance under the most extreme measures consistent with the model.

While the MRE measure, p^*, and the robust allocation, $\hat{b} = b^*(p^*)$, are unique, there is no unique worst-case measure, i.e., there is no unique minimizer

$$\min_{p \in C(M)} K(\hat{b}, p)$$

of the outperformance

$$K(\hat{b}, p) = E_p[U(\hat{b}, \mathcal{O})] - E_p[U(b^*(q^{(0)}), \mathcal{O})]. \tag{11.100}$$

This point is illustrated in Figure 11.4 for an example where $U(W) = \log(W)$, $\mathcal{X}$ has one state, $\mathcal{Y} = \{1, 2\}$, $b = (b_1, 1 - b_1)$, $p = (p_1, 1 - p_1)$, $q^{(0)} = (\frac{1}{2}, \frac{1}{2})$, we obtain $K(b, p) = p_1 \log(b_1) + (1 - p_1) \log(1 - b_1)$, with $\hat{b} = (\frac{1}{2}, \frac{1}{2})$, and $p^* = (\frac{1}{2}, \frac{1}{2})$. For $b_1 = \hat{b}_1 = \frac{1}{2}$, $K(b, p) = \log(\frac{1}{2})$ (the bold line) is independent of p_1, and therefore $\min_{p \in C(M)} K(\hat{b}, p)$ is not unique.

The practical importance of Theorem 11.8 lies in the fact that it leads to efficient numerical solution schemes for finding performance measures for regression models. When one uses the theorem, it is only necessary to solve one measure-related optimization problem, rather than a measure-related optimization problem for each allocation in Definition 11.15. In order to compute the model performance measure, we first have to compute the MRE measure,

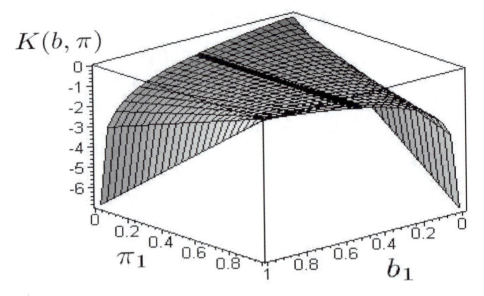

FIGURE 11.4: Outperformance $K(b, p)$ for a particular example. (This figure originally appeared in Friedman and Sandow (2006a).)

p^*, by means of (11.96). This amounts to a minimization of a relative $(U, \mathcal{O})$-entropy under constraints that can be expressed in terms of expectations over the measure we have to find. Having computed p^*, we can proceed to compute the robust allocation, $\hat{b}$, by means of (11.95) and (11.98). Finally, we can compute the relative performance measure, $\Delta_U(M, b^{(0)}, \mathcal{O})$, by means of (11.94) in Definition 11.15.

Theorem 11.8 is also interesting from a theoretical perspective, since it provides an alternative interpretation of our performance measure: The investor constructs a probabilistic model from the regression model by minimizing the information-theoretic distance to his prior over all measures consistent with the regression model. We then evaluate this probabilistic model (with respect to the prior model, which underlies the prior allocation) as in Section 8.3, i.e., by means of Definition 8.4.

11.3.4 Performance Measure for Investors with a Generalized Logarithmic Utility Function

We consider an investor with the generalized logarithmic utility function from Definition 5.4, i.e., with the utility function

$$U(W) = \gamma_1 \, log(W - \gamma B) + \gamma_2 \,, \tag{11.101}$$

with $\gamma < 1$ and $\gamma_1 > 0$. We have seen in Section 8.3.2 that, under certain conditions, a relative performance measure based on a utility function from this family can be used to approximate the relative performance measure for probabilistic models for an investor with a different utility function. Since our performance measure has the same mathematical structure as the one for probabilistic models, we can make the same approximation here.

For utility functions in the generalized logarithmic family, the conditional relative $(U, \mathcal{O})$-entropy, which enters the optimization problem (11.96), is given by

$$D_{U,\mathcal{O}}(p\|q^{(0)}) = \gamma_1 \sum_{x \in \mathcal{X}} \tilde{p}_x \sum_{y \in \mathcal{Y}} p_{y|x} \left[\log \left(\frac{p_{y|x}}{q_{y|x}^{(0)}} \right) \right], \qquad (11.102)$$

which is (up to a constant factor) the Kullback-Leibler conditional relative entropy. From (11.96) in Theorem 11.8, (11.93), and (11.102) we can see that the MRE measure, p^*, is the solution of the following convex optimization problem:

Problem 11.2 *(MRE Problem for U from the Generalized Logarithmic Family)*

$$\text{Find } p^* = \arg \min_{p \in \mathbf{R}^{|\mathcal{X}||\mathcal{Y}|}} \sum_{x \in \mathcal{X}} \tilde{p}_x \sum_{y \in \mathcal{Y}} p_{y|x} \left[\log \left(\frac{p_{y|x}}{q_{y|x}^{(0)}} \right) \right] \qquad (11.103)$$

$$\text{s.t. } p_{y|x} \geq 0 \;, \; \forall x \in \mathcal{X} \,, \; y \in \mathcal{Y} \,, \qquad (11.104)$$

$$\sum_{y \in \mathcal{Y}} p_{y|x} = 1 \,, \; \forall x \in \mathcal{X} \,, \qquad (11.105)$$

$$E_p[f|x] = \overline{f}(x) \,, \; \forall x \in \mathcal{X} \,, \qquad (11.106)$$

$$\text{and } E_p[g] = \overline{g} \,. \qquad (11.107)$$

We show in Section 11.3.5 that the dual of the above problem is the following problem.

Problem 11.3 *(Dual of MRE Problem for U from the Generalized Logarithmic Family)*

$$\text{Find } (\beta^*, \eta^*) = \arg \max_{(\beta, \eta)} h(\beta, \eta) \qquad (11.108)$$

$$\text{with } h(\beta, \eta) = \sum_{x \in \mathcal{X}} \tilde{p}_x \left\{ \beta_x^T \overline{f}(x) - \log Z_x(\beta, \eta) \right\} + \eta^T \overline{g} \,, \quad (11.109)$$

$$\text{and } Z_x(\beta, \eta) = \sum_{y \in \mathcal{Y}} q_{y|x}^{(0)} e^{\beta_x^T f(x,y) + \eta^T g(x,y)} \,, \qquad (11.110)$$

where each of the β_x is an N_f-dimensional real vector, β is the vector with components β_x , $x \in \mathcal{X}$, and η is an N_g-dimensional real vector. The MRE *measure is given by*

$$p^*_{y|x} = Z_x^{-1}(\beta^*, \eta^*) \, q^{(0)}_{y|x} \, e^{(\beta^*_x)^T f(x,y) + (\eta^*)^T g(x,y)} \, . \tag{11.111}$$

Having computed p^* (e.g., by means of solving Problem 11.3), we can use Definition 11.15 and Theorems 11.8 to compute our model performance measure as

$$\Delta_U(M, b^{(0)}) = \gamma_1 \sum_{x \in \mathcal{X}, y \in \mathcal{Y}} \tilde{p}_{x,y} \log \left(\frac{p^*_{y|x}}{q^{(0)}_{y_x}} \right) \, . \tag{11.112}$$

This performance measure is essentially the likelihood ratio between the prior measure, $q^{(0)}$, and the robust measure, p^*.

We can now combine Problem 11.3 with (11.112) to obtain the following theorem:

Theorem 11.9 *For a utility function of the form*

$$U(z) = \gamma_1 \log(z - \gamma B) + \gamma_2 \, , \, \gamma > \max_{x \in \mathcal{X}}(-B_x) \, , \, \gamma_1 > 0 \, , \tag{11.113}$$

the regression model performance measure is given by

$$\Delta_U(M, b^{(0)}) = \gamma_1 \sum_{x,y} \tilde{p}_{x,y} \left[(\beta^*_x)^T f(x, y) + (\eta^*)^T g(x, y) \right.$$
$$\left. - \log(Z_x(\beta^*, \eta^*)) \right] \, , \tag{11.114}$$

where

$$(\beta^*, \eta^*) = \arg\max_{(\beta,\eta)} \left[\sum_{x \in \mathcal{X}} \tilde{p}_x \left\{ \beta_x^T \overline{f}(x) - \log(Z_x(\beta, \eta)) \right\} + \eta^T \overline{g} \right] , \tag{11.115}$$

$$and \; Z_x(\beta, \eta) = \sum_{y \in \mathcal{Y}} q^{(0)}_{y|x} e^{\beta_x^T f(x,y) + \eta^T g(x,y)} \, . \tag{11.116}$$

The above model performance measure is independent of the odds ratios, $\mathcal{O}$, independent of γ and γ_2, and it depends on γ_1 only in a trivial way. Therefore, this performance measure ranks models equivalently under all utility functions in the family. In light of the results from Section 8.3.2, this is, of course, not surprising.

The performance measure from Theorem 11.9 can be used for practical ends. The optimization problem, which must be solved, is a strictly convex problem with no constraints. The dimension of this problem is $N_f \times |\mathcal{X}| + N_g$. For most practical problems, N_f and N_g are small. However $|\mathcal{X}|$ can be quite large. If this is the case, the optimization problem can be rather hard to solve, however, as we shall see below, the problem simplifies for many practical applications.

Models with conditional constraints only

For models that have only conditional constraints, i.e., constraints of the type (11.79), and no unconditional constraints (constraints of type (11.80)), the optimization problem from Theorem 11.9 simplifies considerably. The reason for this is that the β_x-derivative of the objective function from (11.115) does not depend on $\beta_{x'}$ for any $x' \neq x$. Therefore, the condition for the optimal β_x^*, which is $E_{p^*}[f|x] = \overline{f}(x)$, does not depend on $\beta_{x'}^*$ for any $x' \neq x$. Hence, the system of $N_f \times |\mathcal{X}|$ equations we have to solve decouples into $|\mathcal{X}|$ independent systems of N_f equations each. Since for most practical problems N_f is small (typically 1 or 2), each of these independent systems of equations is easy to solve.

For example, Models 11.1 and 11.3 from Section 11.3.1 have only conditional constraints; therefore the computation Δ_U for these models is fairly easy.

Some specific regression models

Next, we specialize Theorem 11.9 to the specific models introduced in Section 11.3.1:

Model 11.1: *(Point estimator)* The regression model performance measure is given by

$$\Delta_U(M, b^{(0)}) = \gamma_1 \sum_{x,y} \tilde{p}_{x,y} \left[\beta_x^* y - \log Z_x(\beta^*)\right], \tag{11.117}$$

$$\text{where} \quad \beta^* = \arg\max_\beta \left[\sum_{x \in \mathcal{X}} \tilde{p}_x \left\{\beta_x \, \mu(x) - \log Z_x(\beta)\right\}\right], \tag{11.118}$$

$$\text{and} \quad Z_x(\beta) = \sum_{y \in \mathcal{Y}} q_{y|x}^{(0)} e^{\beta_x y}. \tag{11.119}$$

The above optimization problem simplifies further if the prior is uniform with $q_{y|x}^{(0)} = \frac{1}{|\mathcal{Y}|}$, $\forall x \in \mathcal{X}, y \in \mathcal{Y}$, and $\mathcal{Y}$ is a uniform grid, i.e.,

$$\mathcal{Y} = \{y_0 + (k-1)\delta \ , \ k = 1, ..., N_y\} \tag{11.120}$$

with constants y_0, δ, and an integer constant N_y. In this case, since (11.119) is a geometric sum, we have

$$Z_x(\beta) = \frac{1}{|\mathcal{Y}|} \frac{e^{\beta_x (N_y \delta + y_0)} - e^{\beta_x y_0}}{e^{\beta_x \delta} - 1}.$$

Hence, the conditions for the optimum,

$$0 = \mu(x) - \frac{1}{Z_x(\beta)} \frac{\partial Z_x(\beta)}{\partial \beta_x} \quad \text{for } x \in \mathcal{X}, \tag{11.121}$$

form a set of $|\mathcal{X}|$ decoupled transcendental equations (see the above discussion of models with conditional constraints only). Each of these equations has a unique solution, which is easy to find numerically by root search.

Model 11.2 *(Point estimator with unconditional variance)* Our performance measure is given by

$$\Delta_U(M, b^{(0)}) = \gamma_1 \sum_{x,y} \tilde{p}_{x,y} \left[\beta_x^* y + \eta^* y^2 - \log Z_x(\beta^*, \eta^*) \right], \qquad (11.122)$$

where

$$(\beta^*, \eta^*) = \arg\max_{(\beta,\eta)} \left[\sum_{x \in \mathcal{X}} \tilde{p}_x \{ \beta_x \, \mu(x) - \log Z_x(\beta, \eta) \} + \eta(\sigma^2 + \bar{y}^2) \right],$$

and

$$Z_x(\beta, \eta) = \sum_{y \in \mathcal{Y}} q_{y|x}^{(0)} e^{\beta_x y + \eta y^2} .$$

Here we have assumed that the model consistency condition $\bar{y} = \sum_{x \in \mathcal{X}} \tilde{p}_x \mu(x)$ holds, and we have rewritten the constraint $E_p[(y - \bar{y})^2] = \sigma^2$ as $E_p[y^2] = \sigma^2 + \bar{y}^2$.

Assuming again that the prior is uniform and $\mathcal{Y}$ has the form (11.120) and, furthermore, $N_y \gg 1$, $\delta \ll \sigma$, $y_0 \ll \mu(x) - \sigma$, and $y_0 + N_y \delta \gg \mu(x) + \sigma$, the above optimization problem can be easily solved by approximating the sums by integrals. The result is

$$\beta_x^* \approx \frac{\mu(x)}{\sigma^2} , \text{ and} \qquad (11.123)$$

$$\eta^* \approx -\frac{1}{2\sigma^2} . \qquad (11.124)$$

We can convince ourselves that this is correct, by using the fact that the MRE measure is $\propto q_{y|x}^{(0)} e^{\beta_x^* y + \eta^* y^2}$ with (β^*, η^*) chosen such that the constraints (11.106) and (11.107) of the primal problem hold. Indeed, the measure $p_{y|x}^* \propto e^{-\frac{(y - \mu(x))^2}{2\sigma^2}}$ with (β^*, η^*) given by (11.123) and (11.124), respectively, approximately satisfies the constraints $E_{p^*}[y|x] = \mu(x)$ and $[E_{p^*}[(y - \bar{y})^2] = \sigma^2$.

Model 11.3: *(Point estimator with conditional variance)* The regression model performance measure is given by

$$\Delta_U(M, b^{(0)}) = \gamma_1 \sum_{x,y} \tilde{p}_{x,y} \left[\beta_{1,x}^* y + \beta_{2,x}^* y^2 - \log Z_x(\beta^*) \right], \tag{11.125}$$

where $\beta^* = \arg\max_\beta \left[\sum_{x \in \mathcal{X}} \tilde{p}_x \left\{ \beta_{1,x} \mu(x) + \beta_{2,x}(\sigma^2(x) + \mu^2(x)) \right. \right.$

$$\left. \left. - \log Z_x(\beta, \eta) \right\} \right], \tag{11.126}$$

and $Z_x(\beta) = \sum_{y \in \mathcal{Y}} q_{y|x}^{(0)} e^{\beta_{1,x} y + \beta_{2,x} y^2}. \tag{11.127}$

We can approximately solve the optimization problem if the prior is uniform, $\mathcal{Y}$ has the form (11.120), $N_y \gg 1$, $\delta \ll \sigma$, $y_0 \ll \mu(x) - \sigma$, and $y_0 + N_y \delta \gg \mu(x) + \sigma$. Following the same logic as for Model 11.2, we obtain

$$\beta_{1,x}^* \approx \frac{\mu(x)}{\sigma^2(x)}, \text{ and} \tag{11.128}$$

$$\beta_{2,x}^* \approx -\frac{1}{2\sigma^2(x)}. \tag{11.129}$$

Example 11.1 Let us consider a numerical example. We assume that $\mathcal{X} = \{1, 2, ..., 10\}$ and $\mathcal{Y} = \{\frac{k}{100}, k = 0, ..., 100\}$. Our test dataset consists of 1,000 observations with the empirical X-probabilities

$$\tilde{p}_x = \frac{1}{10}, \forall x \in \mathcal{X}, \tag{11.130}$$

and the Y-values drawn from the following conditional distribution

$$p_{y|x} = \frac{1}{\hat{Z}_x} e^{-\frac{(y - \hat{\mu}_x)^2}{\hat{\sigma}_x^2}}, \forall x \in \mathcal{X}, y \in \mathcal{Y}, \tag{11.131}$$

$$\text{with } \hat{\mu}_x = 0.1 + 0.05 x, \tag{11.132}$$

$$\hat{\sigma}_x = 0.01 + 0.01 x, \tag{11.133}$$

$$\text{and } \hat{Z}_x = \sum_{y \in \mathcal{Y}} e^{-\frac{(y - \hat{\mu}_x)^2}{\hat{\sigma}_x^2}}. \tag{11.134}$$

This test dataset is depicted in Figure 11.5.

Table 11.2 shows The regression model performance measure, Δ_U, for the specific regression models from Section 11.3.1 as measured on the test sample shown in Figure 11.5, with respect to a uniform prior. The models were specified such that the constraints are consistent with the measure (11.131)-(11.134), from which the test-dataset was sampled. The table also shows the performance of the (test data-generating) probabilistic model (11.131)-(11.134), which was measured by means of the probabilistic model performance measure (8.46). We can see from the table how adding information

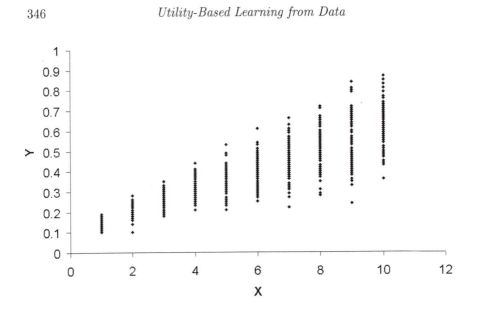

FIGURE 11.5: Test sample drawn from the conditional probability measure (11.131)-(11.134) and x-values chosen according to (11.130). We note that the conditional mean and variance increase with x. (This figure originally appeared in Friedman and Sandow (2006a).)

improves the performance of a model. Model 11.1, for which the MRE measure is exponential with an exponent linear in y, leads to a gain in expected utility of 0.24. By adding the unconditional variance to the model we arrive at Model 11.2, for which the MRE measure is a discretized Gaussian on a finite interval with the correct conditional means and (incorrect) X-independent conditional variances. The gain in expected utility over the prior is 1.14. If we add even more information to the model in terms of the conditional variances, the gain in expected utility over the prior is 1.27. This is the same gain in expected utility an investor would achieve if he knew the measure (11.131)-(11.134) or the empirical measure, $\tilde{p}$, on the test set. (The latter two models perform the same within the precision we have displayed.) The reason for this is that the MRE measure for Model 11.3 is the same as the measure (11.131)-(11.134), which means that the robust allocation is optimal for the measure (11.131)-(11.134). This surprising result is related to the fact that we have sampled from a (truncated and discretized) normal distribution and relies on the assumption that Model 11.3 provides us with the correct (up to some error that is not reflected in Δ_U at the accuracy level we consider) empirical conditional means and variances.

TABLE 11.2: Performance measure Δ_U for the specific models from Section 11.3.1, for the model (11.131)-(11.134), and for the empirical test-set measure, as measured on the test sample from Figure 11.5, with respect to a uniform prior. The performance of the probabilistic model (11.131)-(11.134) was measured by means of (8.46). The constraints in the models are assumed to be consistent with the measure (11.131)-(11.134). We have set $\gamma_1 = 1$.

Model	Δ_U
Model 11.1 (point estimator)	0.24
Model 11.2 (point estimator with unconditional variance)	1.14
Model 11.3 (point estimator with conditional variance)	1.27
model (11.131)-(11.134)	1.27
$\tilde{p}$	1.27

11.3.5 Dual of Problem 11.2

In order to derive the dual of Problem 11.2, we follow the logic from Section 2.2.5. The Lagrangian is

$$\mathcal{L}(p, \beta, \eta, \nu, \xi) = \sum_{x \in \mathcal{X}} \tilde{p}_x \left\{ \sum_{y \in \mathcal{Y}} \left[p_{y|x} \log \left(\frac{p_{y|x}}{q_{y|x}^0} \right) - \nu_{y|x} p_{y|x} \right] \right.$$

$$- \xi_x \left[\sum_{y \in \mathcal{Y}} p_{y|x} - 1 \right]$$

$$\left. - \beta_x^T \left[\sum_{y \in \mathcal{Y}} p_{y|x} f(x, y) - \overline{f}(x) \right] \right\}$$

$$- \eta^T \left[\sum_{x \in \mathcal{X}} \tilde{p}_x \sum_{y \in \mathcal{Y}} p_{y|x} g(x, y) - \overline{g} \right] , \qquad (11.135)$$

where the β_x are N_f-dimensional vector-valued parameters, η is a N_g-dimensional vector-valued parameter, the ξ_x and $\nu_{y|x} \geq 0$ are real parameters, β (ξ) denotes the set of all β_x (ξ_x) with $x \in \mathcal{X}$, and ν denotes the set of all $\nu_{y|x}$ with $x \in \mathcal{X}$, $y \in \mathcal{Y}$.

The dual is the maximization over $\beta, \eta, \nu \geq 0, \xi$ of $\mathcal{L}(\hat{p}(\beta, \eta, \nu, \xi), \beta, \eta, \nu, \xi)$ with

$$\hat{p}(\beta, \eta, \nu, \xi) = \arg \min_p \mathcal{L}(p, \beta, \eta, \nu, \xi). \qquad (11.136)$$

In order to find $\hat{p}$, we solve

$$0 = \left. \frac{\partial \mathcal{L}(p, \beta, \eta, \nu, \xi)}{\partial p_{y|x}} \right|_{p=\hat{p}}$$

$$= \log\left(\frac{\hat{p}_{y|x}}{q^0_{y|x}}\right) + 1 - \nu_{y|x} - \xi_x - \beta_x^T f(x, y) - \eta^T g(x, y)$$

$$\text{(since } \tilde{p}_x \neq 0 \text{).} \tag{11.137}$$

We obtain

$$\hat{p}_{y|x}(\beta, \eta, \nu, \xi) = q^0_{y|x} e^{-1 + \nu_{y|x} + \xi_x + \beta_x^T f(x,y) + \eta^T g(x,y)}. \tag{11.138}$$

From (11.135) and (11.137) we obtain

$$\mathcal{L}(\hat{p}(\beta, \eta, \nu, \xi)), \beta, \eta, \nu, \xi) = \sum_{x \in \mathcal{X}} \tilde{p}_x \left\{ -\sum_{y \in \mathcal{Y}} \hat{p}_{y|x} + \xi_x + \beta_x^T \overline{f}(x) \right\}$$

$$+ \eta^T \overline{g}. \tag{11.139}$$

It follows from complementary slackness and (11.138), which implies the positivity of $\hat{p}$ (assuming that $q^0_{y|x} > 0$), that

$$\nu^*_{y|x} = 0. \tag{11.140}$$

The maximum of the Lagrangian (11.139) with respect to ξ is attained if $\xi = \xi^*$ is such that the $\hat{p}$ is normalized to one. By defining $Z_x(\beta, \eta) = e^{-\xi^* + 1}$, we can rewrite $\hat{p}$ as

$$\hat{p}_{y|x}(\beta, \eta, \nu^*, \xi^*) = Z_x^{-1}(\beta, \eta) q^0_{y|x} e^{\beta_x^T f(x,y) + \eta^T g(x,y)} \tag{11.141}$$

$$\text{with } Z_x(\beta, \eta) = \sum_{y \in \mathcal{Y}} q^0_{y|x} e^{\beta_x^T f(x,y) + \eta^T g(x,y)}, \tag{11.142}$$

and we get for the Lagrangian

$$\mathcal{L}(\hat{p}(\beta, \eta, \nu^*, \xi^*), \beta, \eta, \nu^*, \xi^*) = \sum_{x \in \mathcal{X}} \tilde{p}_x \left\{ \beta_x^T \overline{f}(x) - \log Z_x(\beta, \eta) \right\} + \eta^T \overline{g}, \tag{11.143}$$

i.e., the dual of Problem 11.2 is Problem 11.3. □

Chapter 12

Select Applications

The techniques described in this book can be applied to myriad applications too numerous to list. Suffice it to say that there are applications in virtually all fields in the natural and behavioral sciences. Potential applications exist wherever people are interested in understanding nondeterministic phenomena and need to estimate probabilities. In this chapter, we describe five important applications drawn from financial credit modeling, medicine, and text classification. Mathematically, these problems can all be described as conditional probability estimation problems of the form $p(Y = y|x)$, where $Y \in \mathcal{Y}$, and $\mathcal{Y}$, the set of states, can have two elements, $k > 2$ elements, and a continuum of elements. We apply the methods discussed in this book to each of the applications. We would like to note that each of the applications discussed in this chapter can be approached via different methods not discussed in this book. We also very briefly mention recently developed methods to calibrate, via convex programming methods, financial asset return models that incorporate fat tails while maintaining model flexibility.

12.1 Three Credit Risk Models

Individual consumers, corporations, and governments take on debt to finance various endeavors worldwide. In the aggregate, the amount of debt worldwide is staggering. In the first quarter of 2006, the amount of outstanding public and private bond market debt, *in the U.S. alone*, exceeded $25 trillion (see BondMarkets.com (2006)), with worldwide issuance of debt, equity, and equity-related issuance exceeding $1.8 trillion (see Thompson Financial (2006)). These figures do not include consumer debt, which currently exceeds $2 trillion in the U.S. alone (see Federal Reserve Board (2006)).

Holders of such debt (the lenders) are subject to a number of different types of risk, including interest rate risk and credit risk. Interest rate risk is the risk to the value of a security that results from the possibility of change in the interest rate environment: when rates rise, for example, the holder of a debt instrument that pays fixed coupons will fall, since the present value of the coupons will fall in the new, higher-rate environment. In this section, we focus on credit risk — the risk that the debt issuer will default on the debt. Credit risk can be substantial; even single obligor defaults, such as the \$30 billion default of WorldCom Inc. in 2002 or the \$128 billion default of Lehman Brothers in 2008, can have significant economic impact.[1] Moreover, credit risk can vary from obligor to obligor (based on obligor-specific information) and over the economic cycle: for example, the aggregate global annualized quarterly default rate exceeded 12% in 1991, was less than 2% in 1997, and exceeded 10% in 2002 and was less than 2% in 2005 (see Standard & Poor's Global Fixed Income Research (2007)). The 12-month-trailing global corporate speculative-grade default rate increased by more than a factor of 12 from the 25 year low of 0.79%, recorded in November 2007, to 9.71% in October 2009 (see Standard & Poor's Global Fixed Income Research (2009)). Given the high stakes and lack of homogeneity with respect to obligor and macroeconomic conditions, there is considerable interest in estimating models for credit risk.

Below, we discuss three fundamental aspects of credit risk and review estimation procedures for associated models. In Section 12.1.1, we describe a model of the probability that the debt issuer will default over a fixed time horizon, given a variety of explanatory variables.

Once an obligor has defaulted, the debt issuer and creditor participate in a legal process under which claims against the issuer are resolved. As a result of this process, the holder of the debt may receive a full recovery of the loan amount, no recovery, or something in between.[2] This amount is called the ultimate recovery. In the interim, before the legal process is resolved, the debt might trade in the secondary market at a certain value; this value is referred to as the trading price recovery. In Section 12.1.2, we describe conditional probabilistic models of the ultimate discounted recovery rate.

Investors who are concerned about default risk sometimes rely on credit ratings produced by rating agencies, such as Standard & Poor's, Moody's KMV, Fitch and Dominion Bond Rating Service (DBRS). The value of high quality credit debt, for which the chance of default is remote, is highly correlated with the obligor rating. For this reason, investors are interested in a firm's single period ratings transition probability vector. In Section 12.1.3, we discuss a model of the single period conditional probability of rating transition, given a set of explanatory variables.[3]

---

[1]In fact, shortly after the Lehman Brothers bankruptcy filing, the financial markets grew quite volatile, with the Dow Jones Industrial Average setting records for single day point loss, single day point gain, and intraday point range.

[2]In a small percentage of cases, debt is exchanged for equity that rises sufficiently so that the discounted ultimate recovery exceeds the face value of the debt. It is interesting to note that the data show that a significant percentage of the time, the discounted ultimate recovery takes one of the two values, 0, 1.

[3]We confine our discussion in this book to single period models. We note that the ideas described in this book can be useful in other modeling contexts. For example, in recent work, Friedman et al. (2010a) model transitions to default, as well as credit ratings tran-

12.1.1   A One-Year Horizon Private Firm Default Probability Model

A company's creditworthiness can affect its ability to raise funds, the value of its debt in the secondary markets, and stock price. In addition, regulatory requirements and risk management considerations force holders of debt and loan portfolios to consider the default probabilities of the firms in their portfolios. In fact, there is a large body of literature analyzing default probabilities (see, for example, Schönbucher (2003), and the sources cited therein).

In this section, we review the work of Zhou et al. (2006), who estimate private firm default probabilities over a one year time horizon, given side information, which includes financial ratios, economic indicators, and market prices. They estimate their model via the ℓ_1-regularized maximum-likelihood methods described in Section 9.2.2.

This approach, which can be viewed as a generalization of linear logistic regression, is flexible enough to conform to certain nonlinearities in the data, but avoids overfitting. A comparison with various benchmark models indicates that the model produced by this approach performs well (in the sense of the utility-based performance measure described in Section 8.3.2) with respect to the benchmark models.

12.1.1.1   Modeling Method

Let x denote the vector of explanatory variables (which are listed in Section 12.1.1.4, below). Let the random variable $Y \in \{0, 1\}$ indicate default ($Y = 1$) or survival ($Y = 0$) over the one year time interval starting from the date of observation of $x \in \mathbf{R}^d$. We can imagine a conditional horse race setting, where, given x, there are two horses corresponding to the two states: survival and default, with market makers who will accept bets on either outcome and pay odds ratios, corresponding to the "winning horse." For investors with generalized logarithmic utilities, as we have seen in Chapter 10, these odds ratios drop out. The goal is to estimate the conditional probability measure $p(y|x) = Prob(Y = y|x)$ closest to prior beliefs that is consistent with the feature expectation constraints.

Training data consist of the pairs $\left(x^{(k)}, y^{(k)}\right), k = 1, \ldots, N$. Individual firms are observed annually and therefore can appear several times. All observations over all times and all firms are collected; these observations are indexed by

---

sitions, over various future time horizons. Their approach is built around the notion of the deteriorating value of side information (that is, the value of the information contained in the explanatory variable values declines as the side information "ages" and the model is applied to times further and further in the future). They define the value of side information in economic terms, in the conditional horse race setting described in this book, report that the predictive variables that they examine exhibit pronounced information value decay, and benchmark models that incorporate this decay against alternative approaches; they find that the models that incorporate this decay outperform the benchmark models, on the datasets they examine.

the superscript. The order of the observation does not matter. Under the assumption that the x vectors are unique,[4] these data generate empirical probability measures

$$\tilde{p}(x) = \frac{1}{N} I_{x \in \mathcal{X}} \tag{12.1}$$

and

$$\tilde{p}(y|x) = I_{y=\tilde{y}(x)}, \tag{12.2}$$

where $\mathcal{X} = \{x^{(1)}, \ldots, x^{(N)}\}$, I is the indicator function, and $\tilde{y}$ is the observed Y-value corresponding to x. That is, $\tilde{p}(x)$ represents the observed frequency of a particular x and $\tilde{p}(y|x)$ represents the observed frequency of y, given x.

The model makes use of two fundamental notions:

(i) a prior probability measure (the model that we believe *before* we observe data), p^0, and

(ii) a vector of features, f.

The ℓ_1-regularized likelihood maximization problem is given by

Problem 12.1 *(ℓ_1-Regularized Maximum Likelihood)*

$$\text{Find } \beta^* = \arg\max_{\beta \in \mathbf{R}^J} \left[L(p^{(\beta)}) - \alpha \sum_j |\beta_j| \right], \tag{12.3}$$

$$\text{where } L(p) = \frac{1}{N} \sum_{k=1}^{N} \log p\left(y^{(k)} | x^{(k)}\right) \tag{12.4}$$

is the log-likelihood function,

$$p^{(\beta)}(y|x) = \frac{1}{Z_x} p^0(y|x) e^{\beta^T f(y,x)} \tag{12.5}$$

$$\text{and } Z_x(\beta) = \sum_y p^0(y|x) e^{\beta^T f(y,x)}. \tag{12.6}$$

This method (discussed in Sections 9.2.2 and 10.1.5) combines

(i) a large dictionary of nonlinear features that provides enough flexibility for the model to conform to the data, and

(ii) a consistent method to eliminate unnecessary features and regularize the model to avoid overfitting.

---

[4]It is assumed that each x occurs only once to simplify notation. It is possible to lift this restriction.

We note that linear logistic regression, given by

$$p(1|x) = \frac{1}{1 + e^{-\sum_i \beta_i x_i}},\tag{12.7}$$

where the parameters, β_i, are chosen to maximize the likelihood function, is a special case of Problem 12.1. In this case, there is no regularization (i.e., $\alpha = 0$), the features are of the form $f_j(y,x) = (y - \frac{1}{2})x_j$, and the prior is independent of x.

As has been indicated in Section 9.2.2, ℓ_1-regularized methods often lead to models in which certain features do not contribute to the optimal model at all, i.e., certain elements of the vector β^* are zero.

The x-independent prior probability measure is assumed to be

$$p^0(1|x) = \frac{1}{N} \sum_{k=1}^{N} \tilde{p}(1|x^{(k)}).\tag{12.8}$$

The estimation procedure incorporates four types of features:

(*i*) Linear features

$$f(y,x) = \left(y - \frac{1}{2}\right) x_j \tag{12.9}$$

where x_j denotes the jth coordinate of x, with the convention that $x_0 = 1$.

(*ii*) Quadratic features

$$f(y,x) = \left(y - \frac{1}{2}\right) x_i x_j, \text{ and}\tag{12.10}$$

(*iii*) Cylindrical kernel features

$$f(y,x) = \left(y - \frac{1}{2}\right) g(l(x))\tag{12.11}$$

where $l(x) = x_j$ and $l(x) = x_i \pm x_j$, where $i = 1,2,...d$, $j = 1,2...d$ and $i \neq j$, and $g(x) = e^{-\frac{(x-a)^2}{\sigma^2}}$.

(*iv*) Two-dimensional kernel features

$$f(y,x) = \left(y - \frac{1}{2}\right) g(l(x))\tag{12.12}$$

where $l(x)$ represents a 2-dimensional subvector of $x \in \mathbf{R}^d$, and $g(x) = \frac{1}{2\pi\sigma_j^2} exp[-\frac{(x-\xi_i)^2}{2\sigma_j^2}]$, where ξ_i, $i = 1,2,...m$ are local centers of x and σ_j, $j = 1,2,...n$ are bandwidths.

12.1.1.2 Numerical Procedure

The estimation makes use of the numerical procedure, described in Riezler and Vasserman (2004), that approximately solves Problem 12.1. The optimal hyperparameter value, α^*, was chosen to maximize a 5-fold out-of-sample average performance (see Zhou et al. (2006) for details). Once α^* was determined, all the training data were used to train a final version of the model.

12.1.1.3 Benchmark Models

The model was benchmarked against the linear logit, additive logit, and additive probit models, which are described in this section.

The additive logit and probit models are generalizations of the logit and probit models; these generalizations depend on transformations $\phi_i(x_i)$, such as those described, for example, in Falkenstein (2000); the functions, $\phi_i(x_i)$, are themselves models of default frequency as a function of each univariate explanatory variable. The parameters, β_i, are chosen to maximize the likelihood function.

The additive logit model is given by

$$p(1|x) = \frac{1}{1 + e^{-\sum_i \beta_i \phi_i(x_i)}}. \tag{12.13}$$

The additive probit model is given by

$$p(1|x) = \int_{-\infty}^{\sum_i \beta_i \phi_i(x_i)} \frac{1}{\sqrt{2\pi}} e^{-\frac{z^2}{2}} dz. \tag{12.14}$$

12.1.1.4 Variables and Data

The data were drawn from the Standard and Poor's Credit Risk Tracker North America database, consisting of historical data (about 77,000 observations of 24 explanatory variables and a one-year default indicator, collected between 1995-2002). The explanatory variables that were used[5] were:

1. Net Income,
2. Equity / Total Assets,
3. 4-Quarter Moving Average Industrial Production,
4. Cash over Total Liabilities,
5. Current Liabilities,
6. 4-Quarter Moving Average Delinquency Rates,

---

[5]Zhou et al. (2006) have selected these variables from a much larger set of candidate variables via a step-forward selection process.

7. Year-over-Year Change in Net Sales,

8. EBITDA Margin (EBITDA / Net Sales),

9. Asset Turnover (Net Sales / Average Total Assets),

10. Total Liabilities and Net Worth,

11. Average 3-Month Treasury Yield (quarterly),

12. Total Liability / Equity,

13. 4-Quarter Moving Average Change in Nonrevolving Consumer Credit,

14. Working Capital Ratio, which is defined as (Current Assets-Current Liabilities) / Total Assets,

15. Retained Earnings,

16. Industry Average Year-over-Year Change in Sales,

17. Industry Average S&P Credit Rating,

18. CBOE Volatility Index (VXO),

19. Industry Median Distance to Default,

20. Industry Median One-Year Stock Price Rank, and

four industry-sector indicator variables.[6] All variables were rank transformed.

Defaults in this dataset were consistent with the definition given in Basel Committee on Banking Supervision (2003)).

Features were constructed as indicated in Section 12.1.1.1 on the raw variables described above. The candidate set of features contained 1,379 features. Only 79 were selected by the procedure described in Section 12.1.1.2.

12.1.1.5 Benchmarking

The model produced using the method described in Section 12.1.1.1, and the data and explanatory variables described in Section 12.1.1.4, was compared with the benchmark approaches described in Section 12.1.1.3.

Table 12.1 shows average performance measure results (from 5-fold cross validation on out-of-sample datasets) for three benchmark models.

---

[6]In order to determine the relevant groups of industries, Zhou et al. (2006) categorized the firms under consideration into the following four broad industry groups:

1. Agriculture, mining, oil and gas extraction, manufacturing;
2. Wholesale and retail trade, eating and drinking places, repair services, motion pictures;
3. Utilities, transportation, real estate, hotels; and
4. Printing and publishing, contractors, and other services.

TABLE 12.1: 5-fold cross validation performance measure averages. We express Δ (described in Section 8.3.2) and Model Value (described in Section 8.6) with respect to the prior measure p^0. (We take $S = \$1$ billion and $B = 1.05$.)

Model	ROC	Δ_{log} (bps)	Monetary Value (in Millions of Dollars)
Linear logistic model	0.8512	75.35	3.97
Additive logit model	0.8413	70.98	3.74
Additive probit model	0.8425	69.94	3.68
ℓ_1-Regularized model	0.8622	81.51	4.30

Table 6.1 displays the results of statistical tests (DeLong et al. (1988)), on ROC curves from the two models, which indicate that the ℓ_1-regularized model significantly outperformed, at 5% level, the linear logit model four out of five times.

TABLE 12.2: Statistical test results on 5 pairs of out-of-sample ROC curves.

ℓ_1-Regularized model ROC	Linear Logit ROC	p-value
0.8534	0.8413	0.0190
0.8826	0.8716	0.0336
0.8557	0.8370	0.0067
0.8695	0.8590	0.0325
0.8496	0.8469	0.3580

All one-year default probabilities on a pure hold-out dataset containing 15,207 firm-year observations including 258 defaults were computed. These firm-year observations were not used in the model training. The ROC values for the ℓ_1-regularized model and the linear logit model were 0.8275 and 0.8155, respectively. The statistical test, described in DeLong et al. (1988), on the null hypothesis that the ℓ_1-regularized model is no better than the linear logit model, shows that the ℓ_1-regularized model significantly outperforms the linear logit model, with a p-value of 0.0060 (i.e., the null hypothesis is rejected with a confidence of 99.4%) .

12.1.2   A Debt Recovery Model

The value of a debt security depends to a large extent on the probability of default and the probability distribution of recovery given default (RGD). In this section, we review the work of Friedman and Sandow (2003c), who consider discounted recovery rates of defaulted debt at the time that the obligor emerges from bankruptcy. They estimate the conditional probability distribution of the discounted recovery rate as a function of collateral, debt below class, debt above class, and economy-wide default rates. This estimation was based on the utility-based estimation described in Chapter 10. The model's performance is also measured in economic terms along the lines indicated in Chapter 8. Numerical studies indicate that this model has a clear advantage over certain models under which the recovery rate is β-distributed.

12.1.2.1 Conditional Probabilities

Let x denote a vector of explanatory variables, which are quality of collateral, debt below class, debt above class, and the aggregate (US economy-wide) default rate. Let the random variable R denote the (discounted ultimate) recovery rate of a defaulted debt instrument[7] (with its values denoted by r). R is assumed to take values in the interval $I = [0, r_{max}]$, where $r_{max} = 1.2$ as the data suggest.[8] The conditional probability of the recovery rate, R, is estimated under the condition that the explanatory variables have values x. It is assumed that there is a continuous probability density function (pdf) in the interval $I = [0, r_{max}]$ and positive probabilities at the points $R = 0$ and $R = 1$,[9] i.e., the probability measure is assumed to be of the form

$$p(r|x) = p_I(r|x) + p_0(x)\delta(r) + p_1(x)\delta(r-1) , \qquad (12.15)$$

where δ denotes Dirac's delta function. Here, $p_I(r|x)dr$ is the probability that $R \in (r, r + dr)\backslash\{0, 1\}$ given x, $p_0(x)$ is the probability that $R = 0$ given x, and $p_1(x)$ is the probability that $R = 1$ given x. This measure is normalized as follows:

$$1 = \lim_{\epsilon \to 0^-} \int_{\epsilon}^{r_{max}} p(r|x)dr = \int_{0}^{r_{max}} p_I(r|x)dr + \sum_{\rho=0,1} p_\rho(x) . \qquad (12.16)$$

(Below we will write $\int_I$ instead of $\lim_{\epsilon \to 0^-} \int_{\epsilon}^{r_{max}}$.)

12.1.2.2 Maximum Expected Utility Principle and Dual Problem

The modeling approach used is described in Section 10.2 with a generalized logarithmic utility function, $U(W) = \eta_0 \log(W - \gamma B) + \eta_1$. We can imagine the

---

[7]Here, $R = 0$ corresponds to zero recovery, and $R = 1$ corresponds to complete recovery. The value at emergence from bankruptcy was discounted to the last date on which a cash payment was made, based on the coupon of the debt.

[8]Sometimes, debt is exchanged for equity before emergence; increases in equity values can lead to discounted recoveries greater than 1. This was the case for about 3.5% of the observations.

[9]The point probabilities were introduced to account for the fact that roughly 10% of the observations have $R = 0$, and roughly 20% have $R = 1$.

setting as a continuous horse race with market makers who pay odds ratios if a particular horse (recovery value) "wins." As we have seen in Chapter 10, these odds ratios drop out for the logarithmic family that is used. In this setting, the MRE primal problem arises as the solution to the problem of finding the conditional probability measure closest to prior beliefs that is consistent with the feature expectation constraints. This means that, for each value α, the Pareto optimal measure is found by minimizing, over measures p, the discrepancy (Kullback-Leibler relative entropy) between p and the prior measure $p^0 = p_I^0(r|x) + p_0^0(x)\delta(r) + p_1^0(x)\delta(r-1)$:

$$D(p\|p^0) = \sum_x \tilde{p}(x) \left\{ \int_0^{r_{max}} p_I(r|x) \log \frac{p_I(r|x)}{p_I^0(r|x)} dr \right.$$

$$\left. + \sum_{\rho=0,1} p_\rho(x) \log \frac{p_\rho(x)}{p_\rho^0(x)} \right\}, \qquad (12.17)$$

$$\text{subject to } Nc^T \Sigma^{-1} c \le \alpha , \qquad (12.18)$$

$$\text{with } c = E_p[f] - E_{\tilde{p}}[f] , \qquad (12.19)$$

$$E_p[f] = \sum_x \tilde{p}(x) \int_I p(r|x) f(r,x) \, dr \qquad (12.20)$$

$$\text{and } E_{\tilde{p}}[f] = \sum_x \tilde{p}(x) \int_I \tilde{p}(r|x) f(r,x) \, dr . \qquad (12.21)$$

Here, $f(r,x) = (f_1(r,x), \ldots, f_J(r,x))^T$ is the vector of features, $\tilde{p}$ denotes the empirical distribution, N is the number of observations, and Σ is the empirical covariance matrix of the features. The solutions to the above optimization problems form a family of measures which is parameterized by α. Each of these measures is robust in the sense that it maximizes the worst-case (over all potential true measures) outperformance relative to the prior measure. After computing a number of these measures, the hyperparameter value associated with the best out-of-sample performance, as measured by the expected utility for an investor with a generalized logarithmic utility (see Sections 8.5 and 8.3.2), is selected.

As we have seen in Section 10.2.3, the dual of the above optimization problem is the following:

$$\text{Find } \beta^* = \arg \max_\beta h(\beta) \qquad (12.22)$$

$$\text{with } h(\beta) = \frac{1}{N} \sum_{k=1}^N \log p^{(\beta)}(r_k|x_k) - \sqrt{\frac{\alpha}{N} \beta^T \Sigma \beta} \qquad (12.23)$$

$$\text{with } p^{(\beta)}(r|x) = \frac{1}{Z_x(\beta)} e^{\beta^T f(r,x)} \qquad (12.24)$$

and

$$Z_x(\beta) = \int_0^{r_{max}} p_I^0(r|x)e^{\beta^T f(r,x)}dr + \sum_{\rho=0,1} p_\rho^0(x)e^{\beta^T f(\rho,x)} , \quad (12.25)$$

where the (x_k, r_k) are the observed (x, r)-pairs and $\beta = (\beta_1, \ldots, \beta_J)^T$ is a parameter vector. The optimal (in the sense of the dual and of the original optimization problem) probability measure is then:

$$p(r|x) = p^{(\beta^*)}(r|x)p^0(r|x) . \quad (12.26)$$

The problem (12.22)-(12.25) amounts to a regularized maximum likelihood estimation, or an expected utility maximization, of the parameter vector β for an exponential distribution. The regularization term, $\sqrt{\frac{\alpha}{N}\beta^T\Sigma\beta}$, penalizes large β vectors by an amount proportional to $\sqrt{\alpha}$. This dual perspective on regularization and the role of α is consistent with the primal perspective. We note that the dual formulation of the problem leads to a J-dimensional optimization problem (the primal problem is an infinite dimensional problem). The dual formulation is used to numerically find this optimal measure.[10] The objective function of the maximization problem (12.22)-(12.25) is strictly concave. For this reason, the problem is amenable to a robust numerical solution. In particular, as a consequence of convex optimization theory, the solution depends continuously on the data, and the solution is unique.

For the data and features of this problem, the results depend little on the prior measure; the results described below were obtained for the following convenient choice:

$$p^0(r|x) = \frac{1}{2 + r_{max}} [1 + \delta(r) + \delta(r - 1)] . \quad (12.27)$$

12.1.2.3 Features

The features we use here can be partitioned into two classes:

- *Global features:* $f_j(r, x)$ is defined on the whole interval $I = [0, r_{max}]$ for r.

- *Point features:* $f_j(r, x)$ is defined only for $r = 0$ or for $r = 1$.

12.1.2.4 Global Features

Global features are of the type

$$f_j(r, x) = r^n x_i^m, \text{ where } x_i \text{ is the } i^{th} \text{ component of } x, \quad (12.28)$$

---

[10]For a practical implementation, we can drop the square root in the second term of (12.23); the resulting family, indexed by α, of solutions is the same (see Section 2.2.9, Theorem 2.13).

with $n = 1, 2, 3$ and $m = 0, 1$. Using such features, we obtain

$$E_p[r^n x_i^m] - E_{\tilde{p}}[r^n x_i^m] = c_j. \qquad (12.29)$$

These features force the theoretical (noncentered) covariances of r^n and x_i^m to coincide with the empirical ones — up to the features-error term c_j.

12.1.2.5   Point Features

Point features are of the form

$$f_j(r, x) = \begin{cases} x_i^m & \text{if } r = \rho, \\ 0 & \text{otherwise,} \end{cases} \quad \text{where } x_i \text{ is the } i^{th} \text{ component of } x \qquad (12.30)$$

where $m = 0, 1$ and $\rho = 0, 1$. In this case, the features equation, (12.19), becomes:

$$Pr(r = \rho) E_p[x_i^m | r = \rho] - \tilde{P}r(r = \rho) E_{\tilde{p}}[x_i^m | r = \rho] = c_j , \qquad (12.31)$$

where $Pr(r = \rho)$ $(\tilde{P}r(r = \rho))$ is the probability of finding $r = \rho$ under the model (empirical) measure.

12.1.2.6   Evaluating Model Performance

Model performance was measured via the scaled log-likelihood difference, Δ_{log}, between the model and the (noninformative) prior measure, as estimated on an out-of-sample dataset (see Chapter 8). This performance measure is given by

$$\Delta_{log}(p) = \frac{1}{N} \sum_{k=1}^{N} \log p^{(\beta^*)}(r_k | x_k) , \qquad (12.32)$$

where the (x_k, r_k) are the (x, R)-pairs and N the number of observations of the test sample (recall that $p(r|x) = p^{(\beta^*)}(r|x) p^0(r|x)$ according to (12.26)). As we have seen in Chapter 8, Δ_{log} can be interpreted as the gain in expected logarithmic utility experienced by an investor who uses the model p to design a utility-optimal investment strategy, where the gain is measured with respect to an investor who has no information beyond his prior beliefs and therefore invests according to the prior measure, p^0.

To compute the performance measure, Δ_{log}, the data were randomly split into two parts, one with 75% of the ($\approx$ 1400) observations and another one with the remaining 25%. The model was trained on the first dataset and tested on the second one. This procedure was repeated 120 times to ensure the reliability of the results.

In order to get an idea what kind of values to expect for the performance measure, Δ_{log} was compared with analogous quantities associated with the following simple models:

1. *Naive model:*

$$p_0(x) = w_0 \ , \ p_1(x) = w_1 \ , \ p_I(r|x) = \frac{1 - w_0 - w_1}{r_{max}} \ , \tag{12.33}$$

where w_0 and w_1 are the observed frequencies of finding $R = 0$ and $R = 1$.

2. *Simple β-model:* this model — probably the most widely used method to estimate conditional recovery rates — is built by means of a linear regression of a transformed R on the explanatory variables, where R is transformed such that it becomes approximately normal under the assumption that it is originally β-distributed and has no point probabilities at $R = 0$ and $R = 1$. For the latter problem, there are no point probabilities at $R = 0$ or $R = 1$ as for the ultimate recoveries studied here, so that a β-distribution might be a reasonable approximation.

3. *Generalized β-model:* This model generalizes the β-distribution by including the point probabilities (p_0 and p_1) at $R = 0$ and $R = 1$. It assumes that $p_0(x)$ and $p_1(x)$ are linear logistic functions of x, and that the density $p_I(r|x)$ on the interval $I = [0, r_{max}]$ is a β-distribution, i.e., that $p_I(r|x) \propto \left(\frac{r}{r_{max}}\right)^{\kappa_1(x)} \left(1 - \frac{r}{r_{max}}\right)^{\kappa_2(x)}$, where $\kappa_1(x)$ and $\kappa_2(x)$ are linear in the explanatory variables, x. Model parameters were estimated by maximizing the log-likelihood, i.e., by maximizing the investor's expected utility.

12.1.2.7 Data

The ultimate discounted recovery values from Standard & Poor's US LossStats^{TM} Database (for more details on the database see Bos et al. (2002)) were used. The database contains loans and bonds that have defaulted and emerged since 1988. There are a few recoveries larger than 1.2 which were excluded, leaving a sample of about 1,400 observations. The following factors were used as explanatory variables:

1. Collateral quality. The collateral quality of the debt was classified into 17 categories, ranging from "unsecured" to "all assets," and then ranked (see Table 12.3). The model was based on this rank.

2. Debt below class. This is the percentage of debt on the balance sheet that is inferior to the class of the debt instrument considered.

3. Debt above class. This is the percentage of debt on the balance sheet that is superior to the class of the debt instrument considered.

4. Aggregate default rate. This is the percentage of S&P-rated US bonds that default within the twelve months prior to bankruptcy announcement.

The data for the first three variables were drawn from Standard & Poor's US LossStats^{TM} Database, and the data for the fourth variable were drawn from Standard & Poor's CreditPro®. All data are available at the time of default.

TABLE 12.3: Collateral Ranks.
(This table originally appeared in
Friedman and Sandow (2003c).)

Collateral	Rank
All assets	1
Most assets	2
Cash	3
Inventories/receivables	4
All Noncurrent assets	5
Non-current Assets	6
PP&E	7
Equipment	8
O&G reserves	9
Real estate	10
Guarantee	11
Capital stock of operating units	12
Intercompany debt	13
Second lien	14
Intellectual property	15
Unsecured	16

12.1.2.8 Results

The performance of this model was measured by means of Δ_{log} (see Sections 8.5 and 8.3.2). The result is $\Delta_{log} = 0.63$. We can interpret Δ_{log} as the difference in the expected utilities or, equivalently, in the wealth growth rates, that result from two investment strategies: one based on the target model and another one based on a (noninformative) prior probability measure. This means that the wealth of the investor using the utility-based model grows by $e^{0.63} - 1 \approx 0.88$, i.e., by approximately 88%, faster than the wealth of an investor who knows only the noninformative prior measure. This wealth growth pickup is large; however, we note that 88% is *only a relative* wealth growth rate. This does not mean that the investor realizes this wealth growth.

For comparison, Table 12.4 shows the performance measure Δ_{log} for the simple β-model, the *naive* model, and the generalized β-model from Section 12.1.2.6. The first two of these models perform much worse than the utility-based model. The generalized β-model is closer to the utility-based model; however, the difference of 0.16 (i.e., 16%) in the expected wealth growth rate is substantial.

Figure 12.1.2.8 shows the conditional probability density as a function of each of the explanatory variables and the recovery rate. The shapes of these functions are consistent with our intuition: the better the collateral, the debt superiority, or the economic environment, the more likely is a high recovery. We can also see from the figure that the effect of collateral, debt below class, and debt above class on recoveries is very strong, while the effect of the ag-

TABLE 12.4: Gain in expected utility (wealth growth rate), Δ_{log}, for various models (see Section 12.1.2.6 for the definition of Δ_{log} and the models). (This table originally appeared in Friedman and Sandow (2003c).)

Model	Δ_{log}
Simple β-model	0.07
Naive model	0.25
Generalized β-model	0.47
Utility-based model	0.63

gregate default rates is weaker. One can also observe that the recoveries are highly uncertain. This is consistent with the results from Van de Castle and Keisman (1999) and Bos et al. (2002).

Figure 12.1.2.8 shows some curves of the same probability density together with the actual data. We can see from this picture that the probability densities are consistent with the data: highly concentrated data points correspond to high model densities.

Figure 12.1.2.8 shows the conditional probabilities of finding $R = 0$ or $R = 1$. The model probabilities are consistent with our intuition and with the data. Better collateral, capital structure, and economic environment variables are associated with higher probabilities that $R = 1$ and lower probabilities that $R = 0$.

Figure 12.1.2.8 shows the conditional averages and standard deviations that the model predicts for the recovery rates. They are also consistent with our intuition: the better the collateral, the debt superiority, or the economic environment, the greater the expected recovery. Note that the standard deviation is generally high (mostly in the range of 0.3 to 0.4).

12.1.3 Single Period Conditional Ratings Transition Probabilities

Many investors who are concerned about default risk rely on credit ratings produced by rating agencies, such as Standard & Poor's, Moody's KMV, Fitch, A. M. Best Company, and Dominion Bond Rating Service. The value of debt of high credit quality, for which the chance of default is remote, depends strongly on the obligor rating. For this reason, investors are interested in a firm's ratings transition probability vector. Many risk management tools depend strongly on ratings transition assumptions. Often, historical average rating transition rates are input to such models; however, such average rates

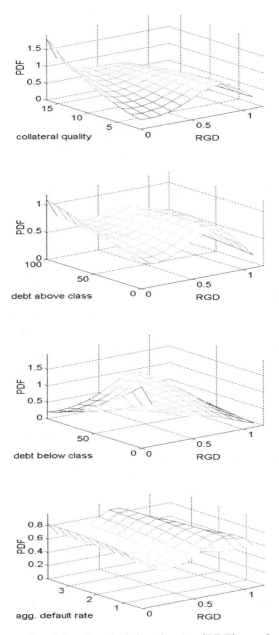

FIGURE 12.1: Conditional probability density (PDF) as a function of each explanatory variable and the recovery given default (RGD). For each plot, the remaining variables are chosen to be in the middle of their observed ranges. (These plots originally appeared in Friedman and Sandow (2003c).)

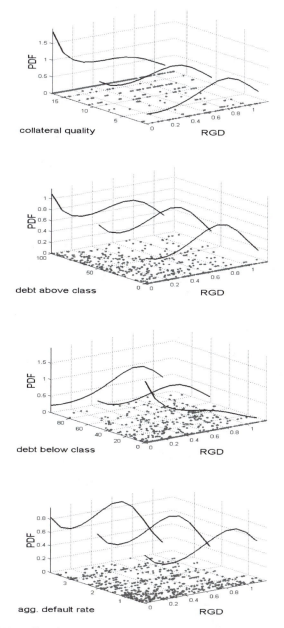

FIGURE 12.2: Conditional probability density (PDF) as a function of each explanatory variable and the recovery given default (RGD), and actual data. For each plot, the remaining variables are chosen to be in the middle of their observed ranges. The lines represent the probability density, and the dots represent observed data. (These plots originally appeared in Friedman and Sandow (2003c).)

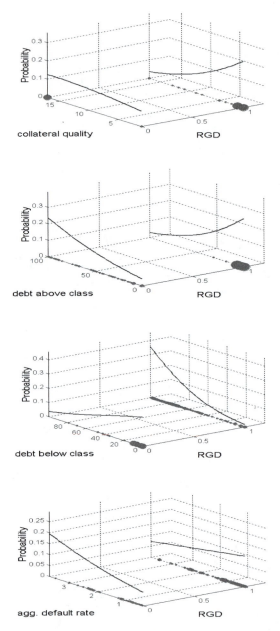

FIGURE 12.3: Conditional point probabilities for recovery given default (RGD) of zero and one. The lines represent the conditional probabilities. The balls represent observed data; the size of the balls is proportional to the number of observations at the same position. (These plots originally appeared in Friedman and Sandow (2003c).)

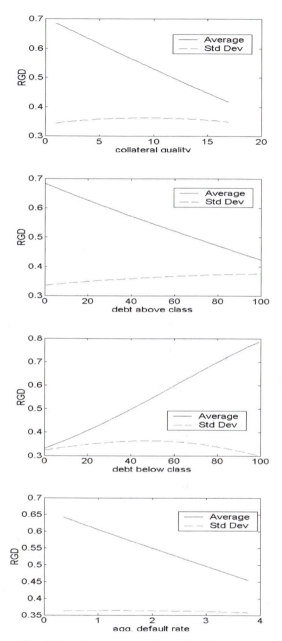

FIGURE 12.4: Conditional average and standard deviation of the recovery given default (RGD) as functions of each explanatory variable. For each plot, the remaining variables are chosen to be in the middle of their observed ranges. (These plots originally appeared in Friedman and Sandow (2003c).)

may not accurately capture the transition probabilities at a given point in time. Moreover, such rates ignore current firm specific side information that could better inform a conditional probability model. For these reasons, it is interesting to estimate the conditional probability model for ratings transition.

In the case of the Standard & Poor's ratings system, there are 20 categories:

$$\{AAA, AA+, AA, AA-, A+, A, A-, BBB+, BBB, BBB-,$$
$$BB+, BB, BB-, B+, B, B-, CCC+, CCC, CCC-, D\}.$$

We discuss a way to estimate the probability distribution over these 20 categories, given the current rating and other explanatory variables, including ratios derived from financial statements and macroeconomic variables.

That is, we seek a model of the form $P(Y = y|x)$, where the random variable Y denotes a rating that the analyst might give at the end of some fixed time horizon, with $Y = y \in \{1, \ldots, 20\}$ and $x \in \mathbf{R}^d$ denotes a vector of explanatory variables, including the current rating. We might imagine a logarithmic utility investor participating in a conditional horse race setting, where, given the explanatory variable values, there are market makers who will pay off odds ratios on the various horses (y-values), depending on which horse (rating) "wins." As we have seen, we can formulate a minimum relative entropy problem formulation or we can solve the dual problem — a maximum-likelihood problem (with or without regularization) over an exponential family. For this MRE formulation, with logarithmic utility, as we have seen, the odds ratios will drop out of the problem formulation. Given the finite number of states, this problem can be viewed as mathematically equivalent to the recovery rate distribution problem of the preceding section, except, since the states are discrete in this context, the integrals are replaced by sums. In the remainder of this section, we describe how such a method could be used to estimate conditional rating transition probabilities.

12.1.3.1 Modeling Method

Our training data would consist of the pairs $\left(x^{(k)}, y^{(k)}\right)$, $k = 1, \ldots, N$. Individual rated firms would be observed annually and could therefore appear numerous times. We would collect all observations, over all times and all firms; these observations are indexed by the superscript. The order of the observation would not matter.[11] Under the assumption, which holds in practical applications, that the x vectors are unique,[12] these data would generate empirical probability measures

$$\tilde{p}(x) = \frac{1}{N} I_{x \in \mathcal{X}} \tag{12.34}$$

---

[11] The model would be sensitive to the economic cycle since some of the explanatory variables are macroeconomic quantities.

[12] We make the assumption that each x occurs only once to simplify notation. It is possible to lift this restriction.

and

$$\tilde{p}(y|x) = I_{y=\tilde{y}(x)}, \tag{12.35}$$

where $\mathcal{X} = \{x^{(1)}, \ldots, x^{(N)}\}$, I is the indicator function, and $\tilde{y}$ is the observed Y-value corresponding to x. That is, $\tilde{p}(x)$ would represent the observed frequency of a particular x and $\tilde{p}(y|x)$ the observed frequency of y, given x.

Our model would make use of two fundamental notions:

(*i*) a prior probability measure (the model that we believe *before* we observe data), p^0,

(*ii*) a vector of features, f.

The ℓ_1-regularized likelihood maximization problem is given by

Problem 12.2 (*ℓ_1-regularized Maximum Likelihood*)

$$Find\ \beta^* = \arg \max_{\beta \in \mathbf{R}^J} \left[L(p^{(\beta)}) - \alpha \sum_j |\beta_j| \right], \tag{12.36}$$

$$where\ L(p) = \frac{1}{N} \sum_{k=1}^{N} \log p\left(y^{(k)}|x^{(k)}\right) \tag{12.37}$$

is the log-likelihood function,

$$p^{(\beta)}(y|x) = \frac{1}{Z_x} p^0(y|x) e^{\beta^T f(y,x)} \tag{12.38}$$

$$and\ Z_x(\beta) = \sum_y p^0(y|x) e^{\beta^T f(y,x)}. \tag{12.39}$$

We take as a prior probability measure the x-independent measure

$$p^0(y|x) = \frac{1}{N} \sum_{k=1}^{N} \tilde{p}(y|x^{(k)}). \tag{12.40}$$

12.1.3.2 Features

We would use features of the form

$$f_j(y, x) = y^n x_i^m, \text{ where } x_i \text{ is the } i^{th} \text{ component of } x, \tag{12.41}$$

with $n = 1, 2, 3$ and $m = 0, 1$. Using such features, we would obtain

$$E_p[y^n x_i^m] - E_{\tilde{p}}[y^n x_i^m] = c_j. \tag{12.42}$$

These features would force the theoretical (noncentered) covariances of y^n and x_i^m to coincide with the empirical ones — up to the features-error term c_j.

12.1.3.3 Numerical Procedure

We could estimate the models via a numerical procedure, described in Riezler and Vasserman (2004), that approximately solves Problem 12.2.

We would seek the hyperparameter value, α, in Problem 12.2 that would maximize performance on an out-of-sample dataset. The procedure is as follows.

1. Randomly partition the entire training dataset into five portions. The splitting is purely random. Each one-year observation of a firm appears in only one partition. If a firm has multiple year observations, the firm may or may not appear in different partitions.

2. For a given candidate α,

    (a) Hold one portion as an "out-of-sample dataset," and build a model on the combined complementary four portions;

    (b) Evaluate the model on the "out-of-sample dataset" and record the value of model performance measure (log-likelihood).

    (c) Repeat the above two steps five times. Each time we hold a different portion as the "out-of-sample dataset;"

    (d) Calculate the average performance measure over the five times of evaluation.

3. Repeat step two for all candidate α values, producing a curve of the average "out-of-sample" performance measure vs. α values.

4. Determine the optimal α which maximizes the average performance measure on the "out-of-sample datasets."

After the optimal α value has been determined, we use all the training data to train a final version of the model.

12.2 The Gail Breast Cancer Model

In this section, we briefly review the Gail breast cancer model described in Gail et al. (1989), which is used to generate risk estimates that allow decision makers to weigh various breast cancer prevention options, including chemoprevention and tamoxifen. Specifically, the model provides an estimate of the probability that a woman will develop breast cancer over a specified interval, given her age, age at menarche, age at first live birth, number of previous biopsies, and number of first degree relatives (parents, siblings, and children) with breast cancer.

Gail et al. (1989) decompose the problem of estimating the probability of developing breast cancer into three subproblems:

(*i*) attribute selection, and relative risk estimation, for a woman at a given age conditioned on risk factors, relative to a woman who does not have those risk factors,

(*ii*) baseline age-specific breast cancer incidence rate estimation for a woman who does not have the risk factors described in (*i*), above, and

(*iii*) estimation of the long-term probability of developing breast cancer, taking into account competing risks, as well as (*i*) and (*ii*) above.

The Gail model was developed using data from the Breast Cancer Detection Demonstration Project (BCDDP). For specific details regarding the selection of data, see Gail et al. (1989). In the next few sections, we briefly describe the above three subproblems.

12.2.1   Attribute Selection and Relative Risk Estimation

Attributes were determined and relative risk was estimated based on an extension of a case-control study. Cases were drawn from white BCDDP participants with *in situ* or invasive cancer incident between 1973 and 1980, but not prevalent at the first screening. Controls were matched from women who did not receive a recommendation for biopsy over the same period. The matching variables were

(*i*) age at entry into the screening program in 5-year age groups,

(*ii*) race,

(*iii*) center at which the participant was screened,

(*iv*) calendar time of entry into the screening program within 6 months, and

(*v*) length of participation in the screening program.

The analysis was based on 2,852 white cases with 3,146 white controls, with 4,496 matched pairs.

A variety of potential explanatory factors were considered individually. Factors that did not affect the risk of developing breast cancer or affected a very limited number of women were eliminated. The five factors on which the model were based, which were identified in earlier studies, were

(*i*) age (AGECAT = 0 for age less than 50 years and AGECAT = 1 for age 50 years or more),

(*ii*) age at menarche (AGEMEN = 0 for age at menarche ≥ 14, 1, for age at menarche = 12 or 13, and 2, for age at menarche < 12),

(*iii*) number of previous breast biopsies (NBIOPS = 0, for no previous biopsies, 1, for 1 previous biopsies, and 2, for 2 or more previous biopsies),

(*iv*) age at first live birth (AGEFLB = 0, for age of first live birth < 20, 1
for age of first live birth $\in [20, 24]$, 2 for age of first live birth $\in [25, 29]$,
and 3, for age of first live birth ≥ 30),

(*v*) and number of first-degree relatives (mother or sisters) with breast cancer (NUMREL = 0, for no first-degree relatives, 1, for 1 first-degree
relative, and 2, for 2 or more first-degree relatives).

A logistic regression (see Section 10.2.3.5) was performed on the case-control
data with the preceding five variables, as well as the quadratic interaction
terms

(*i*) NBIOPS × AGECAT, and

(*ii*) AGEFLB × NUMREL,

to produce an estimate for $p(Y = 1|x)$, where $Y = 1$ denotes an incidence of
cancer over the test period and x denotes the vector

$$(\text{AGECAT}, \text{AGEMEN}, \text{NBIOPS}, \text{AGEFLB}, \text{NUMREL},$$
$$\text{NBIOPS} \times \text{AGECAT}, \text{AGEFLB} \times \text{NUMREL}).$$

This resulted in the model

$$log\frac{p(Y = 1|x)}{1 - P(Y = 1|x)} = -0.74948 + 0.09401(\text{AGEMEN}) + 0.52926(\text{NBIOPS})$$

$$+0.21863(\text{AGEFLB}) + 0.95830(\text{NUMREL})$$
$$+0.01081(\text{AGECAT}) - 0.28804(\text{NBIOPS} \times \text{AGECAT})$$
$$-0.19081(\text{AGEFLB} \times \text{NUMREL}). \qquad (12.43)$$

When viewed from the perspective of a profit-oriented insurance company,
in our horse race context, the two states (horses, $Y = 1$ and $Y = 0$) correspond
to contracting cancer or not. The odds ratios can be thought of as payments
in either state — perhaps a payment from an idealized insurance policy, if
$Y = 1$, or from an idealized annuity if $Y = 0$.[13]

12.2.2   Baseline Age-Specific Incidence Rate Estimation

Next, the authors estimate the baseline incidence rate; that is, the incidence rate for a woman *without* the risk factors (with explanatory variable
values AGEMEN=0, NBIOPS=0, AGEFLB=0, and NUMREL=0). Let $h_1^*(t)$

---

[13] We provide this example for illustrative purposes; we do not suggest that such financial
instruments exist at this time.

denote the incidence rate, over the entire (composite) BCDDP set, associated with breast cancer. In this notation, the superscript $*$ signifies that this is a composite incidence rate; the subscript 1 signifies that this incidence rate is associated with breast cancer — other incidences (competing causes of death) have subscript 2. The goal in this section is to estimate $h_1(t)$, the baseline incidence rate.

To do so, the authors, using the five discrete raw explanatory variables, partition the dataset into $I = 2 \times 3 \times 3 \times 4 \times 3 = 216$ risk groups. They then let $P_i(t)$ denote the proportion of women of age t in risk group i. Then

$$h_1^*(t) = \sum_{i=1}^{I} P_i(t) h_1(t) r_i(t), \qquad (12.44)$$

where $r_i(t)$ denotes the relative risk of the ith risk group, relative to the baseline group. Let $\hat{\rho}_i(t)$ denote the observed proportion of cases aged t in risk group i. Then $\hat{\rho}_i(t)$ is an estimate for

$$\rho_i(t) = \frac{P_i(t) h_1(t) r_i(t)}{h_1^*(t)}. \qquad (12.45)$$

Dividing by $r_i(t)$, multiplying by $h_1^*(t)$, and summing over i, the authors obtain

$$h_1(t) = h_1^*(t) \sum_{i=1}^{I} \frac{\rho_i(t)}{r_i(t)} \equiv h_1^*(t) F(t). \qquad (12.46)$$

The authors estimate $h_1^*(t)$ from the general population, $\hat{\rho}_i(t)$ from the cases, and $\hat{r}_i(t)$ from the case control study, and substitute into (12.46) to obtain an estimate for $h_1(t)$.

12.2.3 Long-Term Probabilities

With age-dependent relative risk given by (12.43) and denoted by $r(t)$, and baseline age-specific incidence rates $h_1(t)$ and $h_2(t)$, the authors apply standard methods in competing risk analysis to compute the probability that a woman of age a who has age-dependent relative risk $r(t)$ will develop breast cancer by age $a + \tau$

$$P\{a, \tau, r(t)\} = \int_a^{a+\tau} h_1(t) r(t) e^{-\int_a^t h_1(u) r(u) du} \frac{S_2(t)}{S_2(a)} dt, \qquad (12.47)$$

where

$$S_2(t) = e^{-\int_0^t h_2(u) du}. \qquad (12.48)$$

12.3 A Text Classification Model

In this section, we review the work of Genkin et al. (2006), which discusses a particular automated approach to text classification — the classification of a document written in natural language into one of two or more categories. Genkin et al. (2006) solve the ℓ_1-regularized maximum-likelihood problem (over an exponential family of models) described in Chapters 9 and 10 and produce models trained and tested on three text categorization collections. Each of these collections consists of a large number of natural language documents that human beings, after an enormous amount of work, have categorized. The documents in each collection contain tens of thousands of unique terms, the frequencies of which serve as possible explanatory variables for the text classification models. Given the huge number of possible explanatory variables, feature selection is a very important aspect of the text categorization problem. The models produced are of the form $p(Y = y|x)$, where $y \in \{-1, 1\}$ denotes membership in a particular category of documents and the explanatory variable x is derived from a particular document as discussed below.

The authors benchmark the ℓ_1-regularized maximum likelihood models against two state-of-the-art text categorization methods: support vector machines (SVMs) and ridge regression. For a general discussion of support vector machines, see Vapnik (1999). In particular Genkin et al. (2006) benchmark against SVM-Lite (see `http://svmlight.joachims.org/`). Ridge regression was discussed in Chapters 9 and 10 of this book. The authors also provide a numerical algorithm suitable for such high dimensional optimization problems. The authors concluded that the models produced via ℓ_1-regularized maximum likelihood were clearly more effective than the ridge regression model and competitive with support vector machine models; moreover, the ℓ_1-regularized maximum likelihood models used only a very small fraction of the set of possible attributes. It should also be noted that the support vector machines do not directly produce probabilities (though probability estimates can be produced by an additional modeling layer).

12.3.1 Datasets

(*i*) ModApte, a subset of the Reuters-21578 collection of news stories, described in Lewis et al. (2004). The Reuters-21578 collection is a test collection that is often used by text categorization researchers. This collection of Reuters, Ltd. news articles was collected and labeled by Carnegie Group, Inc. and Reuters, Ltd. The ModApte subset is a particular training/testing split of the Reuters-21578 collection — with 9,603 training instances and 3,299 test instances. The training set contained 18,978 unique terms (that is, 18,978 possible explanatory variables). For further information, see Lewis (2004).

(*ii*) RCV1-v2, a collection of 804,414 manually categorized newswire stories collected by Reuters, Ltd. The authors used the LYRL2004 split, with 23,149 training documents and 781,265 test documents. According to Lewis (2004),the RCV1 collection is likely to supersede the Reuters-21578 collection. The training set contained 47,152 unique terms (possible explanatory variables).

(*iii*) Medline (the bibliographic database for the U.S. National Library of Medicine) records from 1987 to 1991 — part of the OSHUMED test collection described in Hersh et al. (1994)). The authors used a subset of 233,445 records with nonempty title, abstract, and MeSH (Medical Subject Headings) category fields. The authors used 83,944 documents from 1987 and 1988 to train and 149,501 documents from later years to test. The training set contained 73,269 unique terms (possible explanatory variables).

12.3.2 Term Weights

In order to train a conditional probability model, each document must first be represented as a vector of numerical values. The authors construct such a vector via the following steps:

(*i*) They use the Lemur toolkit (see Lemur Project (2007)) to

- tokenize the text into words (using the TreeParser module),
- discard words from the SMART stopword list of 572 words (available at `ftp://ftp.cs.cornell.edu/pub/smart/english.stop`),
- remove word endings using the Lemur variant of the Porter stemmer.

(*ii*) They convert each unique term to a numerical value related to[14]

$$\log TF \times IDF, \tag{12.49}$$

where TF denotes the term frequency and IDF denotes the inverse document frequency. They then assemble these numerical values into a vector, with one element for each unique term in the document.

(*iii*) Finally, they apply the cosine transformation to the above vector to produce the vector of explanatory variables, where the cosine transformation of the vector x is given by

$$\frac{x}{\|x\|^{\frac{1}{2}}}. \tag{12.50}$$

---

[14] For details, they refer the reader to Lewis et al. (2004).

12.3.3 Models

The authors seek a text classifier $y = f(x)$ from the training examples $D = \{(x_1, y_1), \ldots, (x_n, y_n)\}$, where the components of x are determined as described in Section 12.3.2.

This setup can be viewed as a conditional horse race market (see Section 3.5), where, given x, odds ratios $\mathcal{O}_{y|x}$ are paid by a market maker on "horses" $y \in \{0, 1\}$ (membership in the category under consideration). In this market, suppose that an investor with a generalized logarithmic utility function from Definition 5.4 who allocates according to $p(Y = y|x)$ generated by the authors manages the tradeoff between consistency with the prior distribution consistency with the data by solving the ℓ_1 regularized MRE problem and the associated regularized maximum likelihood dual problem of Section 10.1.4. As we have seen (see Section 10.2.3.5), this setup, with appropriate features, can lead to logistic regression models.

The authors seek logistic regression models of the form

$$p(Y = 1|x) = \frac{1}{1 + exp(-\beta^T x)}, \tag{12.51}$$

where $y = 1$ denotes membership in the category in question. Optimal parameters for maximum likelihood are found by maximizing

$$l(\beta) = \sum_{i=1}^{n} log(1 + exp(-\beta^T x_i y_i)). \tag{12.52}$$

Optimal parameters for maximum likelihood ridge regression are found by maximizing

$$l_{ridge}(\beta) = l(\beta) + \lambda \sum_j \beta_j^2 \tag{12.53}$$

and optimal parameters for maximum likelihood (ℓ_1-regularized) lasso regression are found by maximizing

$$l_{lasso}(\beta) = l(\beta) + \lambda \sum_j |\beta_j|, \tag{12.54}$$

where λ is a hyperparameter selected via cross validation.[15]

---

[15] For the purpose of making a comparative study of models when the number of features was fixed at 6, 51, and 501, the authors sometimes adjusted the value of λ for the lasso model, so that the lasso model would select the desired number of features; for more details, see Genkin et al. (2006).

12.4 A Fat-Tailed, Flexible, Asset Return Model

Fat-tailed distributions seem to be of particular interest,[16] given recent financial market turbulence sometimes attributed to reliance on models that do not adequately capture the likelihood of extreme events.[17] In this section, we briefly discuss the work of Friedman et al. (2010b) who describe

(i) an application of the MRUE method with power-law utility,

$$U(W) = \frac{W^{1-\kappa} - 1}{1 - \kappa}, \qquad (12.55)$$

where κ denotes the investor's (constant) relative risk aversion,[18] for estimating fat-tailed probability distributions for continuous random variables,

(ii) practical numerical techniques necessary for such an undertaking, and

(iii) numerical experiments in which power-law probability distributions are calibrated to asset return data.

They show that, using MRUE methods, even with relatively simple features, it is possible to estimate flexible power law (fat-tailed) distributions. A probability distribution is said to be a power-law distribution[19] if it can be expressed as

$$p(y) \propto L(y)y^{-\alpha}, \qquad (12.56)$$

where $\alpha > 1$ and $L(y)$ is a slowly varying function, in the sense that

$$\lim_{y \to \infty} \frac{L(ty)}{L(y)} = 1, \qquad (12.57)$$

where t is constant.

In particular, the authors have shown that by taking the MRUE approach, with power utility functions and fractional power features, it is possible to

---

[16] See, for example, the following recent New York Times articles: Nocera (2009), Bookstaber (2009), and Safire (2009).

[17] See the the the end of Section 10.3 for a discussion of MRE methods and the calibration of fat-tailed models.

[18] As we have mentioned, power utility functions are used widely in industry (see, for example, Morningstar (2002)). Moreover, power utility functions have constant relative risk aversion and important optimality properties (see, for example, Stutzer (2003)).

[19] Power-law distributions have been proposed for an enormous variety of natural and social phenomena including website popularity, the popularity of given names, conflict severity, the number of words used in a document, and financial asset returns (see Gabaix et al. (2003)). For additional discussion, see, for example, Mitzenmacher (2004), Newman (2005), and Clauset et al. (2009).

obtain a rich family of power-law distributions by solving a continuous alternative version of the convex programming problem, Problem 10.16, with a flat prior distribution, odds ratios set according to (10.150), and a power utility. They note that, given a collection of features, greater relative risk aversion is associated with fatter-tailed distributions. They also note that a number of well-known power-law distributions, including the student-t, generalized Pareto, and exponential distributions, can be obtained as special cases of the connecting equation associated with MRUE approach with power utility and linear or quadratic features; the skewed generalized-t distribution is a special case with power features.

The authors have calibrated such methods to financial asset return data and reported performance superior to that of alternative benchmark models, with respect to log-liklihood, which they attribute to the ability of their models to incorporate fat tails where data are extreme and sparse, with flexibility where data are more plentiful.

References

H. Akaike. Information theory and an extension of the maximum likelihood principle. In B. Petrov and F. Caski, editors, *Proceedings of the Second International Symposium on Information Theory*, page 267. Budapest, 1973.

P. Artzner, F. Delbaen, J. Eber, and D. Heath. Coherent measures of risk. *Mathematical Finance*, 9(3):203–228, 1999.

M. Avellaneda. Minimum-relative-entropy calibration of asset pricing models. *Int. J. Theor. and Appl. Fin.*, 1(4):447, 1998.

G. A. Barnard. Statistical inference (with discussion). *Journal of the Royal Statistical Society, B*, 11:115, 1949.

P. Bartlett, S. Boucheron, and G. Lugosi. Model selection and error estimation. In *COLT '00: Proceedings of the Thirteenth Annual Conference on Computational Learning Theory*, pages 286–297, San Francisco, CA, USA, 2000. Morgan Kaufmann Publishers Inc.

Basel Committee on Banking Supervision. The new Basel capital accord. April:80–81, 2003.

A. Berger, S. Della Pietra, and V. Della Pietra. A maximum entropy approach to natural language processing. *Computational Linguistics*, 22(1):39, 1996.

J. Berger. *Statistical Decision Theory and Bayesian Analysis*. Springer, New York, 1985.

J. Berkovitz. *Convexity and Optimization in R^n*. Wiley, New York, 2002.

J. M. Bernardo. Expected information as expected utility. *Annals of Statistics*, 7:686–690, 1979.

J. M. Bernardo and A. F. M. Smith. *Bayesian Theory*. Wiley, New York, 2000.

D. Bernoulli. Specimen theoriae novae de mensura sortis, Commentarii Academiae Scientiarum Imperialis Petropolitanae (5, 175-192, 1738). *Econometrica*, 22:23–36, 1954, 1738. Translated by L. Sommer.

D. Bertsekas and J. Tsitsiklis. *Introduction to Probability*. Athena Scientific, 2002.

N. Bingham and R. Kiesel. *Risk-Neutral Valuation: Pricing and Hedging of Financial Derivatives, 2nd Edition*. Springer, New York, 2004.

A. Birnbaum. On the foundation of statistical inference (with discussion). *J. Am. Stat. Assoc.*, 69:269, 1962.

C. Bishop. *Pattern Recognition and Machine Learning.* Springer, New York, 2007.

A. Blum and P. Langley. Selection of relevant features and examples in machine learning. *Artificial Intelligence*, 97(1-2):245–271, 1997.

L. Boltzmann. Über die Beziehung zwischen dem zweiten Hauptsatz der mechanischen Wärmetheorie und der Wahrscheinlichkeitsrechnung respective den Sätzen über das Wärmegleichgewicht. *Wiener Berichte*, 76:373–435, 1877.

BondMarkets.com. Outstanding level of public & private bond market debt. http://www.bondmarkets.com/story.asp?id=323, 2006.

R. Bookstaber. The fat-tailed straw man. *New York Times*, March 10, 2009.

R. Bos, K. Kelhoffer, and D. Keisman. Ultimate recovery in an era of record defaults. *Standard & Poor's CreditWeek*, August 7:23, 2002.

S. Boyd and L. Vandenberghe. *Convex Optimization.* Cambridge University Press, Cambridge, 2004.

M. Braun. *Differential Equations and Their Applications.* Springer, New York, 1975.

L. Breiman. Optimal gambling systems for favorable games. *Fourth Berkeley Symposium on Mathematical Statistics and Probability*, pages 65–78, 1961.

L. Breiman. Statistical modeling: The two cultures. *Statistical Science*, 16:199–231, 2001.

S. Browne and W. Whitt. Portfolio choice and the Bayesian Kelly criterion. *Advances in Applied Probability*, 28(4):1145–1176, 1996.

B. Buck and V. A. Macaulay. *Maximum Entropy in Action.* Clarendon Press, Oxford, 1991.

K. Burnham and D. Anderson. *Model Selection and Multimodel Inference.* Springer, New York, 2002.

S. F. Chen and R. Rosenfeld. A Gaussian prior for smoothing maximum entropy models. *Technical Report CMU-CS-99-108, School of Computer Science, Carnegie Mellon University, Pittsburgh, PA*, 1999.

V. Cherkassky and F. Mulier. *Data Mining: Practical Machine Learning Tools and Techniques, Second Edition.* Wiley, New York, 2007.

H. L. Chieu and H. T. Ng. Named entity recognition: A maximum entropy approach using global information. *Proceedings of the 19th International*

Conference on Computational Linguistics, page 190, 2002.

A. Clauset, C. Shalizi, and M. Newman. Power-law distributions in empirical data. *SIAM Review*, 51:661703, 2009.

R. Cont and P. Tankov. Nonparametric calibration of jump-diffusion option pricing models. *Journal of Computational Finance*, 7(3):1–49, 1999.

T. Cover and J. Thomas. *Elements of Information Theory*. Wiley, New York, 1991.

D. R. Cox. Comment on Breiman's 'Statistical modeling: The two cultures'. *Statistical Science*, 16:216, 2001.

G. Cramer. Letter to Bernoulli's cousin (see Bernoulli, 1738). 1728.

I. Csiszár. A class of measure of informitivity of observation channels. *Periodica Mathematica Hungarica*, 2:191–213, 1972.

R. Davidson and J. MacKinnon. *Estimation and Inference in Econometrics*. Oxford University Press, New York, 1993.

B. de Finetti. *Theory of Probabilty. A Critical Introductory Treatment*. John Wiley & Sons, London, 1974.

E. DeLong, D. DeLong, and D. Clarke-Pearson. Comparing the areas under two or more correlated receiver operating characteristic curves: A nonparametric approach. *Biometrics*, 44(3):837–845, 1988.

A. Dreber, C. Apicella, D. Eisenberg, J. Garcia, R. Zamore, J. Lum, and B. Campbell. The 7r polymorphism in the dopamine receptor d4 gene (drd4) is associated with financial risk taking in men. *Evolution and Human Behavior*, 30:85–92, 2009.

M. Dudik, S. Phillips, and R. Shapire. Performance guarantees for regularized maximum entropy density esimation. *Proceedings of the Seventeenth Annual Conference on Computational Learning Theory*, page 472, 2004.

D. Duffie. *Dynamic Asset Pricing Theory*. Princeton University Press, Princeton, 1996.

B. Efron. Comment on Breiman's 'Statistical modeling: The two cultures'. *Statistical Science*, 16:216, 2001.

E. Falkenstein. Riskcalc^{TM} for private companies: Moody's default model. *Moody's Investor Service*, May, 2000.

Federal Reserve Board. Federal Reserve Statistical Release g.19. http://www.federalreserve.gov/releases/g19/current/default.htm, July, 10, 2006.

W. Feller. *An Introduction to Probability Theory and Its Applications*. Wiley,

New York, 1966.

R. A. Fisher. On the mathematical foundation of theoretical statistics. *Philosophical Transactions of the Royal Society of London, A*, 222:309–368, 1922.

R. A. Fisher. Mathematical probability in the natural sciences. *Technometrics*, 1:21–29, 1959.

G. Frenk, G. Kassay, and J. Kolumbán. Equivalent results in minimax theory. *Working paper*, 2002.

M. Friedlander and M. Gupta. On minimizing distortion and relative entropy. *Argonne National Laboratory Preprint*, 2003.

C. Friedman, J. Huang, and S. Sandow. A utility-based approach to some information measures. *Entropy*, 9:1–26, 2007.

C. Friedman, J. Huang, and Y. Zhang. Estimating future transition probabilities when the value of side information decays, with applications to credit modeling. *Working paper*, http://papers.ssrn.com/sol3/papers.cfm?abstract_id=1576673, 2010a.

C. Friedman and S. Sandow. Learning probabilistic models: An expected utility maximization approach. *Journal of Machine Learning Research*, 4: 291, 2003a.

C. Friedman and S. Sandow. Model performance measures for expected utility maximizing investors. *International Journal of Theoretical and Applied Finance*, 6(4):355, 2003b.

C. Friedman and S. Sandow. Ultimate recoveries. *Risk*, August:69, 2003c.

C. Friedman and S. Sandow. Model performance measures for leveraged investors. *International Journal of Theoretical and Applied Finance*, 7(5): 541, 2004.

C. Friedman and S. Sandow. Utility-based performance measures for regression models. *Journal of Banking and Finance*, pages 541–560, 2006a.

C. Friedman and S. Sandow. Utility functions that lead to the likelihood ratio as a relative model performance measure. *Statistical Papers*, pages 211–225, 2006b.

C. Friedman, Y. Zhang, and J. Huang. Estimating flexible, fat-tailed asset return distributions. *Working paper*, http://ssrn.com/abstract=1626342, 2010b.

J. Friedman, T. Hastie, S. Rosset, R. Tibshirani, and J. Zhu. Discussion of three boosting papers. *Annals of Statistics*, 32(1):102, 2004.

M. Frittelli. The minimal entropy martingale measure and the valuation problem in incomplete markets. *Math. Finance*, 10:39, 2000.

X. Gabaix, P. Gopikrishnan, V. Plerou, and H. Stanley. A theory of power-law distributions in financial market fluctuations. *Nature*, 423:267–270, 2003.

M. Gail, L. Brinton, D. Byar, D. Corle, S. Green, C. Schairer, and J. Mulvihill. Projecting individualized probabilities of developing breast cancer for white females who are being examined annually. *Journal of the National Cancer Institute*, 81(No. 24):1879–1886, 1989.

A. Gelman, J. Carlin, H. Stern, and D. Rubin. *Bayesian Data Analysis*. Chapman & Hall/CRC, Boca Raton, 2000.

A. Genkin, D. D. Lewis, and D. Madigan. Sparse logistic regression for text categorization. *Working Paper,* http://www.stat.rutgers.edu/~madigan/mms/loglasso-v3a.pdf, 2006.

A. Globerson, E. Stark, E. Vaadia, and N. Tishby. The minimum information principle and its application to neural code analysis. *Proceedings of the National Academy of Sciences of the United States of America*, February 2009.

A. Globerson and N. Tishby. The minimum information principle in discriminative learning. In M. Chickering and J. Halpern, editors, *Proceedings of the UAI*, page 193200. Assoc for Uncertainty in Artificial Intelligence, 2004.

A. Golan, G. Judge, and D. Miller. *Maximum Entropy Econometrics*. Wiley, New York, 1996.

I. Good. Rational decisions. *Journal of the Royal Statistical Society, Series B*, 14:107114, 1952.

J. Goodman. Exponential priors for maximum entropy models. *Working paper*, 2003.

P. Grünwald. Taking the sting out of subjective probability. In D. Barker-Plummer, D. Beaver, J. van Benthem, and P. S. Di Luzio, editor, *Words, Proofs, and Diagrams*. CSLI Publications, Stanford, CA, 2002.

P. Grünwald and A. Dawid. Game theory, maximum generalized entropy, minimum discrepancy, robust Bayes and Pythagoras. *Annals of Statistics*, 32(4):1367–1433, 2004.

L. Gulko. The entropy theory of bond option pricing. *International Journal of Theoretical and Applied Finance*, 5:355, 2002.

S. F. Gull and G. J. Daniell. Image reconstruction with incomplete and noisy data. *Nature*, 272:686, 1978.

I. Guyon and A. Elisseeff. An introduction to variable and feature selection. *Journal of Machine Learning Research, Special Issue on Variable and Fea-*

ture Selection, 3:1157, 2003.

G. Harańczyk, W. Slomczyński, and T. Zastawniak. Relative and discrete utility maximising entropy. *arXiv:0709.1281v1*, September, 2007.

T. Hastie, R. Tibshirani, and J. Friedman. *The Elements of Statistical Learning: Data Mining, Inference, and Prediction, Second Edition*. Springer, New York, 2009.

W. Hersh, C. Buckley, T. Leone, and D. Hickam. Ohsumed: An interactive retrieval evaluation and new large test collection for research. In *SIGIR*, pages 192–201, 1994.

B. Hoadley. Comment on Breiman's 'Statistical modeling: The two cultures'. *Statistical Science*, 16:216, 2001.

P. Hore. Maximum entropy and nuclear magnetic resonance. In B. Buck and V. A. Macaulay, editors, *Maximum Entropy in Action*. Clarendon Press, Oxford, 1991.

D. Hosmer and S. Lemeshow. *Applied Logistic Regression, Second edition*. Wiley, New York, 2000.

J. Huang. Personal communication. 2003.

J. Huang, S. Sandow, and C. Friedman. Information, model performance, pricing and trading measures in incomplete markets. *International Journal of Theoretical and Applied Finance*, 2006.

J. Ingersoll, Jr. *Theory of Financial Decision Making*. Rowman and Littlefield, New York, 1987.

F. Jamishidian. Asymptotically optimal portfolios. *Mathematical Finance*, 2: 131–150, 1992.

K. Janeček. What is a realistic aversion to risk for real-world individual investors? *Working Paper*, 2002.

E. T. Jaynes. Information theory and statistical mechanics. *Physical Review*, 106:620, 1957a.

E. T. Jaynes. Information theory and statistical mechanics ii. *Physical Review*, 108:171190, 1957b.

E. T. Jaynes. *Probability Theory. The Logic of Science*. Cambridge University Press, Cambridge, 2003.

J. Kallsen. Utility-based derivative pricing in incomplete markets. *Mathematical Finance - Bachelier Congress 2000*, pages 313–338, 2002.

I. Karatzas, J. Lehoczky, S. Shreve, and G. Xu. Martingale and duality methods for utility maximization in an incomplete market. *SIAM J. Control and*

Optimization, 29(3):702–730, 1991.

J. Kazama and J. Tsujii. Evaluation and extension of maximum entropy models with inequality constraints. In *Proceedings of the 2003 Conference on Empirical Methods in Natural Language Processing (EMNLP 2003)*. 2003.

J. Kelly. A new interpretation of information rate. *Bell Sys. Tech. Journal*, 35:917, 1956.

A.N. Kolmogorov. *Grundbegriffe der Warscheinlichkeitsrechnung.* Springer, Berlin, 1933. Translated into English by Nathan Morrison (1950), Foundations of the Theory of Probability, Chelsea, New York. Second English edition 1956.

C. Kuhnen and J. Chiao. Genetic determinants of financial risk taking. *PLoS ONE*, 4(2):e4362, 02 2009.

S. Kullback. *Information Theory and Statistics.* Dover, New York, 1997.

G. Lebanon and J. Lafferty. Boosting and maximum likelihood for exponential models, 2001.

Lemur Project. The lemur toolkit. `http://www.lemurproject.org/`, 2007.

D. Lewis. Reuters-21578 text categorization test collection distribution 1.0 readme file (v 1.3). *http://www.daviddlewis.com/resources/testcollections/reuters21578/readme.txt*, 2004.

D.D. Lewis, Y. Yang, T. Rose, and F. Li. Rcv1: A new benchmark collection for text categorization research. *Journal of Machine Learning Research*, pages 361–397, 2004.

X. Liu, A. Krishnan, and A. Mondry. An entropy-based gene selection method for cancer classification using microarray data. *BMC Bioinformatics*, 6:76, 2005.

R. Luce and H. Raiffa. *Games and Decisions: Introduction and Critical Survey.* Dover, New York, 1989.

D. Luenberger. *Optimization by Vector Space Methods.* Wiley, New York, 1969.

D. Luenberger. *Investment Science.* Oxford University Press, New York, 1998.

G. Lugosi and A. Nobel. Adaptive model selection using empirical complexities. *Annals of Statistics*, 27(6):1830, 1999.

D. MacKay. *Information Theory, Inference, and Learning Algorithms.* Cambridge University Press, Cambridge, 2003.

D. Madan. Equilibrium asset pricing: With non-Gaussian factors and exponential utilities. *Quantitative Finance*, 6(6):455–463, 2006.

P. Massart. Some applications of concentration inequalities to statistics. *Annales de la Faculte des Sciences de Toulouse*, 9(2):245, 2000.

N. Mehra, S. Khandelwal, and P. Patel. Sentiment identification using maximum entropy analysis of movie reviews. *Working paper*, 2002.

R. Merton. Optimum consumption and portfolio rules in a continuous time model. *J. Econ. Theory*, 3:373–413, 1971.

T. Mitchell. *Machine Learning*. McGraw-Hill, New York, 1997.

M. Mitzenmacher. A brief history of generative models for power law and lognormal distributions. *Internet Mathematics*, 1(2):226–251, 2004.

Morningstar. The new Morningstar rating^{TM} methodology. `http://www.morningstar.dk/downloads/MRARdefined.pdf`, 2002.

M. Newman. Power laws, pareto distributions and zipf's law. *Contemporary Physics*, 46(2):323–351, 2005.

J. Neyman and E. S. Pearson. On the problem of the most efficient tests of statistical hypotheses. *Philosophical Transactions of the Royal Society of London, A*, 231:289, 1933.

A. Y. Ng. Feature selection, l_1 vs. l_2 regularization, and rotational invariance. *Proceedings of the 21st International Conference on Machine Learning*, 2004.

J. Nocera. Risk mismanagement. *New York Times*, January 2, 2009.

R. Parker. Discounted cash flow in historical perspective. *Journal of Accounting Research*, 6(1):58–71, 1968.

E. Parzen. Comment on Breiman's 'Statistical modeling: The two cultures'. *Statistical Science*, 16:216, 2001.

S. Perkins, K. Lacker, and J. Theiler. Grafting: Fast, incremental feature selection by gradient descent in function space. *Machine Learning*, 3:1333–1356, 2003.

A. Plastino and A. R. Plastino. Tsallis entropy and Jaynes' information theory formalism. *Brazilian Journal of Physics*, 29(1):50, 1999.

D. Ravichandran, A. Ittycheriah, and S. Roukos. Automatic derivation of surface text patterns for a maximum entropy based question answering system. *Proceedings of the HLT-NAACL Conference*, 2003.

S. Raychaudhuri, J. Chang, P. Sutphin, and R. Altma. Associating genes with gene ontology codes using a maximum entropy analysis of biomedical literature. *Genome Research*, 12, 2002.

A. Rényi. On measures of information and entropy. *Proceedings of the 4th*

Berkeley Symposium on Mathematics, Statistics and Probability 1960, page 547561, 1961.

S. Riezler and A. Vasserman. Incremental feature selection and ℓ_1-regularization for relaxed maximum-entropy modeling. *Working paper*, 2004.

C. Robert. *The Bayesian Choice*. Springer, New York, 1994.

R. Rockafellar, S. Uryasev, and M. Zabarankin. Deviation measures in generalized linear regression. *Research report, University of Florida*, 2002(9), 2002a.

R. Rockafellar, S. Uryasev, and M. Zabarankin. Deviation measures in risk analysis and optimization. *Research report, University of Florida*, 2002(7), 2002b.

R. T. Rockafellar. *Convex Analysis*. Princeton University Press, Princeton, 1970.

B. Roe, M. Tilley, H. Gu, D. Beversdorf, W. Sadee, T. Haab, and A. Papp. Financial and psychological risk attitudes associated with two single nucleotide polymorphisms in the nicotine receptor (chrna4) gene. *PLoS ONE*, 4(8):e6704, 2009.

S. Ross. *A First Course in Probability, 7th Edition*. Prentice Hall, 2005.

M. Rubinstein. The strong case for the generalized logarithmic utility as the premier model of financial markets. *Journal of Finance*, May, 1976.

W. Safire. Fat tail. *New York Times*, February 5, 2009.

P. Samuelson. St. Petersburg paradoxes: defanged, dissected, and historically described. *Journal of Economic Literature*, 15:2455, March 1977.

P.A. Samuelson. The 'fallacy' of maximizing the geometric mean in long sequences of investing or gambling. *Proc. Nat. Acad. Science*, 68:214–224, 1971.

P.A. Samuelson. Why we should not make mean log of wealth big though years to act are long. *Journal of Banking and Finance*, 3:305–307, 1979.

S. Sandow, J. Huang, and C. Friedman. How much is a model upgrade worth? *Journal of Risk*, 10(1):3–40, 2007.

S. Sandow and J. Zhou. Data-efficient model building for financial applications: a semi-supervised learning approach. *Journal of Risk Finance*, 8(2): 133–155, 2007.

L. J. Savage. *The Foundation of Statistics*. John Wiley & Sons, New York, 1954.

W. Schachermayer. Utility maximisation in incomplete markets. *http://www.fam.tuwien.ac.at/ wschach/pubs/*, 2004.

P. Schoemaker. *Experiments on Decisions Under Risk: The Expected Utility Hypothesis*. Kluwer Nijhoff Publishing, Boston, 1980.

P. Schönbucher. *Credit Derivatives Pricing Models: Model, Pricing and Implementation*. Wiley, 2003.

G. Schwarz. Estimating the dimension of a model. *Annals of Statistics*, 6: 461, 1978.

C. E. Shannon. A mathematical theory of communication. *Bell System Technical Journal*, 27:379–423 and 623–656, Jul and Oct 1948.

J. Skilling. Fundamentals of maxent in data analysis. In B. Buck and V. A. Macaulay, editors, *Maximum Entropy in Action*. Clarendon Press, Oxford, 1991.

W. Slomczyński and T. Zastawniak. Utility maximizing entropy and the second law of thermodynamics. *The Annals of Probability*, 32(3A):2261, 2004.

Standard & Poor's Global Fixed Income Research. Global credit trends: Quarterly wrap-up and forecast update, 2007. [Online; accessed 19-March-2007].

Standard & Poor's Global Fixed Income Research. Global corporate default rate rises to 9.71% in October 2009, article says, 2009. [Online; accessed 19-November-2009].

M. Stutzer. Portfolio choice with endogenous utility: A large deviations approach. *Journal of Econometrics*, 116:365–386, 2003.

G. Szekely and D. Richards. The St. Petersburg paradox and the crash of high-tech stocks in 2000. *The American Statistician*, 58:225–231, August 2004.

Thompson Financial. First quarter 2006 managing underwriters debt capital markets review. `http://www.thomson.com/cms/assets/pdfs/financial/league_table/debt_and_equity/1Q2006/1Q06_DE_Debt_Capital_Markets_Review.pdf`, 2006.

E. Thorp. The Kelly criterion in blackjack, sports betting, and the stock market. *The 10th International Conference on Gambling and Risk Taking*, 1997.

R. Tibshirani. Regression shrinkage and seklection via the lasso. *Journal of the Royal Statistical Society, Series B, Methodological*, 58:267, 1996.

F. Topsøe. Information theoretical optimization techniques. *Kybernetika*, 15 (1):8, 1979.

C. Tsallis. Possible generalization of Boltzmann-Gibbs statistics. *Journal of Statistical Physics*, 52:479, 1988.

C. Tsallis and E. Brigatti. Nonextensive statistical mechanics: A brief introduction. *Continuum Mechechanics and Thermodynanmics*, 16:223–235, 2004.

K. Van de Castle and D. Keisman. Recovering your money: Insights into losses from default. *Standard & Poor's CreditWeek*, June:29, 1999.

V. Vapnik. *Statistical Learning Theory*. Wiley, New York, 1998.

V. Vapnik. *The Nature of Statistical Learning Theory*. Springer, New York, 1999.

V. Vapnik and A. Chernovenkis. Uniform convergence of the frequencies of occurrence of events to their probabilities. *Soviet Mat. Doklady*, 9:915, 1968.

V. Vapnik and A. Chernovenkis. On the uniform convergence of relative frequencies to their probabilities. *Theory of Probab. Appl.*, 16(2):264, 1971.

J. von Neumann and O. Morgenstern. *Theory of Games and Economic Behavior*. Princeton University Press, Princeton, 1944.

Wikipedia. Receiver operating characteristic — Wikipedia, the free encyclopedia, 2010. [Online; accessed 19-March-2007].

J. Wilcox. Harry Markowitz and the discretionary wealth hypothesis. *Journal of Portfolio Management*, 29:58–65, 2003.

I. Witten and E. Frank. *Learning from Data: Concepts, Theory, and Methods, Second Edition*. Morgan Kaufmann, 2005.

N. Wu. *The Maximum Entropy Method*. Springer, New York, 1997.

A. Zemanian. *Distribution Theory and Transform Analysis: An Introduction to Generalized Functions with Applications*. Dover Publications, Inc., New York, 1987.

S. Zhong, S. Chew, E. Set, J. Zhang, H. Xue, P. Sham, R. Ebstein, and S. Israel. The heritability of attitude toward economic risk. *Twin Research and Human Genetics*, 12:103–107, 2009.

X. Zhou, J. Huang, C. Friedman, R. Cangemi, and S. Sandow. Private firm default probabilities via statistical learning theory and utility maximization. *Journal of Credit Risk*, 2(1), 2006.

W. Ziemba and L. MacLean. The Kelly capital growth theory and its use by great investors and speculators. In *Conference on Risk Management and Quantitative Approaches in Finance, Gainesville, FL*, 2005.

Index